BGE S1-S3

Mathematics & Numeracy

Fourth Level bridging to National 5

Dr Helen Kelly

Kathleen McQuillan

Dr Alan Taylor

Photo credits

p.8 © pamela_d_mcadams - stock.adobe.com; **p.43** (repeated) © vadim_fl - stock.adobe.com; **p.46** (repeated) © Alexander – stock.adobe.com; **p.55** © blanke1973 - stock.adobe.com; **p.149** (top) © Katherine Johnson/NASA, (middle) © .shock - stock.adobe.com, (bottom) © Monkey Business - stock.adobe.com; **p.183** © Coprid - stock.adobe.com; **p.185** © lady-luck - Shutterstock.com.

Although every effort has been made to ensure that website addresses are correct at time of going to press, Hodder Gibson cannot be held responsible for the content of any website mentioned in this book. It is sometimes possible to find a relocated web page by typing in the address of the home page for a website in the URL window of your browser.

Hachette UK's policy is to use papers that are natural, renewable and recyclable products and made from wood grown in well-managed forests and other controlled sources. The logging and manufacturing processes are expected to conform to the environmental regulations of the country of origin.

Orders: please contact Hachette UK Distribution, Hely Hutchinson Centre, Milton Road, Didcot, Oxfordshire, OX11 7HH. Telephone: (44)01235 827827. Fax: (44)01235 400401. Email education@hachette.co.uk. Lines are open from 9 a.m. to 5 p.m., Monday to Friday. Visit our website at www.hoddereducation.co.uk. If you have queries or questions that aren't about an order, you can contact us at hoddergibson@hodder.co.uk

© Helen Kelly, Kathleen McQuillan and Alan Taylor 2021

First published in 2021 by
Hodder Gibson, an imprint of Hodder Education
An Hachette UK Company
50 Frederick Street
Edinburgh EH2 1EX

Impression number	5	4	3	2
Year	2025	2024		

Cover photo © Alexander – stock.adobe.com

Illustrations by Aptara, Inc.

Typeset in Bliss Regular 12/14 pts. by Aptara, Inc.

Printed and bound by CPI Group (UK) Ltd, Croydon, CR0 4YY

A catalogue record for this title is available from the British Library.

ISBN: 978 1 3983 0881 7

We are an approved supplier on the Scotland Excel framework.

Find us on your school's procurement system as *Hachette UK Distribution Ltd* or *Hodder & Stoughton Limited t/a Hodder Education.*

MIX
Paper | Supporting responsible forestry
FSC™ C104740

Contents

Introduction to BGE Mathematics

Mathematics is the richest language in the world. Learning mathematics is one of the most important things you can do to boost your brain power, now and throughout your life. Mathematicians understand, describe and influence the world around us.

This book covers all the BGE Benchmarks for Mathematics at Fourth Level. It will also extend you where appropriate in preparation for your transition to national qualifications.

The chapters take you on a journey which will both support and challenge you. As you focus and work hard on each section you will gain knowledge, understanding and the ability to solve problems and communicate your solutions to others.

Working through this book will equip you to be the best mathematician you can be. Every topic is explained with rigour, taking no short cuts, and has an abundance of practice and challenge to suit every learner. The book is ambitious, encouraging you to aim high and build the skills you need for future success.

The book is enjoyable and easy to read. Every topic includes a clear, 'straight-to-the-point' method set out using bullet points in concise and simple language. It is full of worked examples and helpful hints and contains proven methods of mastering difficult concepts. This book has been designed both as a classroom aid and for your personal study.

Each lesson starts with an explanation and memorable method to which you can refer.

Each lesson has planned and progressive worked examples which prepare you to tackle the problem set.

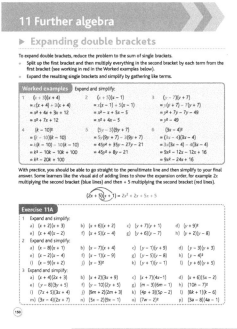

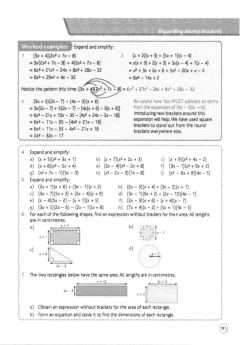

Work through the exercise to practise and consolidate your skills. There are solutions at the end of the book, so check your work as you go to boost your confidence.

Each exercise gets progressively harder. Try to finish the exercises and test your skills with the most challenging problems.

Every chapter has a check-up exercise. Use these to bring your ideas together and boost your memory of key methods.

We hope that you enjoy using this book as much as we have enjoyed creating the content, questions and activities for you!

1 Number work

▶ Significant figures

By considering significant figures (s.f.) we can round whole and decimal numbers to a required accuracy.

- Start counting significant figures at the first non-zero digit from the left.
- Consider the digit after the last significant figure. If it is 5 or above, round the last significant figure up. Otherwise, leave the last significant figure unchanged.
- To keep place value consistent, you may have to add zeros at the end of a whole number.
- Zeros should only appear at the end of a decimal number if they are significant due to rounding.

Worked example

Round **a)** 52 860, **b)** 18·351 and **c)** 0·0031976 to 1, 2 and 3 significant figures.

first significant figure second significant figure third significant figure

a) 5 2 860
= 50 000 to 1 s.f.

52 860
= 53 000 to 2 s.f.

52 8 60
= 52 900 to 3 s.f.

The zeros are required at the end to keep place value consistent i.e. the number approximately 50 000.

b) 1 8·351
= 20 to 1 s.f.

18 351
= 18 to 2 s.f.

18·3 51
= 18·4 to 3 s.f.

Zeros should not be retained to keep the number a decimal or at the end of a decimal unless they are significant due to rounding.

c) 0·003 1976
= 0·003 to 1 s.f.

0·0031 976
= 0·0032 to 2 s.f.

0·00319 76
= 0·00320 to 3 s.f.

The first non-zero digit is 3 so it is the first significant figure. The zeros at the start are required to keep place value consistent, i.e. the number approximately 0·003. For 3 significant figures, the zero is required at the end as it is significant due to rounding.

Exercise 1A

1 Round to 1 significant figure:

    a) 43 000   **b)** 6900   **c)** 261   **d)** 709   **e)** 82

    f) 5·92   **g)** 0·76   **h)** 0·044   **i)** 0·0507   **j)** 0·0089

2 The attendance at an amateur rugby game was 3567. Round the attendance to 1 significant figure.

3 An album is downloaded 718 612 times in the first month after release. Round the number of downloads to 1 significant figure. ➜

4 Round to 2 significant figures:

   a) 63 700   **b)** 4920   **c)** 637   **d)** 805   **e)** 21·3

   f) 8·56   **g)** 0·601   **h)** 0·0924   **i)** 0·309   **j)** 0·004 04

5 A house is valued at £237 500. Round its value to 2 significant figures.

6 An exchange rate is advertised as £1 = AU$1·93597.

   a) What is £100 worth in AU$?

   b) Round your answer to 2 significant figures.

7 Round to 3 significant figures:

   a) 14 930   **b)** 8267   **c)** 932·4   **d)** 51·919   **e)** 2·1553

   f) 0·03461   **g)** 0·7208   **h)** 0·046 28   **i)** 0·7591   **j)** 0·2999

8 On average, a fingernail grows at a rate of 0·003 468 metres per month. Round this rate to 3 significant figures.

9 The 2019 mid-year estimate for the population of Scotland was 5 463 300. Round the population estimate to 1, 2 and 3 significant figures.

10 How many significant figures have the following measurements been rounded to?

   a) 82 cm   **b)** 791 m   **c)** 6 km   **d)** 15·2 cm   **e)** 4·5 m

   f) 3·15 cm   **g)** 6·271 km   **h)** 0·65 m   **i)** 0·005 m   **j)** 0·0080 km

11 The irrational number π can be approximated by 3·14159265.

   a) How many significant figures has π been written to?

   b) Approximate π to:

   i) 3   **ii)** 4   **iii)** 5 significant figures.

12 When a dot appears above a digit (or digits) it means the digit(s) are recurring. For example,

   $0 \cdot \dot{6}$ = 0·666666...   only one dot so only the 6 is repeated

   $0 \cdot \dot{7}\dot{1}$ = 0·7171717171...   two dots so the two numbers with the dots are repeated as a block

   $0 \cdot \dot{2}0\dot{8}$ = 0·208208208...   all digits between the first and last dot are repeated as a block

   By writing the following numbers with a suitable number of digits, round to the accuracy asked for:

   a) $0 \cdot \dot{4}$ to 2 significant figures   **b)** $0 \cdot \dot{8}\dot{3}$ to 2 significant figures

   c) $0 \cdot \dot{5}\dot{7}$ to 3 significant figures   **d)** $0 \cdot \dot{2}6\dot{1}$ to 3 significant figures

   e) $0 \cdot \dot{6}0\dot{9}$ to 4 significant figures   **f)** $0 \cdot \dot{4}7\dot{8}$ to 5 significant figures.

13 A prize fund of £219 500 is to be shared equally amongst 4 people. How much will each person get if the prize fund is:

   a) shared amongst them with no rounding

   b) rounded to 1 significant figure before being shared amongst them

   c) rounded to 2 significant figures before being shared amongst them?

   Consider your answers to **a)**, **b)** and **c)**.

   d) Does it make sense to round the amount before sharing? Explain why not.

   e) What would happen if you rounded a prize fund of £299 500 to 1 significant figure before sharing?

▶ Tolerance

A **tolerance** describes a desired range within which a value is classed as acceptable. The notation $a \pm b$ is used when writing a tolerance where a and b are numbers and $\pm$ is read as 'plus or minus'. For numbers of the form $a \pm b$:

- $a + b$ is the largest acceptable value and $a - b$ is the smallest acceptable value
- any value between $a - b$ and $a + b$ is within the acceptable range
- any value below $a - b$ or above $a + b$ is not within the acceptable range
- brackets are added around $(a \pm b)$ as the units of measurement are for the whole value, not just b.

Worked examples

1 Find the maximum and minimum values of $(6\cdot3 \pm 2\cdot9)$ mm.

Maximum = $6\cdot3 + 2\cdot9 = 9\cdot2$ mm   Minimum = $6\cdot3 - 2\cdot9 = 3\cdot4$ mm

2 An apple must be within a weight range of (200 ± 30) g for a supermarket to sell it. Would they sell an apple which weighs 145 g?

The minimum acceptable weight is $200 - 30 = 170$ g and the maximum acceptable weight is $200 + 30 = 230$ g. The apple weighs 145 g, which is less than 170 g so the supermarket would not sell it.

Exercise 1B

1 For each of the following tolerances, find the maximum and minimum values.
   a) (16 ± 2) cm   b) (11 ± 5) mm   c) (30 ± 16) g   d) $(76 \pm 0\cdot5)$ cm
   e) (240 ± 11) mm   f) $(450 \pm 2\cdot5)$ g   g) $(63 \pm 1\cdot5)$ cm   h) $(2\cdot5 \pm 0\cdot2)$ cm

2 Which lengths from 71 mm, 76 mm, 75 mm and 72 mm are within (73 ± 2) mm?

3 Which lengths from 2·4 cm, 3·6 cm, 3·3 cm and 2·9 cm are within $(3 \pm 0\cdot5)$ cm?

4 A machine produces parts which are used when assembling cars. Only parts with a length satisfying (24 ± 3) mm are accepted. A manager samples 10 parts and the lengths measured in millimetres are shown below. Write down all lengths which are accepted.

   25   24   30   26   24   23   25   29   20   22

5 To make furniture, a joiner requires wood to be cut to an accuracy of (420 ± 3) mm. She measures the following cuts in millimetres:

   417   416   424   420   421   415   422   423   412   422

   a) Which cuts should she reject?
   b) What fraction of cuts are acceptable?

6 Rectangular tiles are manufactured with a length of (580 ± 2) mm and breadth of (290 ± 2) mm.
   a) Find the maximum and minimum possible perimeters of a tile.
   b) Find the difference between the maximum and minimum possible areas of a tile.

7 Sana is a building inspector. She takes measurements to the nearest millimetre and classifies them as acceptable if they are (160 ± 6) mm and unacceptable if they are (148 ± 6) mm. Why is she wrong to do this?

To write a range of values using the tolerance notation $a \pm b$:

● Find half of the difference between the largest and smallest values. This gives you the value of b.

● Add the value for b onto the smallest value (or subtract from the largest one) to get the value of a.

● Write your final answer in the form $a \pm b$ and include brackets and units.

> ## Worked examples
>
> 1 All values within the range 20 mm to 35 mm are considered accurate measurements. Write this range of values using tolerance notation.
>
> Difference = 35 − 20 $b = 15 \div 2$ $a = 20 + 7.5$ The acceptable range of values is
>   = 15   = 7.5   = 27.5 (27.5 ± 7.5) mm
>
> 2 The following lengths of factory components were an acceptable length.
>
> 2.3 mm   4.2 mm   2.6 mm   3.9 mm   2 mm   4.1 mm
>
> Write the range of values shown using tolerance notation.
>
> Difference = 4.2 − 2 $b = 2.2 \div 2$ $a = 2 + 1.1$ The range of acceptable values is
>   = 2.2   = 1.1   = 3.1 (3.1 ± 1.1) mm

8 Write, using tolerance notation, the range of values from:

   a) 10 cm to 14 cm   b) 80 mm to 88 mm   c) 100 mm to 200 mm   d) 73 g to 85 g

   e) 1 mm to 2 mm   f) 12 mm to 15 mm   g) 42.0 g to 42.8 g   h) 6.3 km to 7.1 km

9 Karen accepts the following angle measurements from her design and manufacture pupils.

   28°   25°   27°   25°   24°   30°   27°   26°   28°   27°

   a) Write the range of values shown in tolerance notation.

   b) She decides she wants the measurements to be more consistent. Should she increase or decrease the tolerance?

10 Holes are drilled in a piece of wood and dowels are then inserted in the holes. Holes with the diameters, in millimetres, listed below were acceptable for the dowels to be inserted correctly. Write the acceptable hole diameters using tolerance notation.

   6.3   5.8   6.1   6.2   5.9   6.1   5.9   6.3   5.9

11 Peak flow is a measure of how quickly you can blow air out of your lungs. Simon's doctor asks him to keep a note of his peak flow each morning and night for 5 days. Here are his results in litres per minute:

   430   426   419   432   417   433   416   411   422   426

   a) Write Simon's range of peak flow measurements using tolerance notation.

   b) The following week, Simon's readings were within the tolerance (426 ± 9) litres per minute. What are the smallest and largest values that are within both tolerances?

12 A scientist records the following temperature changes, in degrees Celsius, from an initial temperature during an experiment:

   0.26   0.25   0.1   −0.15   −0.21   −0.34

   Write the range of values recorded using tolerance notation.

▶ Powers and roots

A number written in the form x^n is in index form. The number x is called the base and n the index, or power. For example, for 4^2, 4 is the base and 2 is the index (or power). The value of n tells you how many lots of x must be multiplied together. A number with an index of 2 is squared and an index of 3 is cubed. It is important that you know your square numbers up to 12^2, or even better up to 20^2!

Worked example

Evaluate:

a) 4^2

   $= 4 \times 4$

   $= 16$

b) $(-2)^5$ This is read as negative 2 to the power of 5.

   $= (-2) \times (-2) \times (-2) \times (-2) \times (-2)$

   $= -32$

c) 17^3 Don't be scared to show working.

   $= 17 \times 17 \times 17$

   $= 289 \times 17$

   $= 4913$

$$\begin{array}{r} 17 \\ \times\ 17 \\ \hline 119 \\ +\ 170 \\ \hline 289 \end{array} \qquad \begin{array}{r} 289 \\ \times\ 17 \\ \hline 2023 \\ +\ 2890 \\ \hline 4913 \end{array}$$

Exercise 1C

1. Evaluate:

  a) 7^2 b) 9^2 c) 6^2 d) 12^2 e) 18^2 f) 2^3

  g) 6^3 h) 15^3 i) 3^4 j) 5^4 k) 10^5 l) 2^6

2. Evaluate:

  a) $(-5)^2$ b) $(-8)^2$ c) $(-10)^2$ d) $(-11)^2$ e) $(-20)^2$ f) $(-2)^3$

  g) $(-3)^3$ h) $(-4)^3$ i) $(-10)^4$ j) $(-3)^5$ k) $(-10)^6$ l) $(-1)^7$

3. When the base is a negative number, what must be true about the power if the answer is

  a) positive   b) negative?

4. Evaluate:

  a) $16^2 + 19^2$ b) $14^2 - 2^4$ c) $(-10)^3 + 5^5 - 1^6$

A scientific calculator will have a squared button. It will also have one for other powers which may look like x^y or x^n. You may use a calculator for Q5–Q7.

5. Evaluate:

  a) 43^2 b) 91^2 c) 231^3 d) 46^5 e) 23^5 f) 13^8

6. In binary, one kilobyte is 2^{10} bytes. One megabyte is 2^{10} kilobytes. How many bytes are in one megabyte? Investigate on your calculator what this answer is in index form.

7. Three consecutive numbers multiplied together give 6840.

  a) Which power should you consider to help you to find the numbers quickly?

  b) What are the numbers?

Addition is the inverse operation to subtraction and multiplication is the inverse operation to division. The inverse operation to taking a power is taking a root. The **radical sign** $\sqrt{}$ is used to denote the positive square root (the inverse operation of squaring). For the root of any higher power, the corresponding number is written between the oblique lines in the radical sign ($\sqrt[3]{}, \sqrt[4]{}, \sqrt[5]{}$, etc.).

Worked examples

1 Find the square roots of 64.

To answer this think 'Which numbers give the answer 64 when squared?'

$8^2 = 64$ and $(-8)^2 = 64$ so the square roots of 64 are 8 and –8.

If the question is written as $\sqrt{64}$ then only the positive value is required, $\sqrt{64} = 8$.

2 Find the cube root of 216.

To answer this think 'Which number gives the answer 216 when cubed?' Consider 2^3, 3^3, 4^3, etc.

$6^3 = 216$ so $\sqrt[3]{216} = 6$. Notice the small 3 in the radical sign for the cube root. There is only one solution this time as $(-6)^3 = -216$.

Exercise 1D

1 Find:
   a) the square roots of 25   b) the square roots of 4   c) the square roots of 49
   d) $\sqrt{100}$ e) $\sqrt{81}$ f) $\sqrt{144}$

2 a) Is 10 a cube root of 1000? Explain your answer.
   b) Is –10 a cube root of 1000? Explain your answer.

3 5 is a fourth root of 625 since $5^4 = 625$. Write down a second fourth root of 625.

4 Copy and complete:
   a) $9^3 = 729$ so $\sqrt[3]{729} = ...$ b) $7^3 = 343$ so $\sqrt[3]{343} = ...$ c) $9^4 = 6561$ so $\sqrt[4]{6561} = ...$
   d) $3^2 = 9$ so $\sqrt{...} = 3$ e) $8^3 = 512$ so $\sqrt[3]{...} = 8$ f) $12^4 = 20736$ so $\sqrt[4]{20736} = 12$.

5 Find:
   a) $\sqrt[3]{8}$ b) $\sqrt[3]{125}$ c) $\sqrt[3]{64}$ d) $\sqrt[3]{27}$ e) $\sqrt[4]{16}$ f) $\sqrt[4]{10\,000}$

6 Determine whether the following statements are true or false.
   a) $\sqrt{36} + \sqrt{64} = \sqrt{100}$ b) $\sqrt{16} + 3^2 = 13$ c) $\sqrt{9} = 5^2 - 2^4$
   d) $5 = \sqrt{13^2 - 12^2}$ e) $\sqrt[3]{1000} = \sqrt{100}$ f) $\sqrt[3]{8} = \sqrt[6]{64}$

A scientific calculator will have a square root button. It will also have a button for other roots, and may look like $\sqrt[\square]{\square}$ or $\sqrt[y]{x}$. You may use a calculator for Q7 and Q8.

7 Find, correct to three significant figures:
   a) the square roots of 50   b) the square roots of 116   c) the fourth roots of 2140
   d) $\sqrt[3]{582}$ e) $\sqrt[4]{121}$ f) $\sqrt[5]{6}$ g) $\sqrt[6]{9251}$ h) $\sqrt[8]{87\,000}$ i) $\sqrt[10]{10}$

8 Four consecutive numbers multiplied together give 1 771 560.
   a) Which root should you consider to help you find the numbers quickly?
   b) What are the numbers?

▶ Order of operations

Think about 2 + 3 × 5. Is the answer 5 × 5 = 25 or 2 + 15 = 17?

Now consider   These add to give 17p so 2 + 3 × 5 = 17.

● It is important that calculations are completed in the correct order. The mnemonic **BIDMAS** can help:

Brackets	simplify inside any brackets first
Index	evaluate any index or root
Division	} division and multiplication have the same priority
Multiplication	
Addition	} addition and subtraction have the same priority
Subtraction	

An alternative form of BIDMAS which you may already be familiar with is BODMAS. All roots can be written using an index (for example, the square root is equivalent to an index of one half - try it on your calculator!) so roots are included with index.

Worked example Calculate:

a)
$30 - 40 \div 5 \rightarrow$ divide
$= 30 - 8 \qquad \rightarrow$ subtract
$= 22$

b)
$3 + 4 \times 9^2 \rightarrow$ index
$= 3 + 4 \times 81 \rightarrow$ multiply
$= 3 + 324 \rightarrow$ add
$= 327$

c)
$1 - \frac{1}{2}$ of $\left(\sqrt{144} - 10\right) \rightarrow$ brackets
$= 1 - \frac{1}{2}$ of $(12 - 10)$
$= 1 - \frac{1}{2} \times 2 \qquad \rightarrow$ multiply
$= 1 - 1 \qquad \rightarrow$ subtract
$= 0$

Exercise 1E

1 Calculate:

a) 6 + 4 × 9 b) 8 + 3 × 11 c) 12 − 4 × 2

d) 16 − 3 × 5 e) 12 + 9 ÷ 3 f) 21 − 49 ÷ 7

g) 9 + 3 × 7 − 6 h) 20 + 5 × 4 − 8 i) 30 + 30 ÷ 30 + 30

2 Simplify:

a) 100 − 9 × (2 + 7) b) (53 − 3) × 4 − 9 c) $6 - \frac{1}{5}$ of 100 + 25

d) 6 × (8 × 2 − 1) e) (2 + 3 × 4) ÷ (9 − 2) f) (3 + 4 × 25) − (8 × 7 + 6)

3 Calculate:

a) $8 + 5^2$

b) $10 - 3^2$

c) $6 + 2 \times 4^2$

d) $11^2 - 3 \times 8$

e) $10 + \sqrt{100} \div 2$

f) $(60 - 6^2) \div 6 - 2$

4 Simplify:

a) $18 \div (\sqrt{36} - 3)$

b) $1 + \dfrac{3}{5}$ of $(12 + \sqrt{9})$

c) $15 + \dfrac{1}{2}$ of $(\sqrt{25} + 3 \times 7)$

d) $\dfrac{2}{3}$ of $\left(30 - \dfrac{1}{\sqrt{4}} \text{ of } 18\right)$

e) $\sqrt{400} - 8^2 \div 4$

f) $\left(\dfrac{1}{5} + \dfrac{1}{5}\right)$ of $(12^2 - 4 \times \sqrt{121})$

g) $10^2 - 3 \times 5^2 + \sqrt[3]{8}$

h) $3^3 + 6^2 \div \sqrt[3]{64} - 1$

i) $\left(9^2 - \sqrt[3]{125} \times 2^4\right)^5$

5 Insert one set of brackets to make the following statements true. The first one is completed for you.

a) $(2 + 5) \times 8 - 2 + 3 = 57$ $2 + 5 \times 8 - 2 + 3 = 35$ $2 + 5 \times 8 - 2 + 3 = 37$

b) $6^2 - 3 - 2 \times 5 + 1 = 32$ $6^2 - 3 - 2 \times 5 + 1 = 22$ $6^2 - 3 - 2 \times 5 + 1 = 21$

c) $18 - \sqrt{100} \div 2 - 1 + 9 = 12$ $18 - \sqrt{100} \div 2 - 1 + 9 = 23$ $18 - \sqrt{100} \div 2 - 1 + 9 = 17$

d) $\dfrac{3}{4}$ of $4^3 - 2^3 + 15 - \sqrt{9} = 54$ $\dfrac{3}{4}$ of $4^3 - 2^3 + 15 - \sqrt{9} = 28$ $\dfrac{3}{4}$ of $4^3 - 2^3 + 15 - \sqrt{9} = 51$

When the question is a fraction, simplify the numerator and the denominator of the fraction using BODMAS/BIDMAS before completing the calculation. Remember to always simplify a fraction where possible.

Worked example

Calculate:

a)
$$\frac{6 + 2 \times 3^4}{10 - 3 \times 1} \quad \rightarrow \text{index}$$
$$= \frac{6 + 2 \times 81}{10 - 3 \times 1} \quad \rightarrow \text{multiply}$$
$$= \frac{6 + 162}{10 - 3} \quad \rightarrow \text{add and subtract}$$
$$= \frac{168}{7}$$
$$= 24$$

b)
$$\frac{6 - 50 \div 5}{10 - (-2)^2} \quad \rightarrow \text{index}$$
$$= \frac{6 - 50 \div 5}{10 - 4} \quad \rightarrow \text{divide}$$
$$= \frac{6 - 10}{10 - 4} \quad \rightarrow \text{subtract}$$
$$= \frac{-4}{6}$$
$$= -\frac{2}{3}$$

c)
$$\frac{2 + \sqrt{36} \times 3}{3 + \sqrt{49}} \quad \rightarrow \text{index}$$
$$= \frac{2 + 6 \times 3}{3 + 7} \quad \rightarrow \text{multiply}$$
$$= \frac{2 + 18}{3 + 7} \quad \rightarrow \text{add}$$
$$= \frac{20}{10}$$
$$= 2$$

---

6 Simplify:

a) $\dfrac{12 + 9 \times 8}{3^2 + 5}$

b) $\dfrac{40 \div (3 + 5)}{4^2 - 1}$

c) $\dfrac{8 - 10^2 \div 5}{1 + 6 \div 6}$

d) $\dfrac{1 + \sqrt{25} \times 7}{(1 + 5)^2}$

e) $\dfrac{1 + \sqrt[4]{81}}{2 \times (5 - 1^3)}$

f) $\dfrac{2 + \sqrt{49} \times \sqrt[3]{8} - 1}{\dfrac{2}{3} \text{ of } (10^2 - 4^3)}$

▶ Writing numbers using scientific notation

Scientific notation allows us to write and undertake calculations involving very large and very small numbers easily. Scientific notation is also known as standard form.

- Any number written using scientific notation is of the form $a \times 10^n$.
- The value of a must be greater than or equal to 1 and less than 10 and may be a decimal.
- The value of the index n corresponds to how many times you multiply or divide a by 10 to return to the original number.

Worked example

Write using scientific notation:

a) 8000 $(8 \times 10 \times 10 \times 10)$
   $= 8 \times 10^3$

b) 630 000 $(6.3 \times 10 \times 10 \times 10 \times 10 \times 10)$
   $= 6.3 \times 10^5$

c) 250 700 000 $(2.507 \times 10 \times 10 \times 10 \times 10 \times 10 \times 10 \times 10 \times 10)$
   $= 2.507 \times 10^8$

Exercise 1F

1   Write using scientific notation:

   a) 400   b) 70 000   c) 9000   d) 600 000
   e) 7500   f) 890   g) 62 000   h) 230 000
   i) 590 000   j) 14 000   k) 2900   l) 8 500 000

2   Look at your answers to **Q1**. What is the link between the number of digits after the first digit when the number is written in full and the size of the index when written using scientific notation?

3   The mass of an aeroplane is 43 000 kg. Write the mass of the aeroplane using scientific notation.

4   Write using scientific notation:

   a) 45 100   b) 723 000   c) 1 930 000   d) 231 000 000
   e) 850 600   f) 1 072 000   g) 376 010 000   h) 51 023 000

5   A well-known celebrity has approximately 198 000 000 followers on social media. Write this number using scientific notation.

6   Write using scientific notation:

   a) 1 million   b) 30 million   c) 420 million   d) 1 billion

7   One kilometre is equal to 100 000 centimetres. Write the following distances in centimetres using scientific notation:

   a) 6 km   b) 8.3 km   c) 91 km   d) 507 km

8   Alia receives £400 for every new build flat she sells from plan. If she sells all 320 flats within a new housing development, how much money will she make? Write your answer using scientific notation.

Now consider:

$$5000 = 5 \times 10 \times 10 \times 10 = 5 \times 10^3$$
$$500 = 5 \times 10 \times 10 = 5 \times 10^2$$
$$50 = 5 \times 10 = 5 \times 10^1$$

Every time we divide the starting number by 10, the index when written using scientific notation decreases by 1. Continuing this pattern gives

$$5 = 5 = 5 \times 10^0$$
$$0{\cdot}5 = 5 \div 10 = 5 \times 10^{-1}$$
$$0{\cdot}05 = 5 \div 10 \div 10 = 5 \times 10^{-2}$$
$$0{\cdot}005 = 5 \div 10 \div 10 \div 10 = 5 \times 10^{-3}$$

This demonstrates that to write a number between 0 and 1 using scientific notation, the index is negative.

Worked example   Write using scientific notation:

a) $0{\cdot}01$ $(1 \div 10 \div 10)$

   $= 1 \times 10^{-2}$

b) $0{\cdot}000\,76$ $(7{\cdot}6 \div 10 \div 10 \div 10 \div 10)$

   $= 7{\cdot}6 \times 10^{-4}$

c) $0{\cdot}000\,000\,003\,08$ $(3{\cdot}08 \div 10 \div 10 \div 10 \div 10 \div 10 \div 10 \div 10 \div 10)$

   $= 3{\cdot}08 \times 10^{-9}$

Exercise 1G

1 Write using scientific notation:

   a) $0{\cdot}06$
   b) $0{\cdot}007$
   c) $0{\cdot}00009$
   d) $0{\cdot}0002$

   e) $0{\cdot}0032$
   f) $0{\cdot}041$
   g) $0{\cdot}00067$
   h) $0{\cdot}000\,008\,5$

   i) $0{\cdot}000\,99$
   j) $0{\cdot}0053$
   k) $0{\cdot}000\,008$
   l) $0{\cdot}000\,000\,021$

2 Look at your answers to **Q1**. What is the link between the number of zeros at the start of the number when written in full and the size of the index when written using scientific notation?

3 Write using scientific notation:

   a) $0{\cdot}0153$
   b) $0{\cdot}006\,27$
   c) $0{\cdot}000\,053\,4$
   d) $0{\cdot}000\,848$

   e) $0{\cdot}000\,162$
   f) $0{\cdot}000\,007\,19$
   g) $0{\cdot}003\,05$
   h) $0{\cdot}000\,000\,006\,407$

4 Billy's attempt at writing two numbers using scientific notation is shown below:

   a) $0{\cdot}0062 = 6{\cdot}2 \times 10^3$
   b) $0{\cdot}000\,0309 = 3{\cdot}9 \times 10^{-5}$ ✗

   Billy has made two mistakes. Explain what her two mistakes are.

5 For each question, complete the calculation and write your answer using scientific notation.

   a) A grain of rice has mass $0{\cdot}028\,g$. What is the mass of 5 grains of rice?

   b) A teaspoon holds $0{\cdot}005$ litres of liquid. How many litres will 12 teaspoons hold?

   c) A human hair is $0{\cdot}000\,038\,4\,m$ thick. Find the width in metres of 6 hair strands.

   d) The mass of a Scottish midgie is $0{\cdot}000\,125\,g$. Calculate the mass of 70 Scottish midgies.

   e) Convert $0{\cdot}4\,mm$ into kilometres.

   f) Convert $1\,g$ into tonnes (recall that 1 tonne is equal to $1000\,kg$).

▶ Scientific notation to normal form

To write a number of the form $a \times 10^n$ in full:

- When n is positive, multiply a by 10 n times. When a is a decimal it is sometimes easier to split the multiplication into two steps: the first to multiply until there is no longer a decimal and then the second to complete the multiplication.

- When n is negative, divide a by 10 n times. You should find the number of zeros at the start of the decimal answer matches the size of the power.

Worked example

Write in full:

a) $8 \cdot 105 \times 10^7$

$= 8 \cdot 105 \times 10 \times 10 \times 10 \times 10 \times 10 \times 10 \times 10$

$= 8105 \times 10 \times 10 \times 10 \times 10$

$= 81\,050\,000$

b) $1 \cdot 45 \times 10^{-4}$

$= 1 \cdot 45 \div 10 \div 10 \div 10 \div 10$

$= 0 \cdot 000\,145$

With practice, you should be able to go straight to the final answer.

Exercise 1H

1   Write in full:

   a) 6×10^2 b) 7×10^3 c) $3 \cdot 5 \times 10^5$ d) $8 \cdot 1 \times 10^4$

   e) $4 \cdot 39 \times 10^3$ f) $6 \cdot 82 \times 10^5$ g) $7 \cdot 04 \times 10^6$ h) $9 \cdot 538 \times 10^7$

2   The distance between the Moon and the Earth is approximately $3 \cdot 844 \times 10^5$ km. Write this distance in full.

3   Write in full:

   a) 8×10^{-2} b) 2×10^{-3} c) $4 \cdot 1 \times 10^{-1}$ d) $7 \cdot 9 \times 10^{-4}$

   e) $5 \cdot 62 \times 10^{-3}$ f) $8 \cdot 35 \times 10^{-5}$ g) $2 \cdot 07 \times 10^{-4}$ h) $9 \cdot 34 \times 10^{-6}$

4   A government published that they spent £$1 \cdot 065 \times 10^8$ on education last year.

   a)   Write this number out in full.

   b)   Round your answer in a) to two significant figures.

5   The average growth rate for a child between the age of 2 and 3 is $2 \cdot 44 \times 10^{-2}$ cm per day. Write this number in full.

6   A cell membrane has a thickness of $7 \cdot 3 \times 10^{-9}$ m.

   a)   Write this number in full.

   b)   Convert your answer in a) to cm and write your answer using scientific notation.

   c)   Write down the thickness of the cell membrane in mm using scientific notation.

7   The wavelength interval of violet light is approximately $(4 \cdot 15 \pm 0 \cdot 35) \times 10^{-7}$ m. Write the minimum and maximum wavelength for violet light in full.

► Scientific notation on a calculator

Numbers in scientific notation can be entered into a scientific calculator. The button to use on your calculator may look like $\times 10^x$ or EXP.

Worked example

A factory produces $2\cdot15 \times 10^6$ tins of soup every day. How many tins are produced in 9 days? Write your answer in scientific notation correct to three significant figures.

$2\cdot15 \times 10^6 \times 9$ To enter $2\cdot15 \times 10^6$ on your calculator enter $2\cdot15$ followed by the

$= 19\,350\,000$ $\times 10^x$ button (or EXP button) and then 6.

$= 19\,400\,000$ to 3 s.f.

$= 1\cdot94 \times 10^7$ tins.

Exercise 1I

Throughout this exercise, write your answers using scientific notation correct to three significant figures.

1   Calculate:

a) $9\cdot6 \times 10^7 + 5\cdot8 \times 10^6$ b) $4\cdot5 \times 10^9 - 7\cdot86 \times 10^8$

c) $8\cdot01 \times 10^{-12} + 2\cdot13 \times 10^{-13}$ d) $3\cdot2 \times 10^{-11} \times 7$

e) $(7\cdot34 \times 10^{14}) \times (4\cdot02 \times 10^5)$ f) $(5\cdot39 \times 10^6) \div (6\cdot08 \times 10^{-6})$

2   The gross domestic product (GDP) of a country is a measure of its economy. The GDP of three countries at a given time is shown in the table. Calculate:

Country	GDP (US Dollars)
Canada	$\$1\cdot65 \times 10^{12}$
Netherlands	$\$8\cdot31 \times 10^{11}$
Cyprus	$\$2\cdot21 \times 10^{10}$

a)   the combined GDP of the three countries

b)   the difference in GDP between Canada and Cyprus

c)   how many times greater the GDP of Canada is compared to the GDP of the Netherlands. Write your answer to three significant figures but there is no need for scientific notation.

3   The length of a bacterium is $1\cdot107 \times 10^{-6}$ m. Find the length of:

a)   35 bacteria   b)   half a bacterium.

4   The number of times a website was visited in three months is shown in the table.

Month	Number of visits
November	$1\cdot89 \times 10^9$
December	$2\cdot05 \times 10^9$
January	$9\cdot97 \times 10^8$

a)   How many times was the website visited in total over the three months?

b)   Write the months in order from least to greatest number of visits.

c)   What was the difference between the greatest and least number of visits?

5   One astronomical unit (au) is approximately $1\cdot496 \times 10^{11}$ metres. If Uranus is $19\cdot19$ au from the Sun, how far in metres is it from Uranus to the Sun?

▶ Direct proportion

When quantities are in direct proportion they scale down and up identically. If one goes down, so does the other and if one goes up, so does the other.

- Set the information down under appropriate headings.
- Scale down by dividing each side by the same amount.
- Scale back up by multiplying each side by the same amount.

In this section, apart from Q5, assume the quantities described are in direct proportion.

Worked examples

1 Seven identical books cost £90·65. How much will twelve of these books cost?

Number of books	Cost (£)
7	90·65
1	12·95
12	155·40

÷7 ... ÷7
×12 ... ×12

Twelve books will cost £155·40.

2 Alex paints 6 sections of a fence in 39 minutes. How long will it take her to paint 10 sections of fence?

Number of sections	Time (mins)
6	39
2	13
10	65

÷3 ... ÷3
×5 ... ×5

It will take Alex 65 minutes to paint 10 sections.

Exercise 1J

1 Set down and complete these direct proportion problems.

a)

Number of games	Cost (£)
8	152
1	
5	

b)

Time (h)	Cost (£)
5	185
1	
12	

c)

Episodes	Time (mins)
4	28
2	
10	

2 Five train tickets cost £15. How much will 11 identical tickets cost?

3 William gets his hair cut once a month. He spends £59·50 on haircuts in 7 months. How much will William spend on haircuts in one year?

4 It takes Zena 10 minutes to run 4 lengths of a track. How long will it take her to run 18 lengths if she continues at the same speed?

5 Are the following quantities in direct proportion? Show working to explain your answer.

a) At a concert with 2000 people, 40 t-shirts were purchased. At a concert with 3000 people, 50 t-shirts were purchased.

b) Three loaves cost £2·70. Five loaves cost £4·25.

c) Nine dog walks cost £135. Seven dog walks cost £105.

d) On a bus with 20 passengers, 4 had blue eyes. On a bus with 55 passengers, 10 had blue eyes.

e) Three bookshelves have 129 books on them. Eight bookshelves have 354 books on them.

▶ Inverse proportion

When quantities are in inverse proportion they scale up and down inversely. This means that if one increases, the other decreases.

● Set the information down under appropriate headings.

● Divide one side and multiply the other by the same amount.

● Now multiply one side and divide the other by the same amount (see Worked example below).

In this section, assume the quantities described are in either direct or inverse proportion.

Worked example

It takes 6 people 40 minutes to set out seating for a concert. How long will it take 8 people to set out the same seating?

Number of people	Time (mins)
6	40
1	240
8	30

÷6, ×6, ×8, ÷8

The product of the quantities stays the same.

It will take 240 minutes to set out the seating, no matter how it is shared.

It will take 8 people 30 minutes to set out the seating.

Exercise 1K

1 Five people take 4 hours to paint a building. How long would it take 2 people to paint the same building?

2 A crate contains 25 boxes and each box has 40 jotters in it. Bigger boxes, which hold 50 jotters each, are introduced. How many of the bigger boxes can be filled from one crate?

3 A bag of dog feed will feed 5 dogs for two weeks. How many days will the feed last if there are 2 extra dogs?

4 Hamish makes 240 ml of fruit punch for each of his 6 party guests. An extra 4 guests turn up unexpectedly. How much fruit punch will they get each?

5 A hire-purchase agreement requires 18 monthly payments of £45 per month. How much would have to be paid per month if the same total amount was repaid in one year?

Answer the following questions after you decide whether they are in direct or inverse proportion.

6 Six people take five hours to build a wall. How long will it take six people to build 8 walls?

7 Nine tins of paint cover 45 square metres of canvas. What area of canvas will 7 tins cover?

8 A container fills thirty 250 ml bottles. How many 300 ml bottles can be filled from the container?

9 A machine working at 80% efficiency prints 480 papers. How many will it print at 90% efficiency?

10 The distance between two landmarks on a map with a scale of 1 cm : 25 km is 8 cm. How far apart will the two landmarks be on map with a scale of 1 cm : 10 km?

11 Eight boxes hold 7 bottles each and each bottle has a volume of 500 ml. How many boxes can be filled if the same total amount of liquid fills 400 ml bottles and 10 of these bottles fit in a box?

Check-up 👍

1 Round to one significant figure:

    a) 5800   b) 7·42   c) 0·0031   d) 0·00098

2 The annual turnover of a large company is approximately £6 435 000. Round this number to one significant figure.

3 Round to two significant figures:

    a) 36 700   b) 43·91   c) 8·437   d) 0·001934

4 A hotel recorded 58 607 visitors staying with them in one year. Round the number of visitors to two significant figures.

5 Round to three significant figures:

    a) 62 790   b) 18·987   c) 1·005   d) 0·009 121

6 Find the maximum and minimum values within these ranges:

    a) $(34 \pm 3)\,\text{mm}$ b) $(150 \pm 4)\,\text{cm}$ c) $(185 \pm 2)°$ d) $(284 \pm 19)\,\text{mm}$

7 On a market stall, pears are sorted into small, medium and large categories using the weights shown in the table.

    The stall holder has pears with the following weights, in grams:

Category	Weight range (g)
Small	129 ± 10
Medium	170 ± 30
Large	221 ± 20

    160   190   120   193   200   204
    234   133   199   176   136

    a) Sort the pears into the categories small, medium and large.

    b) The stall holder considers changing their small weight range to be $(130 \pm 10)\,\text{g}$ and keeping the other ranges the same. They always know the weight of each pear to the nearest gram. Explain why the stall holder should not do this.

8 Ava measured the length of six kitchen work surfaces ready for installation. Her measurements in millimetres were:

        713   726   734   725   730   712

    Write the range of values using tolerance notation.

9 Evaluate:

    a) 7^2 b) $(-4)^2$ c) $(-5)^3$ d) $3^4 - 2^6$

10 Find:

    a) $\sqrt{16}$ b) $\sqrt[3]{1000}$ c) $\sqrt[5]{243}$ d) $\sqrt{5^2 + 12^2}$

11 Use a calculator to approximate the following roots to three significant figures.

    a) $\sqrt{200}$ b) $\sqrt[3]{76}$ c) $\sqrt[4]{926}$ d) $\sqrt[5]{8}$

12 Calculate:

    a) $25 + 3 \times 6^2$ b) $18 - 6 \times \sqrt{4} + 5$ c) $3^2 \times (10 - \sqrt{16}) + 4^3$ d) $\dfrac{32 - 16 \div 2^3}{2^4 - 1}$

13 Write using scientific notation:

    a) 57 000   b) 6 490 000   c) 70 800 000   d) 914 300 000

                                                           →

14 There are 604 800 seconds in a week. Write this number using scientific notation.

15 Write using scientific notation:

a) 0·04 b) 0·0037 c) 0·000 009 12 d) 0·000 050 3

16 A spider's silk is measured to be 0·000 007 m wide. Write this number using scientific notation.

17 Write in full:

a) 9×10^3 b) $8·6 \times 10^5$ c) 3×10^{-4} d) $2·05 \times 10^{-6}$

18 The thickness of the central area in a human cornea is $5·6 \times 10^{-4}$ m. Write this thickness in full.

19 Dana attempts to write a number using scientific notation. She writes $52·1 \times 10^{-8}$.

a)   What mistake has Dana made when attempting to use scientific notation?

b)   Write Dana's number correctly using scientific notation.

You may use a calculator for Q20–Q23.

20 Find, writing your answer using scientific notation correct to three significant figures:

a) $2·9 \times 10^{10} + 6·75 \times 10^{11}$ b) $(8·2 \times 10^{-12}) \times (1·8 \times 10^{-13})$

c) $8·59 \times 10^8 - 9·17 \times 10^7$ d) $(6·3 \times 10^{-9}) \div (5·07 \times 10^{-5})$

21 The depth of a sheet of paper is 5×10^{-3} cm. What depth would 45 sheets of the same paper have? Write your answer in full.

22 A famous Parisian museum has an average of $1·02 \times 10^7$ visitors every year. If this continues, how many people will visit the museum over a 15-year period? Write your answer using scientific notation.

23 The table below shows the approximate distance from the Sun and the mass of four planets.

Planet	Distance from the Sun (km)	Mass (tonnes)
Mercury	$5·79 \times 10^7$	$3·30 \times 10^{20}$
Venus	$1·08 \times 10^8$	$4·87 \times 10^{21}$
Earth	$1·52 \times 10^8$	$5·97 \times 10^{21}$
Mars	$2·28 \times 10^8$	$6·42 \times 10^{20}$

a)   How many significant figures has each number been written to?

b)   Write the planets in order from the lightest to the heaviest in terms of their mass.

c)   How much further is Mars from the Sun than Mercury? Write your answer using scientific notation.

d)   How many times heavier is Earth than Mercury? Give your answer to two significant figures.

e)   One kilometre is approximately 0·621 miles. How far is Venus from the Sun in miles? Give your answer in scientific notation correct to two significant figures.

24 The price of a salad box filled at a self-service counter is directly proportional to the weight of the filled box. If a 100 g salad box costs £1·20, how much will a 240 g box cost?

25 The number of tacks required to fit a carpet is directly proportional to the length of the carpet. If a 3·2 m carpet requires 16 tacks, how many tacks will be needed for a 3·8 m carpet?

26 It takes 40 minutes to fill a small swimming pool using 3 identical hoses. How long will it take to fill the pool if there are 5 of the hoses?

27 Every year a company gives their staff a bonus. One year, 27 staff receive £200 each. If the same total amount is shared between 30 staff the following year, how much will they each receive?

2 Fractions, decimals and percentages

▶ Multiplying and dividing decimals

Multiplication	Digits after the decimal point(s) in the question		Digits after the decimal point in the answer	
3 × 7 = 21	None		None	
0·3 × 7 = 2·1	One	0·3	One	2·1
0·3 × 0·7 = 0·21	Two	0·3 × 0·7	Two	0·21
0·03 × 0·7 = 0·021	Three	0·03 × 0·7	Three	0·021

The digits 2 and 1 are common to every answer but their place value changes as the place value in the question changes. Consider 0·3 × 0·7. This is 3 ÷ 10 × 7 ÷ 10 which is 21 ÷ 100 = 0·21. Dividing a whole number by 100 gives two digits after the decimal point. To multiply decimals together:

● Multiply the digits, ignoring any decimal points.

● Consider the place value of your answer. It will have the same number of digits after the decimal point as the two original numbers combined.

Worked example   Calculate:

a) 2·4 × 3

= 7·2

$$\begin{array}{r} 24 \\ \times\ 3 \\ \hline 72 \end{array}$$

b) 6·7 × 0·9

= 6·03

$$\begin{array}{r} 67 \\ \times\ 9 \\ \hline 603 \end{array}$$

c) 8·3 × 0·26

= 2·158

$$\begin{array}{r} 83 \\ \times\ 26 \\ \hline 498 \\ +\ 1660 \\ \hline 2158 \end{array}$$

d) 0·05 × 0·42

= 0·0210

= 0·021

$$\begin{array}{r} 42 \\ \times\ 5 \\ \hline 210 \end{array}$$

As 5 × 42 = 210 we count the last zero as one of the digits after the decimal point. A zero is also required in the tenths column to have four digits in total. Once the place value is correct, there is no need to keep the final zero.

Exercise 2A

1 Calculate:
   a) 0·6 × 8
   b) 0·9 × 3
   c) 0·4 × 0·7
   d) 0·5 × 0·5
   e) 0·2 × 0·6
   f) 0·23 × 0·5
   g) 0·52 × 0·8
   h) 0·7 × 0·41
   i) 0·437 × 0·2
   j) 0·527 × 0·3
   k) 0·03 × 0·08
   l) 0·64 × 0·05

2 Find:
   a) 1·3 × 6
   b) 2·9 × 4
   c) 3·7 × 0·2
   d) 1·8 × 2·9
   e) 4·5 × 6·3
   f) 10·3 × 0·92
   g) 0·51 × 30·8
   h) 0·71 × 9·23

3 Calculate, writing your answer using scientific notation:
   a) 0·002 × 0·004
   b) 0·000 09 × 0·005
   c) 0·000 06 × 0·0003

4 Simplify:
   a) 2 + 0·5 × 0·6
   b) 10 − 0·4 × 0·03
   c) 5 + 0·7 × 0·8 + 1
   d) $\frac{2}{3}$ of (3 + 0·02 × 0·9)
   e) 5·9 + 0·5²
   f) 1·3² − √0·16

How many 50p pieces are there in £6? There are 2 in £1 so there are 12 in £6. If we set this up as a calculation in pounds, it would be $\dfrac{6}{0\cdot5}$. We can make this calculation easier by using equivalent fractions:

$\dfrac{6}{0\cdot5}^{\times10}_{\times10} = \dfrac{60}{5}$. The division is now straightforward, $60 \div 5 = 12$. We also have $\dfrac{6}{0\cdot5}^{\times2}_{\times2} = \dfrac{12}{1} = 12$ but often

multiplying by a multiple of 10 is easier. When dividing decimals:

- Write the question as a fraction.
- Multiply to obtain an equivalent fraction **without a decimal on the denominator.**
- Complete the division from the equivalent fraction – this is your final answer, there is no need to consider the number of digits after the decimal point when dividing.

Worked examples

1 Calculate:

a) $27 \div 0\cdot3$

$$\dfrac{27}{0\cdot3}^{\times10}_{\times10} = \dfrac{270}{3}$$

$$3\overline{)270} = 90$$

$$27 \div 0\cdot3 = 90$$

b) $0\cdot009 \div 0\cdot02$

$$\dfrac{0\cdot009}{0\cdot02}^{\times100}_{\times100} = \dfrac{0\cdot9}{2}$$

$$2\overline{)0\cdot90} = 0\cdot45$$

$$0\cdot009 \div 0\cdot02 = 0\cdot45$$

c) $1\cdot4605 \div 0\cdot005$

$$\dfrac{1\cdot4605}{0\cdot005}^{\times1000}_{\times1000} = \dfrac{1460\cdot5}{5}$$

$$5\overline{)1460\cdot5} = 292\cdot1$$

$$1\cdot4605 \div 0\cdot005 = 292\cdot1$$

2 How many full 0·6 m lengths can be cut from a 2·8 m piece of wire?

$$\dfrac{2\cdot8}{0\cdot6}^{\times10}_{\times10} = \dfrac{28}{6}$$

$$6\overline{)28\cdot{}^40{}^40\ldots} = 4\cdot6\,6\ldots$$

Four full 0·6 m lengths can be cut from a 2·8 m piece of wire.

---

5 Calculate:

a) $8 \div 0\cdot2$ b) $80 \div 0\cdot5$ c) $7\cdot2 \div 0\cdot3$ d) $9\cdot6 \div 0\cdot6$

e) $0\cdot16 \div 0\cdot8$ f) $2\cdot35 \div 0\cdot5$ g) $0\cdot09 \div 0\cdot3$ h) $0\cdot06 \div 0\cdot4$

6 Find:

a) $6 \div 0\cdot04$ b) $12 \div 0\cdot03$ c) $16\cdot2 \div 0\cdot06$ d) $2\cdot61 \div 0\cdot05$

e) $18 \div 0\cdot009$ f) $0\cdot45 \div 0\cdot005$ g) $0\cdot1004 \div 0\cdot002$ h) $0\cdot0207 \div 0\cdot004$

7 How many full 0·4 m lengths can be cut from a 34 m piece of wire?

8 To sew a dress hem it takes 0·9 m of thread. How many full hems can be sewn with 50 m of thread?

9 Complete each calculation and then match those which have the same value when simplified.

a) $1\cdot2 \div 0\cdot4$ $0\cdot96 \div 0\cdot03$ $4\cdot9 \div 0\cdot7$ $6\cdot4 \div 0\cdot2$ $2\cdot7 \div 0\cdot9$ $0\cdot42 \div 0\cdot06$

b) $0\cdot6 \times 0\cdot3$ $0\cdot15 \times 0\cdot8$ $0\cdot02 \times 0\cdot05$ $0\cdot0007 \div 0\cdot7$ $0\cdot06 \div 0\cdot5$ $0\cdot072 \div 0\cdot4$

10 Calculate:

a) $2\cdot3 + 0\cdot4 \times 0\cdot5$ b) $0\cdot95 - 0\cdot7 \times 0\cdot6$ c) $3\cdot2 \times 1\cdot8 + 4\cdot29 \div 0\cdot03$

▶ Expressing one amount as a percentage of another

When given two amounts, A and B, it can be useful to express one as a percentage of the other.
To express A as a percentage of B:

● Construct the fraction $\dfrac{A}{B}$ and then convert to a percentage by either finding an equivalent fraction with a denominator of 100 or multiplying by 100%.

Worked examples

1 Brass is made by combining copper and zinc. 40 g of brass contains 32 g of copper.

What percentage of this brass is copper?

$$\dfrac{32}{40}{\scriptstyle\stackrel{\div4}{\div4}} = \dfrac{8}{10}{\scriptstyle\stackrel{\times10}{\times10}} = \dfrac{80}{100} = 80\%$$

2 Nivedita scores 76 out of 89 in a test. What was her percentage score to the nearest percent?

$$\dfrac{76}{89} \times 100\% = 85 \cdot 3...\%$$

$$= 85\% \text{ to the nearest percent}$$

Exercise 2B

1 Andrea scored 19 out of 25 in a test. What was her score as a percentage?

2 From a year group of 80 pupils, 68 said mathematics was their favourite subject.

What percentage of pupils said mathematics was their favourite subject?

3 Magnalium is a metal alloy comprising two metals: aluminium and magnesium. It is popular with engineers as it is very strong and has a high resistance to corrosion. A 5 kg block of magnalium contains 4750 g of aluminium. Find the percentage of this magnalium which is aluminium.

4 The table below shows the recommended daily allowance (RDA) of calories, total fat, carbohydrate and protein for an adult's daily food intake.

	Calories (kCal)	Carbohydrate (g)	Total fat (g)	Protein (g)
RDA	2000	At least 260	Less than 70	50

Here is some nutritional information for popular foods:

Selecting the calories for a banana (121) and the total RDA of calories (2000) we find that a banana provides $\dfrac{121}{2000} \times 100\% = 6 \cdot 05\%$ of the RDA of calories.

Food	Calories (kCal)	Carbohydrate (g)	Total Fat (g)	Protein (g)
Banana	121	31	0·5	1·4
Egg	71	0·4	5	6
Pasta (100 g)	130	25	1	5
Bagel	360	80	2	14
Cashew nuts (100 g)	553	30	44	18

a) Calculate, to two decimal places where necessary, the percentage of the RDA each food type contains for calories, carbohydrate, total fat and protein.

b) Design a food label for one of the foods from the table, displaying the percentages calculated in a).

▶ Percentage increase and decrease

We can use a decimal multiplier to summarise the effect of a percentage increase (or decrease). This will make our calculations more efficient. To do this:

- Consider the original amount to be 100%. To increase an amount, add to 100% to find the new percentage. To decrease, subtract from 100% to find the new percentage.

- Convert the new percentage into a decimal multiplier by dividing by 100.

- Multiply to carry out the calculation.

Worked examples

1   Increase 15 kg by 2·5%
    New percentage = 100% + 2·5%
                   = 102·5%

    multiplier = 102·5 ÷ 100
               = 1·025

    15 × 1·025 = 15·375 kg

2   Decrease 1200 ml by 18%
    New percentage = 100% − 18%
                   = 82%

    multiplier = 82 ÷ 100
               = 0·82

    1200 × 0·82 = 984 ml

Exercise 2C

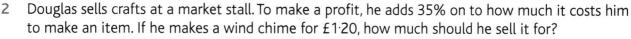

1   Create a decimal multiplier and increase:

   a)   £200 by 20%   b)   700 km by 8%   c)   800 miles by 67%
   d)   £325 by 19%   e)   62 m by 8·5%   f)   £14 by 2·5%

2   Douglas sells crafts at a market stall. To make a profit, he adds 35% on to how much it costs him to make an item. If he makes a wind chime for £1·20, how much should he sell it for?

3   A research company currently pays £250 000 per year for raw materials. Due to expansion, their next order will increase by 14%. How much will their next order cost?

4   Create a decimal multiplier and decrease:

   a)   £600 by 40%   b)   300 mg by 75%   c)   900 km by 23%
   d)   6 kg by 9%   e)   450 km by 88%   f)   £10 by 16·8%

5   In January, Caleb makes £497 commission. In February, he makes 18% less commission. How much commission does he make in February?

6   Layla's mortgage will decrease by 0·5% next month. If she currently pays £725 a month for her mortgage, how much will her next monthly payment be?

7   A house is purchased for £135 500. After one year, the house increases in value by 6% and in the second year it increases again by 8%. In the third year, the house falls in value by 14%.

   a)   What is the value of the house after:

       i)   one year   ii)   two years   iii)   three years?

   b)   Explain why the house is not worth its original value after three years.

▶ Repeated percentage calculations

When the **same** percentage change is repeatedly applied we can calculate the final amount quickly rather than complete separate calculations. To do this:

- Find the new percentage by adding to or subtracting from 100%.
- Convert the new percentage into a decimal multiplier by dividing by 100.
- Use the formula

$$\text{final amount} = \text{initial amount} \times (\text{decimal multiplier})^n,$$

where the initial amount is what you start with and n is the number of times the percentage is applied (see Worked examples below).

This technique is used to calculate compound interest in savings accounts. If you save £100 in an account paying 3% p.a. (per annum or year) you will have £103 at the end of the first year. The next year, assuming no money is withdrawn, you will earn 3% of £103 = £3·09 and you will have £106·09 in total. You earn interest on your interest! Over time the interest accrued (earned) grows faster and faster. Compound interest is a very important technique in finance: it makes savings grow and can make debt (money owed) grow too.

Worked examples

1 £100 is placed in a new savings account with an interest rate of 2% p.a.
How much is in the account after 3 years?

New % = 100% + 2%   Year 1 balance: 100 × 1·02 = £102

   = 102%   Year 2 balance: 102 × 1·02 = £104·04

multiplier = 1·02   Year 3 balance: 104·04 × 1·02 = £106·12

To obtain the balance after three years we have multiplied by $1·02 \times 1·02 \times 1·02 = 1·02^3$. We can do the calculation more efficiently using the formula:

$$\text{final amount} = \text{initial amount} \times (\text{decimal multiplier})^n$$

$$= 100 \times 1·02^3 \quad \leftarrow n = 3 \text{ for three years}$$

$$= 106·120...$$

$$= £106·12 \qquad \leftarrow \text{money, so final answer to 2 decimal places}$$

2 £500 is placed in a new savings account with an interest rate of 3·1% p.a.
How much interest is accrued after 5 years?

New % = 100% + 3·1%   final amount = $500 \times 1·031^5$

   = 103·1%   = 582·456...

   = 1·031   = £582·46

Interest accrued after 5 years is
582·46 − 500 = £82·46.

Questions using this technique can include many contexts, including profit and loss, appreciation and depreciation. If something appreciates in value its value increases. If it depreciates, its value decreases. ➡

3   A new car depreciates by 1·3% every month for 6 months.

If the car originally cost £24 000, how much is it worth at the end of 6 months?

New % = 100% − 1·3%   final amount = 24 000 × 0·987^6 The car is worth £22 187·80

= 98·7%   = 22 187·795...   at the end of 6 months.

= 0·987   = £22 187·80

Exercise 2D

1   Calculate how much money is in each savings account at the end of the term stated if no money is withdrawn:

	Amount initially deposited (£)	Interest rate (% per annum)	Term (years)
a)	200	4	3
b)	700	2	4
c)	1500	3	5
d)	2000	5	3
e)	2948	6	4
f)	3450	2·8	3
g)	5000	3·9	4

2   For each account in **Q1**, calculate the amount of interest accrued over the full term.

3   A caravan which cost £25 000 depreciates in value at a rate of 7% per annum. How much is the caravan worth after 4 years?

4   A motorbike which cost £8000 depreciates in value at a rate of 1·5% per month.
How much is the motorbike worth after 8 months?

5   Farm machinery is estimated to depreciate at a rate of 14% per annum. How much would £300 000 worth of machinery be worth after 3 years?

6   The number of bacteria in a petri dish increases at 12% per minute. If the initial sample contained 224 bacteria, how many are present after half an hour?

7   Scientists have been studying the reproduction rates of a breed of insect. They have found that the insect population increases on average by 10% every 5 days. If there are 260 insects initially, how many will there be after 15 days?

8   The global population is expected to increase by 1·1% annually.
If the population is approximately 7·3 billion, to two significant figures, how many people will there be in:

a)   12 years time   b)   15 years time?

9   Fraser has just started his first job and his annual salary is £28 000. His contract states that he will receive a 2·5% pay rise every year for the first 5 years and then a 4·1% pay rise every year thereafter to a maximum annual salary not exceeding £55 000.

a)   What is Fraser's annual salary after:

i)   5 years   ii)   8 years?

b)   After how many years from starting his job will Fraser no longer get a pay rise?

10 A car is expected to depreciate in value by 6% each year. At the end of how many years will the car first fall below 75% of its original value?

▶ Finding the percentage change

If £200 increases to £220 this is an increase of 10% from the original amount of £200. We can calculate the percentage change between two values by firstly considering the fractional change from the original amount. Here, we have a change of 20 from the original amount of 200, or $\dfrac{20}{200}$.

Converting this to a percentage we get $\dfrac{20}{200} \times 100\% = 10\%$, as expected. When calculating a percentage change we find the size of the change (i.e. the positive value of the change) and use:

- $\dfrac{\text{change}}{\text{original amount}} \times 100\%$.

Communicate whether the change is an increase or a decrease in words.

Worked examples

1   Calculate as a percentage of the original amount an increase from £160 to £216.

The original amount is £160.

The change is 216 − 160 = 56.

$$\dfrac{\text{change}}{\text{original amount}} \times 100\%$$

$$= \dfrac{56}{160} \times 100\%$$

$$= 35\%$$

An increase from £160 to £216 is an increase of 35%.

2   Daniel monitors his blood glucose level. At noon, his blood glucose level was 6 mmol/l and at 2 pm his reading was 4·75 mmol/l. Calculate Daniel's percentage decrease in blood glucose over the two readings to three significant figures.

The original amount is 6 mmol/l

The change is 6 − 4·75

   = 1·25 mmol/l

$$\dfrac{\text{change}}{\text{original amount}} \times 100\%$$

$$= \dfrac{1·25}{6} \times 100\%$$

$$= 20·8\dot{3}\%$$

$$= 20·8\% \quad \text{to 3 s.f.}$$

Daniel's blood glucose fell by 20·8% over the two-hour period.

Exercise 2E

Throughout this exercise, give your answers to three significant figures where necessary.

1   Calculate the increase as a percentage of the original amount for each of the following:

   a) £10 to £12   b) 600 ml to 900 ml   c) 35 g to 49 g   d) 16 cm to 20 cm
   e) 9 kg to 10 kg   f) 60 mm to 94 mm   g) £4·50 to £6·20   h) 9·3 m to 9·8 m

2   A train fare has increased from £4·25 to £5·10.

   What is the increase in the fare as a percentage of the original cost?

3   A shop owner buys rolls from the baker at 20p each and sells them for 42p each.

   Calculate the shop owner's profit per roll as a percentage of her buying price.

➡

4   A painting increases in value from £800 to £890. Calculate the increase in value as a percentage of the original value.

5   Calculate the percentage decrease for each of the following:

   a) £100 to £90   b) 400 g to 180 g   c) 75 kg to 45 kg   d) 80 cm to 60 cm

   e) 250 km to 182 km   f) 99p to 82p   g) 20 m to 7 m   h) 615 ml to 483 ml

6   Alain bought a tenor horn for £520. He didn't enjoy playing it, so he sold it for £400. Calculate Alain's loss as a percentage of the original price.

7   A jacket normally costs £75. In a sale, it is advertised for £50.

   Calculate the sale saving as a percentage of the original cost.

Attempt Q8–Q11 without a calculator. Simplify fractions where possible.

8   The number of pupils who want to go on a school trip increases from 120 to 150.

   Calculate the percentage increase in the number of pupils who want to go on the trip.

9   Due to its popularity, a retailer increases the cost of a book from £40 to £46.

   Calculate the increase in price as a percentage of the original cost.

10 A workforce decreased from 200 to 182. Calculate the reduction in workforce as a percentage of the original workforce.

11 After heavy rain, the number of people staying at a campsite fell from 60 to 42. Calculate the decrease in people camping as a percentage of the original number.

12 A farmer stops spraying her crops with a fertiliser. When the crop was harvested and weighed, the total yield was 500 kg. The year previously, with the fertiliser, the total yield was 630 kg. Calculate the percentage decrease in yield.

13 A 2 m cable has 40 cm cut from the end of it.

   Calculate the percentage decrease in length of the original cable.

14 The table shows the sales of diesel and electric cars over the first two years of a garage trading.

	Year 1	Year 2	Total
Electric	12	35	47
Diesel	259	370	629
Total	271	405	676

   a) What percentage of the total number of cars sold in the first year were electric?

   b) What percentage of the total cars sold over the two years were diesel?

   c) Calculate the percentage increase in sales from the first year to the second year of:

      i)   electric cars   ii) diesel cars   iii) total car sales.

15 A newsreader makes the following claim during a broadcast: 'The company's profit has increased from £320 000 to £640 000. This is a 50% increase in profits'.

   Is the claim correct? Justify your answer.

16 A car depreciates in value by 5% in the first year and then by 6% in the second year. Joe thinks this means the car has depreciated in value by 11% over 2 years.

   a) Explain why Joe is wrong.

   b) What overall percentage has the car depreciated in value by over the 2-year period?

▶ Reverse percentages

Sometimes we are given an amount **after** a percentage has been added on or taken off from the original amount. In this case:

- Find the percentage given by adding to or subtracting from 100%.
- Equate the new percentage to the amount given in the question.
- Find 1% (or any other equally useful percentage).
- Multiply to obtain 100%. This is the original amount.

Worked examples

1 A house appreciates in value by 10% and is now worth £165 000.

How much was the house originally?

New % = 100% + 10% = 110%

Percent %	Value (£)
110	165 000
÷110 ↷ ↷ ÷110	
1	1500
×100 ↷ ↷ ×100	
100	150 000

The house was originally worth £150 000.

2 To meet healthy eating guidelines a chocolate bar is reduced in size by 28%. It now weighs 45 g.

How much did the original bar weigh?

New % = 100% − 28% = 72%

Percent %	Weight (g)
72	45
÷72 ↷ ↷ ÷72	
1	0·625
×100 ↷ ↷ ×100	
100	62·5

The original bar weighed 62·5 g.

3 A car depreciates by 20% and is now worth £8800.

Without using a calculator, find the original cost of the car.

New % = 100% − 20% = 80%

Percent %	Value (£)
80	8800
÷8 ↷ ↷ ÷8	
10	1100
×10 ↷ ↷ ×10	
100	11 000

Here we can find 10% as 8800 is easily divisible by 8.

Alternatively, find 1%. Writing the division as a fraction and simplifying can help with non-calculator questions.

$$1\% = \frac{8800}{80}$$
$$= 110$$
$$100\% = 11000$$

The car originally cost £11 000.

Further strategies include converting the new percentage to a decimal multiplier and dividing by it, or converting to a fraction and completing the arithmetic from there. Find what works best for you!

Exercise 2F

1 Write down the new percentage associated when the following changes have already taken place:

a) an increase of 20%   **b)** an increase of 45%   **c)** an increase of 19%

d) a decrease of 10%   **e)** a decrease of 30%   **f)** a decrease of 9·5%

2 The price of a football shirt increases by 15% after the team win a trophy. The shirt now costs £46. Find the original price of the shirt before they won the trophy.

3 A shampoo bottle on special offer has 20% extra free and contains 900 ml. What is the capacity of a regular bottle of the same shampoo?   ➜

4 Piper gets a pay rise of 5% and her new annual salary is £25 200.

What was her salary before her pay rise?

5 After a new housing estate was built, the population of a village increased by 8% to 2700 people. How many people were in the village before the new estate was built?

6 The cost of a jacket in a 10% off sale is £81.

How much did the jacket cost before it was in the sale?

7 A mobile phone fell in price by 15% after a newer model was introduced. The phone now costs £195·50. How much did it cost originally?

8 A valuer tells homeowners that their house has decreased in value by 5% and is now worth £142 500. How much was the house worth before the fall in price?

9 Students receive a 20% discount in a restaurant. Barny is a student and he paid £12·80 for his meal with this special offer. How much would Barny have to pay if he didn't get the student discount?

10 A successful company have increased their workforce by 12% to 28 workers.

How many workers were there before the increase?

11 When buying a new sofa, the sellers add a 3% administration fee to the price of the sofa. The bill for a sofa, including the fee, is £406·85. What price is the sofa without the fee?

12 The average attendance at a rugby ground decreased by 60% after the club were relegated to a lower league. The average attendance is now 3120 people. What was the average attendance before the team were relegated?

13 Alex makes floral displays. He adds 30% on to the cost of the flowers and materials to make a profit. If Alex sells a display for £45·50, how much did he spend on flowers and materials?

Attempt Q14–Q17 without a calculator. Simplify fractions where possible.

14 A TV costs £240 in a 40% off sale. How much is the TV when not in the sale?

15 A music venue failed to sell 30% of the tickets available for a concert. If 490 tickets were sold, how many tickets were available to begin with?

16 A bill which includes a tip of 20% added on is £150. How much is the bill without the tip?

17 Laura earns £30 000 a year. This is 25% more than her partner. How much does her partner earn in a year?

18 a) Sort the following questions into those that require the reverse percentage technique you have been using in this exercise and those that do not.

A: A pair of boots in a 30% off sale cost £66·50. How much were they before the sale?

B: A pair of boots which cost £70 are put into a 30% off sale. How much do they cost in the sale?

C: Viren's current wage is £14·60 an hour. He is about to receive a 5% pay rise. What will his wage be after the pay rise is added on to his current wage?

D: Jack gets a 5% pay rise and his hourly wage is now £12·20. How much did he get per hour before the pay rise?

E: A bottle of juice on special offer advertises 25% extra free. A normal bottle holds 500 ml. How much will the bottle on special offer hold?

F: A bottle of juice on special offer advertises 25% extra free. The bottle on special offer holds 600 ml. How much does a normal bottle hold?

b) Calculate the answer to each of the questions in a).

▶ Adding and subtracting mixed numbers

Recall that a mixed number has a whole number and fraction part. To add or subtract mixed numbers:

● Add or subtract any whole numbers.

● Write the fraction parts with a common denominator.

● Complete the addition or subtraction to obtain a final answer.

● Make sure that, where possible, you remember to simplify any final answer which contains a fraction.

Worked examples

1
$$3\frac{1}{8}+6\frac{3}{8}$$
$$=9\frac{1}{8}+\frac{3}{8}$$
$$=9\frac{4}{8}$$
$$=9\frac{1}{2}$$

2
$$5\frac{1}{4}+\frac{5}{9}$$
$$=5\frac{1}{4}^{\times9}+\frac{5}{9}^{\times4}$$
$$=5\frac{9}{36}+\frac{20}{36}$$
$$=5\frac{29}{36}$$

3
$$9\frac{7}{15}-2\frac{1}{5}$$
$$=7\frac{7}{15}-\frac{1}{5}^{\times3}$$
$$=7\frac{7}{15}-\frac{3}{15}$$
$$=7\frac{4}{15}$$

Exercise 2G

1 Calculate:

a) $5\frac{1}{3}+1\frac{1}{3}$

b) $3\frac{1}{5}+2\frac{3}{5}$

c) $8\frac{2}{9}+3\frac{5}{9}$

d) $10\frac{1}{7}+5\frac{3}{7}$

e) $7\frac{4}{5}-3\frac{1}{5}$

f) $8\frac{6}{7}-1\frac{1}{7}$

g) $10\frac{8}{9}-5\frac{4}{9}$

h) $12\frac{2}{3}-11\frac{1}{3}$

2 Find:

a) $2\frac{1}{4}+5\frac{3}{8}$

b) $4\frac{2}{3}+7\frac{1}{9}$

c) $1\frac{1}{2}+2\frac{1}{3}$

d) $5\frac{1}{3}+3\frac{2}{5}$

e) $6\frac{7}{8}-4\frac{1}{2}$

f) $5\frac{9}{10}-1\frac{3}{5}$

g) $8\frac{3}{4}-2\frac{2}{3}$

h) $3\frac{5}{6}-2\frac{1}{7}$

i) $6\frac{3}{10}+8\frac{1}{10}$

j) $15\frac{5}{6}-7\frac{1}{6}$

k) $6\frac{1}{4}+8\frac{3}{10}$

l) $14\frac{8}{9}-3\frac{5}{6}$

3 Hope studied for $2\frac{1}{2}$ hours in the morning and then a further $1\frac{1}{4}$ hours in the afternoon.

How long, in hours, did she study for altogether?

4 To raise money for charity, pupils were asked to complete laps of a running track. Darci completed $20\frac{5}{9}$ laps and Grace completed $15\frac{1}{6}$ laps. How much further did Darci run than Grace?

Worked examples

1 $1\frac{5}{8}+3\frac{7}{8}$

$=4\frac{5}{8}+\frac{7}{8}$

$=4\frac{12}{8}$

$=4\frac{3}{2}$

$=5\frac{1}{2}$

When adding, the fraction parts can add to more than one whole.

Change the improper fraction into a mixed number and add to the whole number:

$4\frac{3}{2}=4+\frac{3}{2}=4+1\frac{1}{2}=5\frac{1}{2}$

2 $10-3\frac{2}{15}$

$=7-\frac{2}{15}$

$=6\frac{15}{15}-\frac{2}{15}$

$=6\frac{13}{15}$

When subtracting, the fraction parts can subtract to less than zero.

Convert one whole to an improper fraction:

$7=6+1=6+\frac{15}{15}=6\frac{15}{15}$

3 $9\frac{2}{9}-6\frac{3}{4}$

$=3\frac{2}{9}^{\times4}-\frac{3}{4}^{\times9}$

$=3\frac{8}{36}-\frac{27}{36}$

$=2\frac{44}{36}-\frac{27}{36}$

$=2\frac{17}{36}$

$3\frac{8}{36}=2\frac{36}{36}+\frac{8}{36}$

$=2\frac{44}{36}$

Alternatively, we can convert the mixed numbers to improper fractions at the start.

$1\frac{5}{8}+3\frac{7}{8}$

$=\frac{13}{8}+\frac{31}{8}$

$=\frac{44}{8}$

$=\frac{11}{2}$

$=5\frac{1}{2}$

$10-3\frac{2}{15}$

$=\frac{150}{15}-\frac{47}{15}$

$=\frac{103}{15}$

$=6\frac{13}{15}$

$9\frac{2}{9}-6\frac{3}{4}$

$=\frac{83}{9}^{\times4}-\frac{27}{4}^{\times9}$

$=\frac{332}{36}-\frac{243}{36}$

$=\frac{89}{36}$

$=2\frac{17}{36}$

5 Calculate:

a) $1\frac{2}{3}+5\frac{2}{3}$

b) $3\frac{4}{5}+7\frac{3}{5}$

c) $2\frac{5}{9}+6\frac{8}{9}$

d) $1\frac{1}{2}+4\frac{5}{8}$

e) $3\frac{1}{2}+5\frac{2}{3}$

f) $6\frac{3}{4}+8\frac{2}{5}$

g) $4\frac{2}{3}+2\frac{6}{7}$

h) $1\frac{5}{6}+3\frac{7}{8}$

6 Find:

a) $8-1\frac{1}{2}$

b) $6-2\frac{1}{5}$

c) $9-4\frac{2}{3}$

d) $7-1\frac{3}{4}$

e) $3\frac{1}{3}-1\frac{2}{3}$

f) $8\frac{2}{5}-4\frac{4}{5}$

g) $6\frac{1}{8}-2\frac{1}{4}$

h) $9\frac{2}{7}-3\frac{3}{4}$

7 Find:

a) $8\frac{5}{6}+4\frac{2}{3}$

b) $5\frac{3}{4}+6\frac{8}{9}$

c) $5-3\frac{1}{6}$

d) $10\frac{4}{5}-1\frac{7}{8}$

8 A garden is rectangular with the dimensions shown in the diagram.

a) Calculate the perimeter of the garden.

b) How much longer is the garden than it is wide?

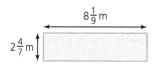

▶ Multiplying and dividing mixed numbers

Consider $2\frac{1}{2} \times 3\frac{1}{5}$. Using decimals, $2 \cdot 5 \times 3 \cdot 2 = 8$. We need two and a half lots of all of three and one fifth.

When multiplying mixed numbers **do not** treat the whole numbers separately. Instead, change mixed

numbers to improper fractions before multiplying $\dfrac{\overset{1}{\cancel{5}}}{\underset{1}{\cancel{2}}} \times \dfrac{\overset{8}{\cancel{16}}}{\underset{1}{\cancel{5}}} = 8$. The answer is not $6\frac{1}{10}$!

To multiply mixed numbers:

- Convert mixed or whole numbers to improper fractions.
- If possible, simplify numbers on the numerators with numbers on the denominators. If you do not simplify at this stage the arithmetic will be more difficult to get to the final answer.
- Multiply numerators together and multiply denominators together.

Worked examples

1 $3\frac{1}{2} \times 6\frac{2}{3}$

$= \dfrac{7}{\underset{1}{\cancel{2}}} \times \dfrac{\overset{10}{\cancel{20}}}{3}$

$= \dfrac{70}{3}$

$= 23\frac{1}{3}$

2 $2\frac{2}{5} \times 5\frac{5}{6}$

$= \dfrac{\overset{2}{\cancel{12}}}{\underset{1}{\cancel{5}}} \times \dfrac{\overset{7}{\cancel{35}}}{\underset{1}{\cancel{6}}}$

$= \dfrac{14}{1}$

$= 14$

3 $4 \times 2\frac{7}{9}$

$= \dfrac{4}{1} \times \dfrac{25}{9}$

$= \dfrac{100}{9}$

$= 11\frac{1}{9}$

Simplifying the fractions in the first two examples makes the arithmetic easier.

Exercise 2H

Throughout this exercise, write any answer which is an improper fraction as a mixed number.

1 Simplify:

a) $1\frac{1}{2} \times 6\frac{1}{5}$

b) $2\frac{1}{3} \times 2\frac{1}{4}$

c) $1\frac{4}{5} \times 4\frac{2}{9}$

d) $3\frac{5}{6} \times 1\frac{1}{2}$

e) $2\frac{2}{5} \times 3\frac{1}{3}$

f) $6\frac{1}{4} \times 3\frac{7}{15}$

g) $4\frac{8}{9} \times 1\frac{5}{16}$

h) $5\frac{1}{3} \times 1\frac{1}{8}$

2 Simplify:

a) $6 \times 2\frac{3}{4}$

b) $3\frac{1}{6} \times 8$

c) $9\frac{4}{5} \times 10$

d) $6 \times 8\frac{1}{4}$

3 A rectangle has length $2\frac{3}{4}$ cm and breadth $3\frac{1}{5}$ cm. Calculate the area of the rectangle.

4 A panoramic photo is printed so that its length is $3\frac{1}{3}$ times as long as its height.

If the photo is 30 cm high when printed, how long is it?

5 Simplify:

a) $2 + 3\frac{1}{2} \times 4\frac{2}{7}$

b) $1\frac{4}{11} \times \left(2\frac{1}{3} + 1\frac{4}{5}\right)$

c) $\left(3\frac{1}{6} + 2\frac{7}{8}\right) \times \left(4\frac{1}{4} - 3\frac{4}{5}\right)$

To divide mixed numbers:

- Convert mixed and whole numbers to improper fractions.

- Keep the first fraction unchanged and multiply by the reciprocal of the dividing fraction.

- Complete the multiplication.

- Do not try to simplify until you have written it as a multiplication.

Worked examples

1 $5\frac{2}{3} \div 2\frac{1}{2}$

$= \frac{17}{3} \div \frac{5}{2}$

$= \frac{17}{3} \times \frac{2}{5}$

$= \frac{34}{15}$

$= 2\frac{4}{15}$

2 $3\frac{3}{4} \div 4\frac{1}{6}$

$= \frac{15}{4} \div \frac{25}{6}$

$= \frac{\cancel{15}^{3}}{\cancel{4}_{2}} \times \frac{\cancel{6}^{3}}{\cancel{25}_{5}}$

$= \frac{9}{10}$

3 $6 \div 7\frac{1}{8}$

$= \frac{6}{1} \div \frac{57}{8}$

$= \frac{\cancel{6}^{2}}{1} \times \frac{8}{\cancel{57}_{19}}$

$= \frac{16}{19}$

Exercise 2I

Throughout this exercise, write any answer which is an improper fraction as a mixed number.

1 Simplify:

a) $1\frac{1}{2} \div 4\frac{1}{3}$

b) $1\frac{2}{5} \div 2\frac{1}{4}$

c) $2\frac{1}{3} \div 5\frac{1}{2}$

d) $2\frac{4}{9} \div 1\frac{4}{7}$

e) $3\frac{1}{2} \div 1\frac{2}{3}$

f) $1\frac{7}{8} \div 2\frac{3}{8}$

g) $5\frac{5}{6} \div 2\frac{1}{3}$

h) $6\frac{2}{9} \div 2\frac{2}{3}$

2 Simplify:

a) $5 \div 3\frac{1}{4}$

b) $8 \div 6\frac{7}{8}$

c) $3 \div 5\frac{5}{6}$

d) $9 \div 8\frac{1}{7}$

3 The area of a parallelogram is $3\frac{3}{5}$ m².

If the base is $2\frac{1}{4}$ m long, find the height of the parallelogram.

4 A large container holds 7·5 litres of liquid. When emptied, $6\frac{3}{4}$ small containers can be filled. How much liquid does a small container hold?

5 Luke runs $1\frac{7}{8}$ miles at an average speed of $4\frac{1}{2}$ mph.

Calculate how long Luke was running for. Give your answer in minutes.

6 Simplify:

a) $1\frac{3}{4} - 1\frac{7}{8} \div 1\frac{1}{2}$

b) $8 \div \left(\frac{3}{5} + 2\frac{1}{3}\right)$

c) $\left(10\frac{1}{3} + 7\frac{1}{6}\right) \div \left(2\frac{7}{9} - 2\frac{1}{2}\right)$

Check-up 👍

1 Calculate:
   a) 0.2×4
   b) 0.7×0.8
   c) 0.32×0.9
   d) 0.62×0.81
   e) $5 \div 0.1$
   f) $1.88 \div 0.2$
   g) $4.8 \div 0.03$
   h) $0.0072 \div 0.004$

2 Find:
   a) $10 + (2.3 + 1.2) \times 0.08$
   b) $7 \div 0.5 - 0.9 \times 0.02$

3 In a group of 40 adults, 12 have a food allergy. What percentage of the group have a food allergy?

You may use a calculator for Q4–Q20.

4 A conductor counts the players in each part of her 105-piece orchestra: 70 string, 16 woodwind, 17 brass and 2 percussionists. To the nearest percent, what percentage are playing:
   a) a string instrument
   b) percussion
   c) brass?

5 Increase:
   a) 92 g by 7%
   b) 480 kg by 34%
   c) 125 mm by 81%

6 Decrease:
   a) 7 kg by 4%
   b) 810 cm by 61%
   c) 58 miles by 38%

7 After a football team were relegated, the number of season tickets bought by supporters fell by 13%. If they sold 2000 season tickets prior to relegation, how many did they sell after they were relegated?

8 A newspaper currently costs 75p. Due to an increase in production costs, the newspaper is going to increase in price by 8%. How much will the newspaper cost after the increase?

9 Calculate how much money is in each savings account at the end of the term stated:

	Amount initially deposited (£)	Interest rate (% per annum)	Term (years)
a)	500	3	5
b)	1200	4	2
c)	3000	2	3
d)	720	5	4
e)	900	1.8	2
f)	4200	2.9	6
g)	650	0.7	4

10 For each part in Q9, calculate the amount of interest accrued over the full term.

11 It is expected that a computer will depreciate in value by 7% every year. If the computer cost £2400 when bought, how much will it be worth after 3 years?

12 During an experiment, scientists approximate that 11% of a liquid evaporates every minute. If there was 100 ml of liquid at the start of the experiment, to the nearest millilitre, how much was left after 5 minutes?

13 Myra designs her own cards. It costs her £1.50 per card for the materials and she sells each card for £2.25. How much profit does Myra make per card as a percentage of the cost price?

14 A trampoline is reduced from £130 to £95. Calculate, to three significant figures, the reduction in price of the trampoline as a percentage of its original cost. ➜

15 Robert bought a wood-burning stove for £900. It was too big so he sold it to his sister for £765. Calculate Robert's loss on the stove as a percentage of the cost price.

16 Over the course of a week, a plant increases in height from 61 cm to 69 cm. Calculate, to three significant figures, the percentage growth of the plant over the week.

17 A large suitcase has a capacity of 100 litres. This is 25% more than the capacity of a medium suitcase. What capacity does a medium suitcase have?

18 A furniture showroom is advertising 20% off all dining tables in a sale. If the table costs £176 in the sale, how much would it cost when not in the sale?

19 Sales of umbrellas in a shop fell by 87·5% between April and May. If 3 umbrellas were sold in May, how many were sold in April?

20 The cost of an online gaming subscription increased by 2·5% and now costs £8·20 per month. How much was the monthly subscription before the increase?

Attempt the remainder of this check-up without a calculator.

21 A fridge in a 25% off sale costs £225. How much did the fridge cost originally?

22 Driving lessons increase in price by 20% to cover rising fuel costs. A lesson now costs £30. How much did a lesson cost before the price increase?

23 Calculate, writing your answer as a mixed number:

a) $1\frac{2}{3} + 5\frac{1}{4}$

b) $6\frac{1}{3} + 4\frac{2}{9}$

c) $2\frac{3}{4} + 5\frac{1}{2}$

d) $4\frac{2}{3} + 1\frac{3}{5}$

e) $7\frac{5}{6} - 4\frac{3}{8}$

f) $9\frac{1}{6} - 2\frac{2}{3}$

g) $8\frac{3}{7} - 6\frac{1}{2}$

h) $5\frac{7}{8} - 3\frac{9}{10}$

24 Find, writing your answer as a mixed number where possible:

a) $1\frac{1}{2} \times 3\frac{2}{5}$

b) $2\frac{3}{4} \times 5\frac{1}{3}$

c) $6\frac{4}{5} \times 4\frac{3}{8}$

d) $3 \times 7\frac{8}{9}$

e) $2\frac{1}{4} \div 1\frac{1}{3}$

f) $4\frac{1}{5} \div 3\frac{1}{2}$

g) $8\frac{4}{9} \div 2\frac{8}{15}$

h) $8 \div 1\frac{3}{7}$

25 One set of golf clubs weighs $13\frac{3}{5}$ kg and a second one weighs $12\frac{5}{6}$ kg. Calculate:

a) the total weight of both sets

b) the difference in weight between the sets.

26 Simplify, writing your answer as a mixed number where possible:

a) $2\frac{3}{7} + 4\frac{2}{7} \times 2$

b) $\frac{2}{9}$ of $\left(8\frac{3}{5} - 1\frac{2}{5}\right)$

c) $4\frac{2}{5} + 6\frac{2}{3}$ of $7\frac{1}{2}$

d) $4\frac{2}{3} \times 3\frac{6}{7} - 5\frac{5}{8} \div 3\frac{3}{4}$

27 A rectangular table has length 60 inches and breadth $22\frac{1}{2}$ inches.

a) Calculate the area of the table.

A second rectangular table has the same area and a breadth of $16\frac{2}{3}$ inches.

b) Calculate the length of the second table.

3 Time

▶ Time intervals

When calculating a time interval:

● Count on to the next/required hour.

● Count on to the required minutes.

● Add all time intervals together to form your answer.

Worked examples

1   Calculate the length of each time interval in hours and minutes.

    a) 5:15 am to 3:10 pm

    b) 08:40 to 13:52

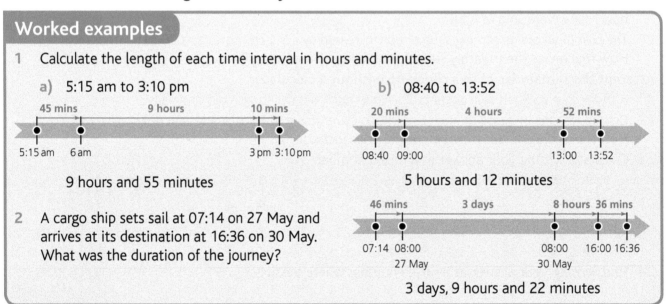

    9 hours and 55 minutes

    5 hours and 12 minutes

2   A cargo ship sets sail at 07:14 on 27 May and arrives at its destination at 16:36 on 30 May. What was the duration of the journey?

    3 days, 9 hours and 22 minutes

Exercise 3A

1   Calculate the time interval, in hours and minutes, from:

    a) 5:35 am to 11:40 am   b) 2:55 pm to 11:17 pm   c) 9:45 am to 3:30 pm

    d) 01:19 to 12:00   e) 06:34 to 15:18   f) 04:25 to 14:19

2   Write the following durations in hours.

    a) 4 days and 7 hours   b) 8 days and 3 hours   c) 2 weeks

3   Write the following durations in weeks, days and hours.

    a) 50 hours   b) 157 hours   c) 197 hours   d) 344 hours

4   The Apollo 11 mission to the Moon lifted off at 13:32 on 16 July, 1969 and returned to Earth at 16:50 on 24 July. Calculate the duration of the mission.

5   Calculate the duration of each holiday below, giving your answers in days, hours and minutes.

	Destination	Date and time of departure	Date and time of return
a)	Venice	7 July, 4:40 am	14 July, 7:15 pm
b)	Budapest	13 August, 2:30 pm	16 August, 4:25 pm
c)	Paris	28 March, 18:30	31 March, 13:10
d)	Dublin	26 February 2022, 06:55	2 March 2022, 14:40

The globe is split up into different **time zones**. This means that although it may be 9 am where you are, in other countries the hour will be quite different. When travelling between two countries we need to take account of the **time difference** as well as the length of journey and the time we set off.

When calculating a duration that spans different time zones:

- Convert the time zone in one location to match the other.
- Calculate the time interval as normal.

Worked example

New York is 5 hours behind Glasgow. A flight leaves Glasgow Airport at 07:40 local time and arrives in New York at 12:05 local time. What is the duration of the flight?

First, we calculate what the local time is in New York when the flight leaves Glasgow.

5 hours before 07:40 is 02:40

Next, we count on the required amount.

Finally, we write our answer.

The duration of the journey is 9 hours, 25 minutes.

---

6   Tokyo is 8 hours ahead of the UK. What time is it in Tokyo if the time in the UK is:

    a) 6:00 am   **b)** 11:15 am   **c)** 5:30 pm on Friday   **d)** 20:05 on Monday?

7   Edinburgh is 5 hours ahead of Havana. What is the time in Havana if the time in Edinburgh is:

    a) 09:00   **b)** 8:15 pm   **c)** 3:45 am on Tuesday   **d)** 23:40 on Sunday?

8   Stockholm is one hour ahead of the UK. A flight leaves Edinburgh airport at 07:18, local time, and takes 2 hours and 13 minutes to reach its destination.

    a) What is the local time in Stockholm when the plane leaves Edinburgh?

    b) What is the local arrival time in Stockholm?

9   Calculate the duration of the following journeys.

	Origin	Local departure time	Destination	Time difference (hours)	Local arrival time
a)	Glasgow	15:40	Paris	+1	18:29
b)	Glasgow	06:30	Reykjavik	−1	07:45
c)	Athens	13:15	Glasgow	−2	16:50
d)	Lima	11:50	Buenos Aires	+2	17:55
e)	Reykjavik	23:15, 1 August	Paris	+2	04:45, 2 August
f)	Glasgow	02:15, 4 July	Lima	−6	03:15, 5 July
g)	Athens	16:17, 20 June	Buenos Aires	−6	04:47, 21 June
h)	Berlin	14:50, 10 January	Sydney	+10	00:10, 12 January

▶ Decimal time

Time does not follow the decimal system (since there are 60 minutes in an hour, not 100).

- To convert decimal hours to minutes, multiply the decimal part by 60.
- To convert minutes to hours, divide by 60.

Worked examples

1 Convert the following times from decimal hours to hours and minutes.

a) 2·3 hours

$$0·3 \times 60$$
$$= 0·3 \times 10 \times 6$$
$$= 3 \times 6$$
$$= 18$$

2 hours 18 minutes

b) 12·8 hours

$$0·8 \times 60$$
$$= 0·8 \times 10 \times 6$$
$$= 8 \times 6$$
$$= 48$$

12 hours 48 minutes

c) 1·65 hours

$$0·65 \times 60$$
$$= 0·65 \times 10 \times 6$$
$$= 6·5 \times 6$$
$$= 39$$

1 hour 39 minutes

2 Convert the following times from hours and minutes to decimal hours.

a) 4 hours 15 minutes

$$\frac{15}{60} = \frac{1}{4}$$
$$= 0·25$$

4·25 hours

b) 3 hours 24 minutes

$$\frac{24}{60} = \frac{4}{10}$$
$$= 0·4$$

3·4 hours

c) 177 minutes

177 minutes = 2 hours 57 minutes

$$57 \div 60$$
$$= 5·7 \div 6$$

$$6)\overline{5·^{5}7^{3}0} = 0·9\,5$$

2·95 hours

Exercise 3B

1 Convert the following times from decimal hours to hours and minutes.

a) 4·5 hours
b) 3·25 hours
c) 6·75 hours
d) 12·25 hours
e) 0·5 hours
f) 6·1 hours
g) 10·8 hours
h) 7·2 hours
i) 7·4 hours
j) 9·6 hours
k) 3·05 hours
l) 12·85 hours

2 Convert the following times from hours and minutes to decimal hours.

a) 2 hours 30 minutes
b) 3 hours 15 minutes
c) 6 hours 45 minutes
d) 1 hour 12 minutes
e) 9 hours 24 minutes
f) 12 hours 36 minutes

3 Convert the following times from minutes to decimal hours.

a) 30 minutes
b) 42 minutes
c) 18 minutes
d) 21 minutes
e) 66 minutes
f) 132 minutes
g) 168 minutes
h) 234 minutes
i) 333 minutes
j) 408 minutes
k) 156 minutes
l) 426 minutes

▶ Calculations with decimal time

When performing distance, speed, time calculations it is crucial that the units we are using 'match up'. This means that we may have to convert between hours and minutes before substituting into the formula.

≫ Calculating distance

To calculate distance, use the formula distance = speed × time or $D = S \times T$.

> ### Worked examples
>
> 1 A runner maintains a speed of 4 mph for 90 minutes. What distance do they cover?
>
>   The speed is in miles per hour so we must first write the time in hours.
>
> $$90 \text{ mins} = \frac{90}{60}$$
> $$= \frac{3}{2} \text{ hours}$$
>
>   Calculate distance $D = S \times T$
> $$= 4 \times \frac{3}{2}$$
> $$= 6 \text{ miles}$$
>
> 2 Whilst training for a race, a cyclist has an average speed of 36 km/h. They cycle for 1 hour and 24 minutes. What distance do they cover?
>
>   Change minutes to decimal hours
>
> $$\frac{24}{60} = \frac{2}{5}$$
> $$= 0\cdot4 \text{ hours}$$
>
>   1 hour 24 minutes = 1·4 hours
> $$D = S \times T$$
> $$= 36 \times 1\cdot4$$
> $$= 50\cdot4 \text{ km}$$
>
>    ```
>          36
>       ×  14
>         144
>       + 360
>         504
>    ```

Exercise 3C

1 Calculate the distance of each journey, giving your answer with appropriate units.

	a)	b)	c)	d)	e)	f)
Speed	50 mph	16 km/h	20 mph	25 km/h	24 mph	35 mph
Time	1 hour, 30 minutes	3 hours, 15 minutes	4 hours, 45 minutes	1 hour, 36 minutes	5 hours, 48 minutes	2 hours, 12 minutes

2 A motorist drives at a speed of 80 km/h. How far would they drive in 1 hour and 15 minutes?

3 A jogger has an average speed of 5 miles per hour. How far would they cover in 2 hours and 30 minutes?

4 An Edinburgh tram has a top speed of 70 km/h. How far could the tram drive in 24 minutes?

5 A garden snail moves at an average speed of 0·047 km/h. How far could it move in 15 minutes? Give your answer in metres.

6 A hare runs at an average speed of 35 mph. How far would it cover in 45 minutes?

7 A peregrine falcon can reach a speed of approximately 240 mph. How far could it fly in 12 minutes?

›› Calculating speed

To calculate speed, use the formula $\text{speed} = \dfrac{\text{distance}}{\text{time}}$ or $S = \dfrac{D}{T}$.

> ### Worked examples
>
> **1** A greyhound runs 10 miles in 15 minutes. Calculate its average speed in miles per hour.
>
> First, convert units
> $$15 \div 60$$
> $$= 0 \cdot 25 \text{ hours}$$
>
> Calculate speed
> $$S = \dfrac{D}{T}$$
> (multiply both numbers by 100)
> $$= \dfrac{10}{0 \cdot 25}$$
> $$= \dfrac{1000}{25}$$
> $$= 40 \text{ mph}$$
>
> **2** A train travels 240 km in 66 minutes. Calculate the speed of the train, giving your answer in km/h correct to 2 decimal places.
>
> First, convert units
> $$66 \div 60$$
> $$= 11 \div 10$$
> $$= 1 \cdot 1 \text{ hours}$$
>
> Calculate speed
> $$S = \dfrac{D}{T}$$
> $$= \dfrac{240}{1 \cdot 1}$$
> $$= 218 \cdot 181 \ldots$$
> $$= 218 \cdot 18 \text{ km/h (to 2 d.p.)}$$

Exercise 3D

1 Calculate the speed of each journey, giving your answer in appropriate units per hour.

	a)	b)	c)	d)	e)	f)
Distance	75 miles	90 km	64 miles	88 km	87 miles	63 km
Time	1 hour, 30 minutes	2 hours, 15 minutes	15 minutes	4 hours, 24 minutes	2 hours, 54 minutes	4 hours, 12 minutes

2 The driving distance from Glasgow to Edinburgh is approximately 75 kilometres. Calculate the average speed in km/h of a car that makes this journey in 90 minutes.

3 Angus and Betty set out to cycle round Arran, a journey of 57 miles. Angus completes the journey in 6 hours.

a) Calculate Angus' average speed in miles per hour.

b) Betty completes the cycle in 4 hours and 45 minutes. How much faster is Betty than Angus?

Try Q4 and Q5 without a calculator.

4 A golden eagle travels 60 miles in 24 minutes. Calculate the eagle's speed in miles per hour.

5 A cyclist covers 64 km in 3 hours 12 minutes. Calculate their speed in km/h.

6 Brigid Kosgei's world record time for the women's marathon is 2 hours, 14 minutes and 4 seconds. The marathon is 42·195 km in length.

a) Convert the time to seconds and the distance to metres.

b) Calculate her average speed in metres per second correct to 3 significant figures.

c) Hence, find her average speed in km/h correct to three significant figures.

≫ Calculating time

To calculate time, ensure your units match and use the formula $\text{time} = \dfrac{\text{distance}}{\text{speed}}$ or $T = \dfrac{D}{S}$.

Worked examples

1 How long would it take a car to travel 156 km at an average speed of 60 km/h? Give your answer in hours and minutes.

$$T = \frac{D}{S}$$
$$= \frac{156}{60}$$
$$= \frac{13}{5}$$
$$= 2\frac{3}{5}$$

The time taken is $2\frac{3}{5}$ hours, but we must now convert this into hours and minutes.

$$\frac{3}{5} \times 60$$
$$= 3 \times 12$$
$$= 36$$

The car takes 2 hours and 36 minutes.

2 An aircraft flies 682 miles at an average speed of 575 mph. How long will the flight take? Give your answer in hours and minutes to the nearest minute.

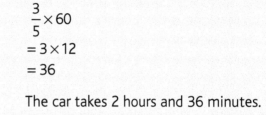

Write the formula, substitute and evaluate.

$$T = \frac{D}{S}$$
$$= \frac{682}{575}$$
$$= 1 \cdot 186 \ldots \text{hours}$$

Subtract the 1 hour and convert the decimal part to minutes by multiplying by 60.

$0 \cdot 186 \ldots \times 60$

$= 11 \cdot 16 \ldots$

$= 11$ minutes to the nearest minute

The flight takes 1 hour and 11 minutes (to the nearest minute).

Exercise 3E

1 Calculate the time taken for each journey, giving your answers in hours and minutes where possible.

	a)	b)	c)	d)	e)	f)
Distance	60 miles	80 km	3 miles	18 km	62 miles	360 km
Speed	40 mph	64 km/h	0·8 mph	15 km/h	20 mph	75 km/h

2 A horsefly flies at approximately 90 mph. How long would it take to cover 15 miles? Give your answer in minutes.

3 How long does it take a mako shark to swim 54 miles at a speed of 45 mph? Give your answer in hours and minutes.

4 A goods vehicle weighing over 7·5 tonnes is restricted to a speed of 60 mph on a motorway. How long will it take to complete a journey of 216 miles?

5 The Concorde airliner can reach a speed of 1350 mph. How long would it take to fly 350 miles (approximately the distance from Glasgow to London)? Give your answer to the nearest minute.

6 The distance from the Earth to the Moon is approximately 384 400 km. How long would it take a spacecraft to make this journey at a speed of 3280 km/h? Give your answer in days and hours, to the nearest hour.

Check-up 👍

1 Calculate the interval, in hours and minutes, from:

   a) 6:15 am to 9:40 pm
   b) 5:45 am to 12:58 pm
   c) 09:17 to 18:43
   d) 11:38 to 23:12
   e) 7:42 am to 6:18 pm
   f) 8:51 am to 4:59 pm

2 A bus service operating along the West Highland Way is introduced. A section of the timetable is shown opposite. The time difference between each stop stays the same for each journey.

Milngavie	07:20	09:43	11:57	14:08
Rowardennan	08:08	10:31	12:45	14:56
Tyndrum	09:40	12:03	14:17	16:28
Kingshouse	10:04	12:27	14:41	
Fort William	10:42	13:05	15:19	

   a) How long is the journey from Milngavie to Rowardennan?

   b) How long is the journey from Tyndrum to Fort William?

   c) How long is the entire journey from Milngavie to Fort William?

   d) Adam is catching the 14:08 bus from Milngavie and intends to meet his friends as they finish the walk in Fort William. They expect to finish at 5:20 pm. Do you think Adam will be on time to meet them? Give a reason for your answer.

3 Iman walks the West Highland Way to raise money for charity. She aims to complete it in 4 days. Ten company sponsors agree to pay £2 each for every minute under the 4-day mark her final time is. Iman sets off at 7:15 am on 29 April and finishes at 10:50 pm on 2 May.

   a) How long did the walk take in total?

   b) How much sponsorship money will Iman receive for the charity?

4 A plane leaves Las Vegas at 11:40, local time, and arrives in Miami at 19:25, local time. Miami is two hours ahead of Las Vegas. What is the duration of the flight?

5 A plane leaves Heathrow airport in London at 05:40 local time. It arrives in Istanbul, Turkey at 12:35 local time. Turkey is two hours ahead of the UK.

   a) What is the local time in Turkey when the plane takes off from London?

   b) Calculate the duration of the journey.

6 A cruise ship sets off from Cyprus at 8:15 am local time. It arrives in Venice, Italy at 10:20 am local time. Italy is 1 hour behind Cyprus.

   a) What is the local time in Venice when the plane takes off from Cyprus?

   b) Calculate the duration of the journey.

7 For each of the journeys in the table below, find:

   i)   the local time at the destination when the flight departs

   ii) the duration of the journey.

City	Country	Time difference (hours)
Montreal	Canada	−5
Glasgow	Scotland	0
Helsinki	Finland	+2

	Origin	Local departure time	Destination	Local arrival time
a)	Glasgow	09:20	Montreal	18:55
b)	Glasgow	04:55	Helsinki	16:20
c)	Montreal	19:20, 3 April	Glasgow	13:05, 4 April
d)	Helsinki	16:30, 31 January	Montreal	00:25, 1 February

→

8 Melbourne, Australia is nine hours ahead of the UK. A flight leaves Melbourne Airport at 11:50 on 13 December, local time, and takes 22 hours and 50 minutes to reach Heathrow Airport in London.

    a) What is the local time in London when the plane leaves Melbourne?

    b) What is the local time when the plane arrives in London?

9 A cross-country train journey goes from Paris to Istanbul, with stopovers in Munich, Vienna and Bucharest. The train leaves Paris on 3 June at 10:30 local time and arrives in Istanbul on 6 June at 20:15 local time.

    a) If Istanbul is one hour ahead of Paris, how long was the journey in total, including stopovers?

    b) The stopover in Munich was twice as long as the one in Bucharest, and one third the length of the one in Vienna. If the total duration of the stopovers was 36 hours, how long did the train spend in each city?

10 Convert the following times from decimal hours to hours and minutes.

    a) 3·2 hours   **b)** 6·4 hours   **c)** 12·85 hours   **d)** 0·65 hours

11 Convert the following times from hours and minutes to decimal hours.

    a) 2 hours 45 minutes   **b)** 5 hours 15 minutes

    c) 3 hours 6 minutes   **d)** 9 hours 27 minutes

You may use a calculator for Q12–Q14.

12 Calculate the distance of each journey, giving your answer in units per hour.

    a) speed = 90 km/h, time = 30 minutes

    b) speed = 20 mph, time = 1 hour, 12 minutes

    c) speed = 25 km/h, time = 3 hours, 36 minutes

13 Calculate the speed of each journey, giving your answer in units per hour.

    a) distance = 60 km, time = 45 minutes

    b) distance = 85 miles, time = 2 hours, 30 minutes

    c) distance = 36 km, time = 1 hour, 12 minutes

14 Calculate the time of each journey, giving your answer in hours and minutes where possible.

    a) speed = 40 mph, distance = 60 miles   **b)** speed = 30 km/h, distance = 72 km

    c) speed = 20 km/h, distance = 16 km   **d)** speed = 80 mph, distance = 376 miles

15 Two friends cycle the same route. Callum took 2 hours and 15 minutes to complete the route at 40 km/h. How long did it take Deirdre to complete it at a speed of 50 km/h?

16 The Ben Nevis Race takes place annually. Eilidh completes the 14 km course maintaining a steady speed of 10 km/h.

    a) How long does it take Eilidh to complete the race? Give your answer in hours and minutes.

    b) Fergus is 12 minutes slower than Eilidh. What is Fergus' average speed?

17 The London Marathon is 42·195 km in length and the women's record time set by Paula Radcliffe in 2003 was approximately 2 hours and 15 minutes. Calculate her average speed, giving your answer correct to three significant figures.

4 Measurement

▶ Similarity

Shapes are **congruent** if they are identical in shape and size. Shapes are **similar** if they have the same angles and the ratios of the lengths of their corresponding sides are equal. We will use big and small to refer to the corresponding sides on the big shape and the small shape, respectively.

To calculate unknown lengths in similar shapes:

● If the unknown side is in the larger shape, find the **enlargement scale factor**, esf $= \dfrac{\text{big}}{\text{small}}$

● If the unknown side is in the smaller shape, find the **reduction scale factor**, rsf $= \dfrac{\text{small}}{\text{big}}$

● Multiply a corresponding side by the scale factor to calculate the length of the unknown side.

Worked example

For each pair of similar shapes identify the scale factor and find the unknown side.

a)
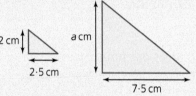

The unknown length is in the larger shape. We calculate the enlargement scale factor.

$$\text{esf} = \dfrac{\text{big}}{\text{small}}$$

Every side is 3 times larger.

$$= \dfrac{7 \cdot 5}{2 \cdot 5}$$
$$a = 2 \times 3$$
$$= 6 \text{ cm}$$
$$= 3$$

b)
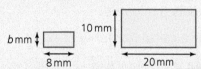

The unknown length is in the smaller shape. We calculate the reduction scale factor.

$$\text{rsf} = \dfrac{\text{small}}{\text{big}}$$

Every side is $\dfrac{2}{5}$ of the original.

$$= \dfrac{8}{20}$$
$$b = \dfrac{2}{5} \times 10$$
$$= \dfrac{2}{5}$$
$$= 4 \text{ mm}$$

Exercise 4A   1   For each pair of similar shapes, identify the enlargement or reduction scale factor and calculate the unknown dimension.

a)

b)
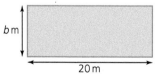

c)
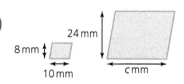

d)

e)
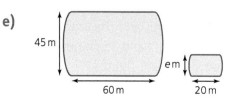

f)
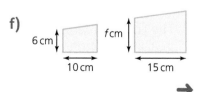

➡

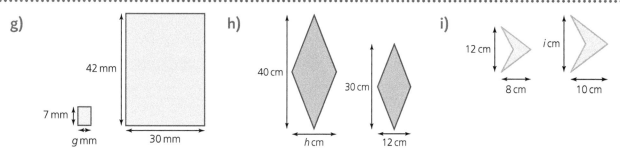

g) 42 mm   30 mm   7 mm   *g* mm

h) 40 cm   30 cm   *h* cm   12 cm

i) 12 cm   *i* cm   8 cm   10 cm

2   The dimensions of this portrait are shown. Calculate the dimensions of each reproduction.

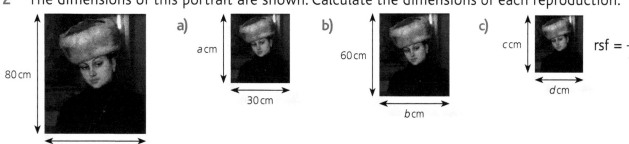

80 cm   60 cm

a) *a* cm   30 cm

b) 60 cm   *b* cm

c) *c* cm   *d* cm   rsf = $\frac{1}{8}$

d) James has a frame with dimensions 50 cm by 25 cm. Would it be a good idea to put a print of this picture in the frame? Explain your answer.

3   A manufacturer makes shorts in 3 sizes: small, medium and large. Each size is 1·25 × larger than the size below. Calculate the unknown measurements.

waist 28 inches
length _____

waist _____
length _____

waist _____
length 25 inches

4   Two friends attempt the similarity problem shown. Both are wrong!
Read through each answer and explain what is wrong with it.

6 cm   12 cm   *x* cm   4 cm

a) Attempt 1

rsf = $\frac{small}{big}$ *x* = 6 × 0·3 = 1·8 cm

= $\frac{4}{12}$ X

= 0·3

b) Attempt 2

esf = $\frac{big}{small}$ *x* = 6 × 3 = 18 cm

= $\frac{12}{4}$ X

= 3

c) Calculate the correct length of the side marked *x*.

5   A heavy machinery company makes full size vehicles.

They also make $\frac{1}{6}$ scale 'ride on' toy vehicles and $\frac{1}{32}$ scale toy models which are both similar in shape to the full size vehicles.

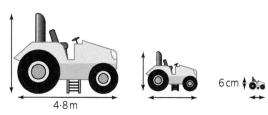

4·8 m   6 cm

a) Calculate the length and height of each model.
b) Calculate the enlargement scale factor from the toy to the ride on vehicle.

▶ Similar triangles

Similar triangles form around parallel lines. Look for corresponding angles and sides.

Worked example

Calculate the unknown lengths in each pair of similar triangles:

a)

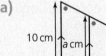

Look carefully to identify the smaller triangle embedded in the larger.

$$rsf = \frac{small}{big} = \frac{16}{20} = \frac{4}{5}$$

$$a = \frac{4}{5} \times 10 = 8 \, cm$$

b)

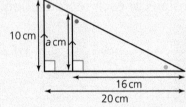
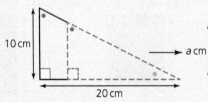
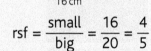

Look carefully, b is not a full side. Call the larger side $b + 10$.

$$esf = \frac{40}{16} = \frac{5}{2} \quad b + 10 = \frac{5}{2} \times 10$$

$$b + 10 = 25, \quad b = 15 \, cm$$

Exercise 4B

1   Find the length of the sides marked with a letter:

a)

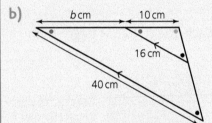

12 cm, a cm, 10 cm, 15 cm

b)

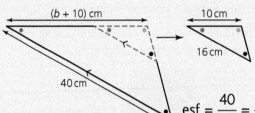

35 mm, 28 mm, 30 mm, b mm

c)

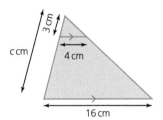

3 cm, c cm, 4 cm, 16 cm

d)

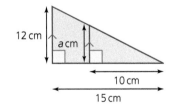

40 cm, 35 cm, 120 cm, d cm

e)

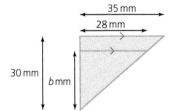

60 cm, 18 cm, 20 cm, e cm

f)

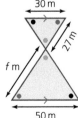

30 m, 27 m, f m, 50 m

g)

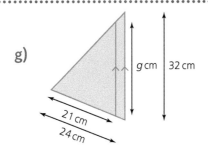

h)

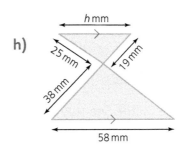

i)

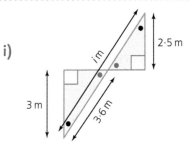

2   A folding table has the dimensions shown. Calculate the width of the table top.

3   Just before sunset the Titan crane in Clydebank casts a shadow 115 m long.

At the same time, a nearby bollard 1 m high casts a shadow 2·5 m long. What is the height of the Titan crane?

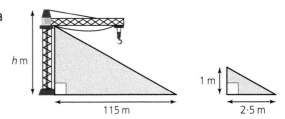

4   Find the length of the parts marked with a letter.

a)

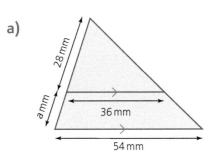

b)

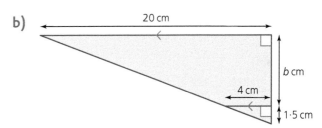

c)

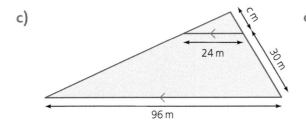

d)

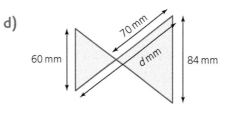

e)

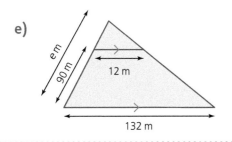

f)

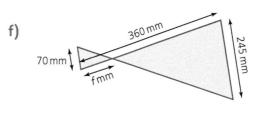

45

▶ Area scale factor

In general, area scale factor = (linear scale factor)², that is   asf = (esf)² or asf = (rsf)²

To calculate the unknown area of similar shapes:

- Remember to square the linear scale factor. See the examples below.

Worked example   Calculate each unknown area.

a) $esf = \dfrac{big}{small} = \dfrac{36}{9} = 4$ $a = 4^2 \times 30 = 480\,cm^2$

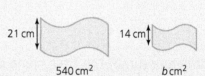

36 cm
9 cm
30 cm²
a cm²

b) $rsf = \dfrac{small}{big} = \dfrac{14}{21} = \dfrac{2}{3}$ $b = \left(\dfrac{2}{3}\right)^2 \times 540 = 240\,cm^2$

21 cm   14 cm
540 cm²   b cm²

Exercise 4C

1   Find each unknown area:

a)
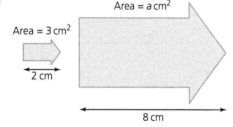
Area = a cm²
Area = 3 cm²
2 cm
8 cm

b)
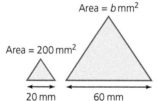
Area = b mm²
Area = 200 mm²
20 mm   60 mm

c)
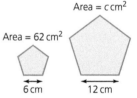
Area = c cm²
Area = 62 cm²
6 cm   12 cm

d)
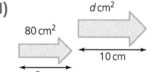
d cm²
80 cm²
10 cm
8 cm

e)
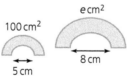
e cm²
100 cm²
5 cm   8 cm

f)
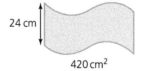
24 cm   12 cm
420 cm²   f cm²

2   A shop prices art prints in proportion to their area.
The smallest print costs £9.

Calculate the price of the medium and large prints.

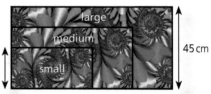

large
medium
small
30 cm   15 cm   45 cm

3   Use linear scale factor = $\sqrt{\text{area scale factor}}$ to calculate each unknown length.

a)

40 cm
Area = 2000 cm²
a cm
Area = 500 cm²

b)
 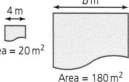
4 m   b m
Area = 20 m²
Area = 180 m²

c)

c cm
1 cm
Area = 4 cm²   Area = 9 cm²

▶ Volume scale factor

In general, volume scale factor = (linear scale factor)³, that is   vsf = (esf)³ or vsf = (rsf)³

To calculate unknown volumes of similar shapes:

● Remember to cube the linear scale factor. See the example below.

Worked example

The smaller cup holds 40 ml. Calculate the volume of the larger cup.

$esf = \dfrac{10}{4} = \dfrac{5}{2}$

larger volume $= \left(\dfrac{5}{2}\right)^3 \times 40 = 625$ ml

4 cm   10 cm

Exercise 4D

1   Calculate the unknown volumes:

a)

2 cm   6 cm
Volume = 30 ml   Volume = a ml

b)

6 cm   12 cm
Volume = 40 ml   Volume = b ml

c)
Volume = 20 m³   Volume = c m³

4 m

10 m

d)
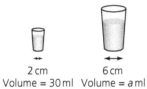

Volume = 540 cm³   Volume = d cm³

15 cm   10 cm

e)

28 cm   21 cm

Volume = 960 cm³   Volume = e cm³

f)
45 mm

22·5 mm

Volume = 1 litre   Volume = f ml

2   The smaller jug measures up to 200 ml and the large jug measures up to 3125 ml..

Use $esf = \sqrt[3]{vsf}$ to find the height of the larger jug.

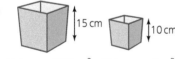

8 cm   h cm

3   Alan sees two similar water bottles for sale. The smaller costs £3·20 and the larger costs £9·50. Is the price proportional to the volume of the bottles? Show working to justify your answer.

12 cm   18 cm

4   Perfume is sold in 3 similar bottles. The prices are proportional to the volume of the bottle.

The smallest bottle costs £25. Calculate the cost of the medium and large bottles.

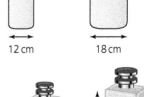

15 cm   21 cm   27 cm

▶ Areas of quadrilaterals

The area a shape occupies is measured in square units. To calculate the area of these quadrilaterals:

- State the correct formula.
- Substitute the correct dimensions and calculate.

Areas of Quadrilaterals	
Parallelogram	$A = b \times h$
Rhombus or Kite	$A = \frac{1}{2}d_1 \times d_2$
Trapezium	$A = \frac{1}{2}(a + b)h$

Worked example   Calculate the area of each shape:

a)
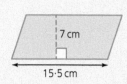

Area = base × height

$A = b \times h$

$= 15.5 \times 7$

$= 108.5 \, cm^2$

b)
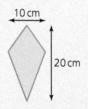

Area = $\frac{1}{2}$ diagonal$_1$ × diagonal$_2$

$A = \frac{1}{2}d_1 \times d_2$

$= \frac{1}{2} \times 10 \times 20 = 100 \, cm^2$

A trapezium has one pair of parallel sides.

c)
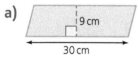

Area = $\frac{1}{2}$ × (sum of parallel sides) × height

$A = \frac{1}{2}(16 + 12)9$

$= \frac{1}{2} \times 28 \times 9 = 126 \, cm^2$

Exercise 4E

1 Calculate the area of each parallelogram.

a)
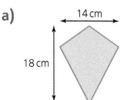

b)

c)
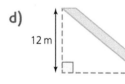

d)

2 Calculate the area of each kite or rhombus.

a)

b)

c)
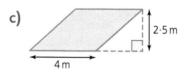

d)

3 Calculate the area of these trapeziums.

a)
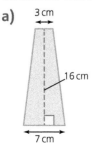

b)

c)
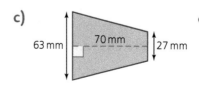

d)
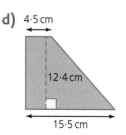

▶ Area of compound shapes

To calculate the area of compound shapes:

- Break the shapes down into simpler shapes. Take extra care to find the correct dimensions.
- Calculate the simpler areas. Add or subtract to calculate the required area.

> **Worked example** Calculate the total area of the shape.
>
> height of trapezium = 2 − 0·8 = 1·2 m
>
> 0·3 m
>
> A_2
>
> 2 m
>
> 0·8 m A_1
>
> 2·4 m
>
> $A_1 = l \times b$
> $= 2 \cdot 4 \times 0 \cdot 8$
> $= 1 \cdot 92\, m^2$
>
> $A_2 = \frac{1}{2}(a+b)h$
> $= \frac{1}{2}(2 \cdot 4 + 0 \cdot 3)1 \cdot 2$
> $= 1 \cdot 62\, m^2$
>
> Total area $= A_1 + A_2$
> $= 1 \cdot 92 + 1 \cdot 62$
> $= 3 \cdot 54\, m^2$

4   Calculate the area of each shape.

   a)   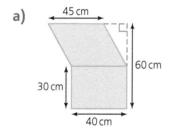   45 cm, 60 cm, 30 cm, 40 cm

   b)   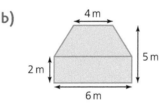   4 m, 5 m, 2 m, 6 m

   c)   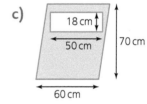   18 cm, 50 cm, 70 cm, 60 cm

   d)   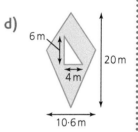   6 m, 4 m, 20 m, 10·6 m

5   A design for a conference centre carpet is shown.

   The design is a kite inside a parallelogram.

   Show that the kite occupies one quarter of the parallelogram.

   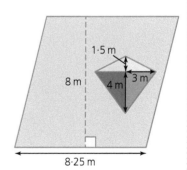

   1·5 m, 8 m, 4 m, 3 m, 8·25 m

6   The dimensions of a design are shown.

   The white rhombus is a half-size reduction of the overlapping yellow rhombuses.

   Calculate the area which is:

   a) white   b) yellow   c) grey.

   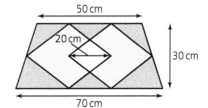

   50 cm, 20 cm, 30 cm, 70 cm

▶ Circumference of a circle

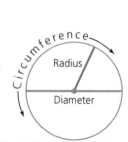

The **circumference** is the distance around the circle. A **diameter** is a line through the centre of the circle joining two points on the circumference. A **radius** is half of a diameter, from the centre to a point on the circumference.

Investigation

Continue this table with measurements from your own circular objects. Use a tape measure or string to measure the circumference.

How many times bigger than the diameter is the circumference?

Object	Diameter (cm)	Circumference (cm)	C ÷ D
tin base	7	22	
pencil pot	5	15·7	
guitar hole	7·5	23·6	
bin base	22	69	

We might guess 3 times bigger but it is always a little more than that. To calculate the ratio complete the last column in the table (circumference ÷ diameter). Round your answers to 2 decimal places.

The ratio of the circumference to the diameter is always π, $C = π × D$

Pi, π = 3·14159265358979323846... is an *irrational number*. Irrational numbers cannot be written as exact fractions. The digits go on and on forever without any pattern developing.

To calculate the circumference of a circle:

- Set down the formula $C = π × D$ or $C = πD$
- Substitute the correct length for the diameter and calculate. Use units of length.

Worked example

Find the circumference of each circle a) to 2 d.p. b) no calculator.

a)

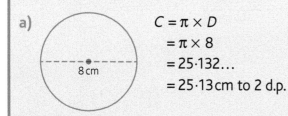

8 cm

$C = π × D$
$= π × 8$
$= 25·132...$
$= 25·13$ cm to 2 d.p.

Use the [π] button on your calculator for an exact answer.

Write down the **truncated** (cut off/ not rounded) answer to 3 d.p. and then round to 2 d.p.

b)

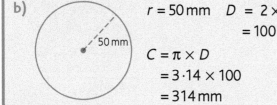

50 mm

$r = 50$ mm $D = 2 × 50$
        $= 100$ mm

$C = π × D$
$= 3·14 × 100$
$= 314$ mm

Remember: **the circumference formula uses the diameter.** Double the given radius.

For non-calculator work use 3·14 as a reasonable approximation.

Exercise 4F

Give your answers correct to 2 decimal places throughout unless told otherwise.

1   Calculate the circumference of each circle or circular object shown.

a)

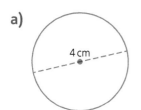

4 cm

b)

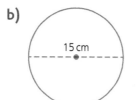

15 cm

c)

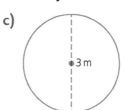

3 m

d)

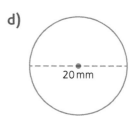

20 mm

e)
60 cm

f)
19 inches

g)
I ♥ Maths
34 mm

h)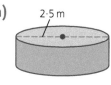
2·5 m

2   The table shows the diameter of some parts of a drum kit. Calculate the circumference of each drum.

Snare drum	Kick drum	Tom tom
14 inch	22 inch	12 inch

3   The diameter of the Earth is approximately 12 700 km. Calculate the approximate circumference.

4   A circular party cake has a diameter of 32 cm. The cake maker has 2 metres of red ribbon. Do they have enough ribbon to wrap around the cake twice?

5   Calculate the circumference of each circle. Be careful to calculate the **diameter** first.

a)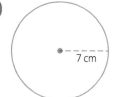
7 cm

b)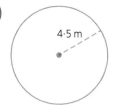
4·5 m

c)
18 cm

d)
14 inches

Worked example   Calculate the perimeter of this semicircular window correct to 2 decimal places.

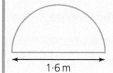
1·6 m

$C = \pi \times D$
$= \pi \times 1\cdot6$
$= 5\cdot026...$

length of semicircle
$= 5\cdot026 \div 2$
$= 2\cdot513...$

total perimeter
$= 2\cdot513 + 1\cdot6$
$= 4\cdot113...$
$= 4\cdot11$ m to 2 d.p.

---

6   Calculate the perimeter of each shape. Be careful not to round too soon.

a)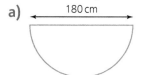
180 cm

b)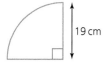
19 cm

c)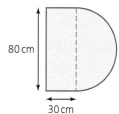
80 cm
30 cm

d)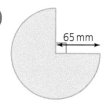
65 mm

7   Use $D = \dfrac{C}{\pi}$ to calculate the size of the diameter needed to draw a circle with circumference:

a) $C = 31\cdot4$ cm   b) $C = 25$ cm   c) $C = 60$ m   d) $C = 2$ km

8   Go back to Q1 a) – d) and find the answers without a calculator, using $\pi = 3\cdot14$. Compare your answers to those you obtained with a calculator. What differences do you notice?

▶ Area of a circle

Think about cutting and unwrapping a circle and stacking it into a triangle as shown. Every layer is a little shorter than the last and the more layers we make the more perfect the triangle would be.

$A = \pi r^2$

Area of a circle equals pi multiplied by the radius squared.

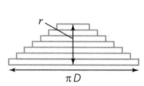

$$A = \frac{1}{2}bh$$
$$= \frac{1}{2} \times \pi \times D \times r$$
$$= \frac{1}{2} \times \pi \times r \times 2 \times r$$
$$= \pi r^2$$

To calculate the area of a circle:

● Set down the formula $A = \pi r^2$. Substitute the correct value for the radius.

● Calculate using the π button ($\pi = 3\cdot14$ for non-calculator work).

Worked example

Calculate the area of each circle or circular object. Round your answers to 2 d.p.

a)
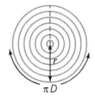
3 cm

$A = \pi r^2$
$= \pi \times 3^2$
$= 28\cdot274...$
$= 28\cdot27$ cm² to 2 d.p.

Non-calculator
$A = \pi \times 3^2$
$= 3\cdot14 \times 9$
$= 28\cdot26$ cm²

The 4th significant figure is different.

b)
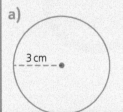
1·5 m

$D = 1\cdot5$m
$r = 1\cdot5 \div 2$
$= 0\cdot75$m

$A = \pi r^2$
$= \pi \times 0\cdot75^2$
$= 1\cdot767...$
$= 1\cdot77$m² to 2 d.p.

Remember that the **area** formula uses the **radius**.

First, find half of the diameter.

Exercise 4G   In questions 1–4 round your answers to 2 decimal places.

1 Calculate the area of each circle or circular object.

a)
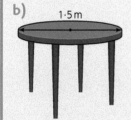
2 cm

b)
5 cm

c)
10 m

d)
20 mm

e)

45 cm

f)

11 inches

g)

I ♥ Maths
17 mm

h)

18 cm

2 Calculate the area of each circle. Be careful to calculate the **radius** first.

a)
24 cm

b)
7 m

c)
49 m

d)
120 mm

3 A shop sells pizzas with 8 inch and 12 inch diameters.

Joe thinks the area of the larger pizza is twice the area of the smaller. Is he correct? Show working to explain your answer.

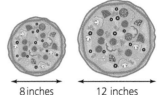
8 inches   12 inches

4 Three options for garden trampolines are shown below.

a) Calculate the area of each trampoline.

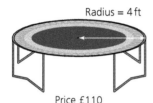
Radius = 4 ft

Price £110

Diameter = 9 ft

£229·99

6 ft
12 ft
£240

b) Which trampoline offers the biggest surface area for the money?

Worked example   A window is in the shape of a rectangle and semicircle. Calculate the area of the window. Give your answer correct to 3 significant figures.

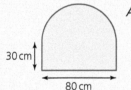
30 cm
80 cm

$A_{rectangle} = l \times b$
$= 80 \times 30$
$= 2400 \text{ cm}^2$

$r = 80 \div 2 = 40 \text{ cm}$

$A_{circle} = \pi r^2$
$= \pi \times 40^2$
$= 5026...$

$A_{semicircle} = 5026... \div 2$
$= 2513... \text{ cm}^2$

$A_{window} = A_{rectangle} + A_{semicircle}$
$= 2400 + 2513...$
$= 4913...$
$= 4910 \text{ cm}^2$ to 3 s.f.

Be very careful not to round too soon. If we had rounded the full circle to 3 s.f. (5030) and then halved (2515) our final answer would be wrong (4915 = 4920 to 3 s.f.).

---

5 Calculate the area of each window shape. Give your answer correct to 3 significant figures.

a)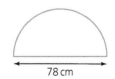
78 cm

b)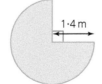
1·4 m

c)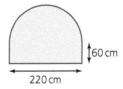
60 cm
220 cm

d) Careful!
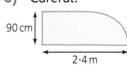
90 cm
2·4 m

6 A racetrack has two straight sections joined by two semicircles with dimensions shown.

a) Calculate the area enclosed by the track correct to 3 s.f.

b) Calculate the distance around the track correct to 3 s.f.

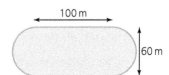
100 m
60 m

▶ Surface area: cuboid and triangular prism

A cuboid has six faces. To calculate the surface area of a cuboid:

● Calculate the total area of all six faces. Use squared units for area.

Worked example Find the total surface area of the cuboid.

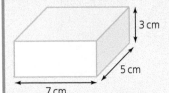

$A(\text{base}) = l \times b$ $A(\text{side}) = l \times b$ $A(\text{front}) = l \times b$

$\qquad\qquad = 7 \times 5 \qquad\qquad\quad = 5 \times 3 \qquad\qquad\quad = 7 \times 3$

$\qquad\qquad = 35\,\text{cm}^2 \qquad\qquad = 15\,\text{cm}^2 \qquad\qquad = 21\,\text{cm}^2$

$35 \times 2 = 70$
$15 \times 2 = 30$
$21 \times 2 = 42$
Total SA $= 142\,\text{cm}^2$

A cuboid has base and top, left side and right side, front and back.

Exercise 4H   1   Calculate the total surface area of each cuboid or cube.

a)

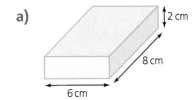

b)

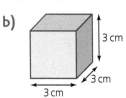

c)

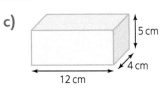

A triangular prism has three rectangular faces and two triangular faces. To calculate the surface area of a triangular prism:

● Calculate the total area of all five faces. Use squared units for area.

Worked example Calculate the total surface area of this triangular prism.

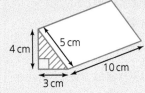

$A(\text{base}) = l \times b$ $A(\text{back}) = l \times b$ $A(\text{slope}) = l \times b$ $A(\text{end}) = \frac{1}{2}bh$

$\qquad\qquad = 10 \times 3 \qquad\qquad = 10 \times 4 \qquad\qquad = 5 \times 10$

$\qquad\qquad = 30\,\text{cm}^2 \qquad\qquad = 40\,\text{cm}^2 \qquad\qquad = 50\,\text{cm}^2$

$\qquad\qquad\qquad\qquad\qquad\qquad\qquad\qquad\qquad\qquad\qquad\qquad = \frac{1}{2} \times 3 \times 4$

$\qquad\qquad\qquad\qquad\qquad\qquad\qquad\qquad\qquad\qquad\qquad\qquad = 6\,\text{cm}^2$

The shape has base, back, slope and two ends. Total surface area = 30 + 40 + 50 + 6 + 6 = 132 cm²

2   Calculate the total surface area of each triangular prism.

a)

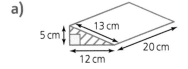

b)

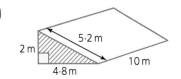

c)

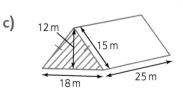

3   Francis is designing dishwasher tablets. Each tablet should be a cuboid with volume 72 cm³.

   a)   Sketch three cuboids which have volume 72 cm³. Label the *l*, *b*, and *h* clearly.

   b)   Calculate the surface area of your designs. Investigate which designs have the largest and smallest surface areas.

▶ Surface area of a cylinder

A cylinder has two circular faces and a curved surface. Think about the label on the curved surface of a tin of soup. What shape would it be if you laid it flat?

To calculate the surface area of a cylinder:

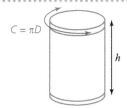

$C = \pi D$

$C = \pi D$

$A = l \times b$
$= \pi D \times h$
$= 2\pi rh$

h

- Set down the formula for curved surface area, CSA = $2\pi rh$, substitute radius and height and calculate.
- Add the area of the base and top using $A = \pi r^2$. Use squared units for area.

Worked example   Calculate the total surface area of the cylinder correct to 3 significant figures.

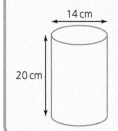

14 cm

20 cm

$A(\text{base}) = \pi r^2$
$= \pi \times 7^2$
$= 153 \cdot 9 \dots \text{cm}^2$

CSA $= 2\pi rh$
$= 2 \times \pi \times 7 \times 20$
$= 879 \cdot 6 \dots \text{cm}^2$

Total surface area
$= (153 \cdot 9 \dots \times 2) + 879 \cdot 6 \dots$
$= 1187 \cdot 5 \dots$
$= 1190 \, \text{cm}^2$ to 3 s.f.

A closed cylinder has base, top and curved surface area.

Exercise 4I   Throughout this exercise, round your answers to 3 significant figures.

1. Find the total surface area of each cylinder.

   a)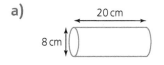
   20 cm
   8 cm

   b)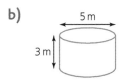
   5 m
   3 m

   c)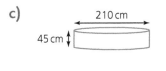
   210 cm
   45 cm

   d)
   13 mm
   24 mm

2. A flower vase is in the shape of an open cylinder. This means it has no top. Calculate the outer surface area of the vase.

25 cm

30 cm

3. The roller of a soil compactor is 3·5 m long with diameter 1·4 m. Calculate the area flattened by ten turns of the roller if the soil compactor travels in a straight line.

4. Beth keeps a note of formulas for surface area but she has smudged ink on her work. Copy out the table and fill in the missing formulae.

Shape	Total surface area
Cuboid	$A = 2lb + $ ⬤ $ + 2lh$
Closed cylinder	$A = 2 \times$ ⬤ $ + 2\pi r \times h$

▶ Volume of a prism

A **prism** is a 3D shape with a constant **cross-section**. We can imagine cutting the shape into identical slices.

The space a prism occupies depends on the area of its cross-section (slice) and the height of the prism.

The formula is:

Volume = area of cross-section × height

$V = A \times h$

To find the volume of a prism:

- Clearly state the formula: $V = A \times h$ or $V = Ah$.
- Substitute the correct values for the cross-sectional area A and height h.
- Calculate and include cubic units.

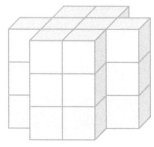

area (cross-section) = 8 cm²
$V = A \times h$
   = 8 × 3
   = 24 cm³

Worked example   Calculate the volume of each prism and give its name.

a)

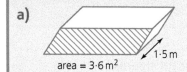

area = 3·6 m²   1·5 m

$V = A \times h$
$= 3·6 \times 1·5$
$= 5·4 \text{ m}^3$

This is a parallelogram-based prism.

The 'height' of a prism is the length between opposite identical faces.

b)

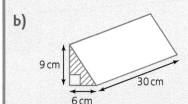

9 cm   30 cm   6 cm

First, calculate the area of the triangular cross-section.

$A = \frac{1}{2}bh$

$= \frac{1}{2} \times 6 \times 9$

$= 27 \text{ cm}^2$

This is a triangular prism.

Now calculate the volume of the prism.

$V = A \times h$
$= 27 \times 30$
$= 810 \text{ cm}^3$

Exercise 4J

1   Calculate the volume of each prism.

a)

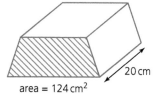

area = 124 cm²   20 cm

b)

30 mm

area = 175 mm²

c)

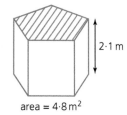

2·1 m

area = 4·8 m²

d)

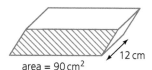

area = 90 cm²   12 cm

e)

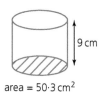

9 cm

area = 50·3 cm²

f)

30 mm

area = 1450 mm²

➡

2   Calculate the volume of each prism by first calculating the area of the cross-section.

a)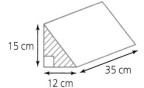
15 cm
35 cm
12 cm

b)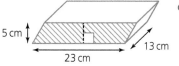
5 cm
23 cm
13 cm

c)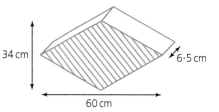
34 cm
6·5 cm
60 cm

d)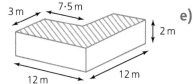
3 m   7·5 m
2 m
12 m   12 m

e)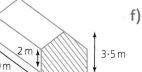
2 m
10 m
4 m
3·5 m

f)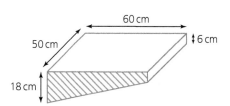
60 cm
50 cm
6 cm
18 cm

3   A hotel has three pools in the shape of prisms. The internal dimensions of the pool walls are shown.

a)   Calculate the volume of each pool.

It takes 90 minutes to drain the hexagonal pool and all the pools drain at the same rate.

b)   Calculate the time it will take to drain each of the other pools.

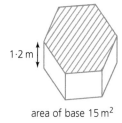
1·2 m
area of base 15 m²

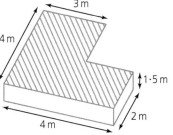
3 m
4 m
1·5 m
4 m
2 m

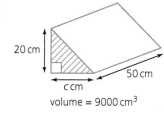
1 m
1 m
5 m

A square pool with a square island.

4   Calculate each unknown dimension.

a)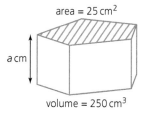
area = 25 cm²
a cm
volume = 250 cm³

b)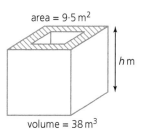
area = 9·5 m²
h m
volume = 38 m³

c)
20 cm
50 cm
c cm
volume = 9000 cm³

5   The volume of this building is 72 m³. Calculate the width of the building, *w*.

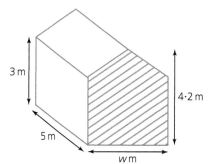
3 m
4·2 m
5 m
w m

▶ Volume of a cylinder

A cylinder is a circle-based prism. To find the volume of a cylinder:

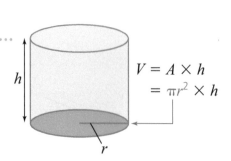

$$V = A \times h$$
$$= \pi r^2 \times h$$

- Clearly state the formula $V = \pi r^2 h$.
- Substitute the correct values for the radius r and perpendicular height h of the cylinder.
- Calculate and include cubic units.

Worked example

Calculate the volume of each cylinder, correct to 3 significant figures.

a)

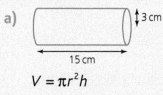

3 cm
15 cm

$V = \pi r^2 h$

$= \pi \times 3^2 \times 15$

$= 424 \cdot 1...$

$= 424 \text{ cm}^3$ to 3 s.f.

b) A water storage tank

$D = 5\,m$

$r = 5 \div 2$

$= 2 \cdot 5\,m$

$V = \pi r^2 h$

$= \pi \times 2 \cdot 5^2 \times 2 \cdot 8$

$= 54 \cdot 97...$

$= 55 \cdot 0\,m^3$ to 3 s.f.

5 m
2·8 m

Exercise 4K

Throughout this exercise round your answers to three significant figures.

1 Calculate the volume of each cylinder.

a)

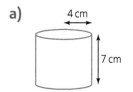

4 cm
7 cm

b)

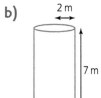

2 m
7 m

c)

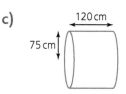

120 cm
75 cm

d)

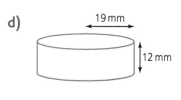

19 mm
12 mm

2 Calculate the volume of each cylinder. Take care to calculate the radius first.

a)

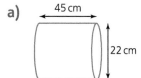

45 cm
22 cm

b)

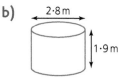

2·8 m
1·9 m

c)

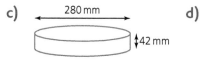

280 mm
42 mm

d)

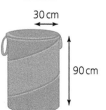

90 cm
1·8 m

3 Cameron has two bins for sorting his laundry. The bins are different shapes: one is a cuboid and one is a cylinder.

Which bin holds the most laundry?

80 cm
45 cm
60 cm

30 cm
90 cm

4 The diagram shows a cylindrical fish tank. The tank has diameter 48 cm and the water is 22 cm deep.

a) Calculate the volume of the water in the tank.

b) Calculate the volume of the water in the tank in litres.

Remember: $1\,cm^3 = 1\,ml$ and $1000\,ml = 1$ litre

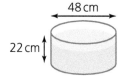

48 cm
22 cm

Worked example

The diagram shows a section of hollow concrete pipe.

The diameter of the pipe is 10 cm and the concrete is 2 cm thick all the way around. Calculate the volume of concrete used to make the pipe.

Give your answer correct to 3 significant figures.

Look carefully at the pipe. Do you see the concrete part is a large cylinder with a smaller cylinder taken away from the inside?

D_{large} = 10 cm

r_{large} = 10 ÷ 2

= 5 cm

$V_{large} = \pi r^2 h$

= $\pi \times 5^2 \times 50$

= 3926·9... cm³

r_{small} = 5 – 2

= 3 cm

$V_{small} = \pi r^2 h$

= $\pi \times 3^2 \times 50$

= 1413·7... cm³

$V_{pipe} = V_{large} - V_{small}$

= 3926·9... – 1413·7...

= 2513·2...

= 2510 cm³ to 3 s.f.

Using the prism formula, $A_{cross\text{-}section}$ = $\pi \times 5^2 - \pi \times 3^2$ and $V_{prism} = Ah$

= 50·26... cm²   = 50·26... × 50 = 2510 cm³ to 3 s.f.

---

5   Calculate the volume of each object.

a)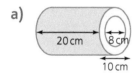

20 cm   8 cm

10 cm

b)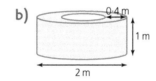

0·4 m

1 m

2 m

c)

2·3 m

140 cm

We have been using the calculator and π button to find answers. Our rounded answers are correct to 3 significant figures. We can also state the **exact value** of the answer as a multiple of π.

Worked example

For the concrete pipe example above:

V_{large} = $\pi \times 5^2 \times 50$

= 1250π

V_{small} = $\pi \times 3^2 \times 50$

= 450π

$V_{pipe} = V_{large} - V_{small}$

= 1250π – 450π

= 800π cm³

The exact answer is

800π cm³.

Check: 800π = 2510 to 3 s.f.

---

6   Match each of these shapes up with the exact values of their volumes. Work without a calculator.

A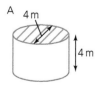

4 m

4 m

B

6 m

2 m

C

4 m

5 m

D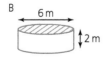

2 m   8 m

10π

16π

8π

18π

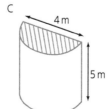

▶ Volume of a pyramid

Pyramids do not have a constant cross-section. If we slice a pyramid the slices would get smaller and smaller as we reached the **apex** or highest point. Pyramids are named after their base shape.

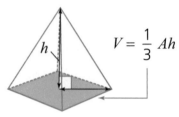
$V = \frac{1}{3}Ah$

The volume of any pyramid is $\frac{1}{3}$ of the volume of a prism with the same base and height. To find the volume of a pyramid:

square-based pyramid

- Clearly state the formula $V = \frac{1}{3}Ah$.

- Substitute the correct values for the area of the base A and the perpendicular height h.

- Calculate and include cubic units.

Worked example   Calculate the volume of each pyramid.

a)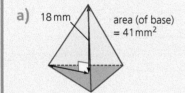
18 mm
area (of base) = 41 mm²

$V = \frac{1}{3}Ah$

$= \frac{1}{3} \times 41 \times 18$

$= 246\,\text{mm}^3$

b)
5 cm
6 cm
4 cm

$A = l \times b$
$= 6 \times 4$
$= 24\,\text{cm}^2$

$V = \frac{1}{3}Ah$

$= \frac{1}{3} \times 24 \times 5$

$= 40\,\text{cm}^3$

Exercise 4L

1   Calculate the volume of each pyramid.

a)
9 cm
area = 50 cm²

b)
15 cm
area = 210 cm²

c)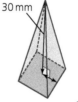
30 mm
area = 225 mm²

d)
12 cm
area = 36 cm²

e)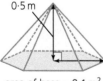
0·5 m
area of base = 8·1 m²

f)
0·9 m
area of base = 1·8 m²

g)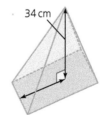
34 cm
area of base = 240 cm²

h)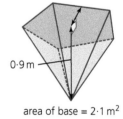
0·9 m
area of base = 2·1 m²

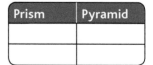

2   Which family of shapes do each of these objects belong to? Copy the headings and make a list, add more prism and pyramid shaped objects to each list.

Prism	Pyramid

 →

3 Alan eats one cube of fudge with length 15 mm. Martin has the same fudge in the shape of square-based pyramids with length 15 mm and height 15 mm.

How many of his pyramid sweets can Martin eat to have the same amount of fudge as Alan?

4 Calculate the volume of each pyramid or pyramid shaped object by first calculating the area of the base shape.

a)

19 cm

13 cm

6 cm

b)

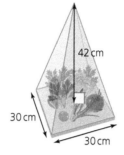

42 cm

30 cm

30 cm

c)

23 cm

18 cm   18 cm

Worked example

Calculate the volume of the gift box shown.

Can you see that the box is a cuboid with a rectangle-based pyramid on top?

We will find the volume of each and then add them together.

Height of pyramid = 24 − 15 = 9 cm

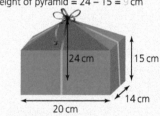

24 cm   15 cm

20 cm   14 cm

$$V_{cuboid} = l \times b \times h$$
$$= 20 \times 14 \times 15$$
$$= 4200 \text{ cm}^3$$

$$A_{pyramid\ base} = l \times b$$
$$= 20 \times 14$$
$$= 280 \text{ cm}^2$$

$$V_{pyramid} = \frac{1}{3} Ah$$
$$= \frac{1}{3} \times 280 \times 9$$
$$= 840 \text{ cm}^3$$

$$V_{total} = V_{cuboid} + V_{pyramid}$$
$$= 4200 + 840$$
$$= 5040 \text{ cm}^3$$

---

5 Calculate the volume of the gift box correct to 3 significant figures.

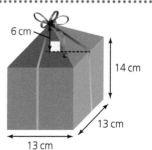

6 cm

14 cm

13 cm

13 cm

6 The diagram shows a candle in the shape of a **truncated** square-based pyramid. Can you see a smaller pyramid has been sliced off the top?

a) Calculate the volume of the smaller pyramid that has been removed.

b) What was the original height of the larger pyramid?

c) Calculate the volume of the original large pyramid.

d) Calculate the volume of the candle.

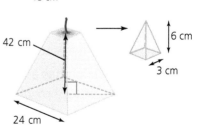

42 cm

6 cm

3 cm

24 cm

▶ Volume of a cone

A cone is a circle-based pyramid. To find the volume of a cone:

- Clearly state the formula $V = \frac{1}{3}\pi r^2 h$.
- Substitute the correct values for the radius of the base r and the perpendicular height h of the cone.
- Calculate and include cubic units.

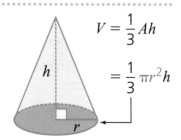

$$V = \frac{1}{3}Ah$$
$$= \frac{1}{3}\pi r^2 h$$

Worked example

Calculate the volume of the cone. Give your answer correct to three significant figures.

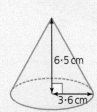

$$V = \frac{1}{3}\pi r^2 h$$
$$= \frac{1}{3} \times \pi \times 3\cdot6^2 \times 6\cdot5$$
$$= 88\cdot21\ldots$$
$$= 88\cdot2\ \text{cm}^3 \text{ to 3 s.f.}$$

Make use of the fraction button on your calculator.

Exercise 4M

Throughout this exercise give your answers correct to three significant figures.

1 Calculate the volume of each cone.

a)  6 cm, 4 cm

b) 9 cm, 5 cm

c) 18 mm, 10 mm

d) 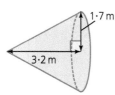 1·7 m, 3·2 m

e) Be careful. 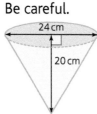 24 cm, 20 cm

f) 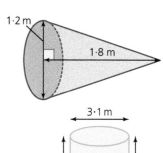 1·2 m, 1·8 m

2 The diagram shows a silo for ready-mixed mortar on a building site.

a) Find the volume of the cylinder to 2 decimal places.

b) Find the height of the cone.

c) Calculate the total volume of the silo.

3·1 m, 2·8 m, 5 m

3 A water station has a 5-litre bottle of water and small paper cone cups with the dimensions shown.

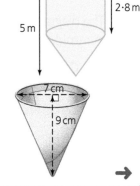
7 cm, 9 cm

Two friends attempt to calculate the number of full cups that can be filled from the bottle. Both have made a big mistake and both are wrong. Read through each attempt and describe the mistakes.

a) Attempt 1:   5 litre $= 5000\,\text{ml} = 5000\,\text{cm}^3$ **b)** Attempt 2:

$V = \dfrac{1}{3}\pi r^2 h$

$= \dfrac{1}{3} \times \pi \times 7^2 \times 9$

$= 461\cdot81\ldots$

$5000 \div 461\cdot81$
$= 10\cdot82$
$= 10\,\text{cups}$
(can't round up to fill 11)

X

$V = \dfrac{1}{3}\pi r^2 h$

$= \dfrac{1}{3} \times \pi \times 3\cdot5^2 \times 9$

$= 115\cdot45\ldots$

$= 100\,\text{cm}^3\ \text{to 1 s.f.}$

$5000 \div 100$
$= 50\,\text{cups}$

X

c) Show that the correct answer is 43 full cups.

4 Dominic is making jelly for a party. He wants to fill 20 identical cone-shaped moulds each with radius 3 cm and height 5·5 cm. Will 1 litre of jelly be enough? Show working to explain your answer.

Remember 1 litre = 1000 cm³.

5 A plastic pencil sharpener is made from a cube with a cone removed.

Calculate the volume available for the pencil sharpenings.

You may ignore the blade for sharpening the pencils.

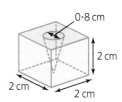

6 Two designs for a tent are shown. One is a cone and the other is a rectangle-based pyramid.

Calculate the difference between the volume of the tents.

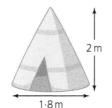
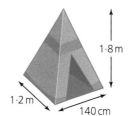

7 The structure of an ornament is shown. It is a cylinder enclosing two cones.

Calculate the volume of the cylinder which is not occupied by the cones.

8 This diagram shows a truncated cone. A smaller cone has been removed from the top. Shapes like these, where a solid shape is cut with a line parallel to the base, are also called **frustrums**.

Calculate the volume of the frustrum.

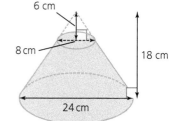

9 Libbie forms the similar triangle shown to help find the volume of this frustrum.

a) Find the length marked x on the similar triangle.

b) Hence calculate the volume of the frustrum.

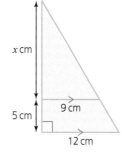
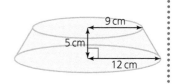

▶ Volume of a sphere

The volume of a sphere depends only on its radius. To find the volume of a sphere:

- Clearly state the formula: $V = \frac{4}{3}\pi r^3$.
- Substitute the correct value for the radius r of the sphere.
- Calculate and include cubic units.

Worked example

Calculate the volume of each object correct to 3 significant figures.

a)

6 cm

$$V = \frac{4}{3}\pi r^3$$

$$= \frac{4}{3} \times \pi \times 6^3$$

$$= 904 \cdot 7 \ldots$$

$$= 905 \text{ cm}^3 \text{ to 3 s.f.}$$

b) A toy

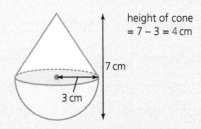

height of cone
$= 7 - 3 = 4$ cm

7 cm

3 cm

$V_{cone} = \frac{1}{3}\pi r^2 h$	$V_{sphere} = \frac{4}{3}\pi r^3$	$V_{hemisphere} = 113 \cdot 097\ldots \div 2$	$V_{toy} = V_{cone} + V_{hemisphere}$
$= \frac{1}{3} \times \pi \times 3^2 \times 4$	$= \frac{4}{3} \times \pi \times 3^3$	$= 56 \cdot 548\ldots$	$= 37 \cdot 699\ldots + 56 \cdot 548\ldots$
$= 37 \cdot 699\ldots$	$= 113 \cdot 097\ldots$		$= 94 \cdot 24\ldots = 94 \cdot 2$ to 3 s.f.

Using exact values $V_{cone} = 12\pi$, $V_{hemisphere} = 18\pi$, $V_{toy} = 18\pi + 12\pi = 30\pi = 94 \cdot 2$ to 3 s.f.

Exercise 4N

In Q1–4 give your answers correct to 3 significant figures.

1 Calculate the volume of each sphere.

a)

10 cm

b)

2 m

c)

5·4 m

d)

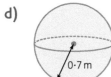
0·7 m

e)

38 cm

f)

4·5 cm

2 The table lists the diameters of different sports balls.

Sport	Diameter (cm)
Squash	4
Tennis	6·7
Football	22
Basketball	23·8

a) Calculate the radius of each ball.

b) Calculate the volume of each ball.

3   A cake cover is in the shape of a hemisphere with diameter 19 cm. Calculate the volume enclosed by the cover.

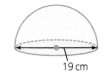

19 cm

4   A playground bollard is formed from a cylinder with radius 20 cm and height 50 cm, topped by a hemisphere which also has radius 20 cm.

Calculate the total volume of the bollard.

5   The diagram shows a paperweight made from a cone and hemisphere.

   a)   What is the height of the cone?

   b)   Calculate the volume of the paperweight. Give your answer correct to **2** significant figures.

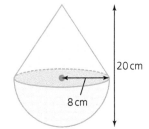

20 cm

8 cm

6   A vitamin pill is a cylinder with two hemisphere ends.

   a)   What is the radius of the hemispherical ends?

   b)   What is the height of the cylinder?

   c)   Calculate the volume of the pill. Give your answer correct to **4** significant figures.

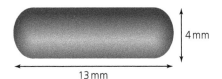

4 mm

13 mm

7   Try this question without a calculator, using π = 3·14 or comparing the exact values as multiples of π.

   The dimensions of two candles are shown. Anas thinks the spherical candle will burn for longer than the cylindrical candle. Is he correct? You can assume the candles burn at the same rate.

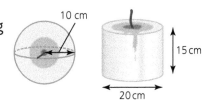

10 cm

15 cm

20 cm

8   Three identical tennis balls fit perfectly in a cylindrical tube.

   a)   Calculate the volume of one tennis ball. Leave your answer as an exact multiple of π.

   b)   Calculate the volume of the tube. Leave your answer as an exact value of π.

   c)   What fraction of the volume of the tube do the three balls occupy?

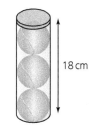

18 cm

9   Graeme is making bath bombs. He has two moulds which have the same radius. The height of the cylindrical mould is the same as its radius. Graeme has a set amount of ingredients and wants to make as many bath bombs as possible.

r

r

r

   a)   Which shape should Graeme choose? Justify your answer.

   b)   If Graeme can make 300 of the shape with the smaller volume, how many of the shape with the larger volume can he make?

▶ Volume problems

To solve for unknown dimensions:

- Set down the formula for the shape and substitute all known values.
- Use balancing steps to solve the equation and find the unknown value.

Worked examples

1   Find the length of the base of a square-based pyramid with height 7 cm and volume 84 cm³.

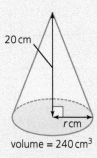

7 cm

a cm

volume = 84 cm³

$$V_{pyramid} = \frac{1}{3}Ah$$

$$84 = \frac{1}{3}A \times 7$$

$$\times 3 \quad \times 3$$

$$252 = A \times 7$$

$$\div 7 \quad \div 7$$

Area of base = 36 cm²

$$A_{square} = l \times b$$

$$36 = a \times a$$

$$36 = a^2$$

$$a = \sqrt{36}$$

$$a = 6 \text{ cm}$$

2   Calculate the radius of a cone with height 20 cm and volume 240 cm³ correct to 2 decimal places.

20 cm

r cm

volume = 240 cm³

$$V_{cone} = \frac{1}{3}\pi r^2 h$$

$$240 = \frac{1}{3}\pi r^2 20$$

$$\times 3 \quad \times 3$$

$$720 = \pi r^2 \times 20$$

$$\div 20 \quad \div 20$$

$$36 = \pi \times r^2$$

$$36 = \pi \times r^2$$

$$\div \pi \quad \div \pi$$

$$r^2 = \frac{36}{\pi}$$

$$r = \sqrt{\frac{36}{\pi}}$$

$$r = 3 \cdot 385...$$

$$r = 3 \cdot 39 \text{ cm to 2 d.p.}$$

Exercise 4O

1   Calculate the unknown dimension in each shape.

a)

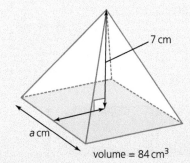

12 cm

a cm

volume = 100 cm³

b)

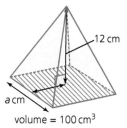

b cm

30 cm

18 cm

volume = 2700 cm³

c)

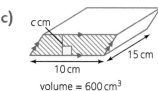

c cm

10 cm

15 cm

volume = 600 cm³

2 Use a calculator to find the unknown dimension in each shape. Give your answers correct to 2 decimal places.

a)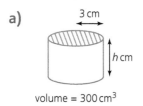

3 cm

h cm

volume = 300 cm³

b)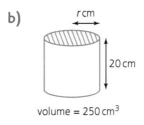

r cm

20 cm

volume = 250 cm³

c)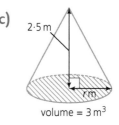

2·5 m

r m

volume = 3 m³

3 These two boxes have the same volume.

Find the height of the triangular prism.

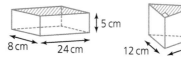

5 cm

8 cm   24 cm

h cm

12 cm   16 cm

4 A manufacturer sells 21-litre fish tanks in two shapes. Calculate the height of each tank correct to 2 significant figures.

21 litres = 21 000 cm³

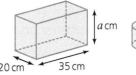

a cm

20 cm   35 cm

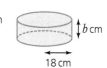

b cm

18 cm

5 In a gift pack of 3 candles all the candles have the same volume. Find the dimensions marked with a letter. Give your answers correct to 3 significant figures.

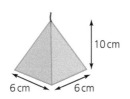

10 cm

6 cm   6 cm

a cm

8 cm

b cm

6 Laura is designing a pipe to carry waste-water. The pipe will be 1·5 m long and must have a minimum volume of 1·42 m³.

Calculate the minimum diameter of the pipe to the nearest centimetre.

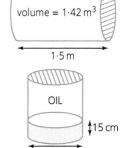

volume = 1·42 m³

1·5 m

7 A kitchen has a storage drum for oil to be recycled with the depth of oil shown. Michael tips 10 litres of waste oil into the drum.

Find the new depth of the oil to the nearest centimetre.

OIL

15 cm

40 cm

8 A jeweller smelts the small block of silver shown and uses it to make a sphere. We will assume that no silver is lost in the process.

Calculate the radius of the largest possible sphere to the nearest mm.

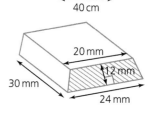

20 mm

12 mm

30 mm

24 mm

9 In this question you may find it easier to work with exact value multiples of π.

Lily is making chocolates. She melts a cylindrical block of chocolate with radius 5 cm.

In the process Lily will lose 10% of the chocolate.

Lily uses all the chocolate that is left to make 25 hemisphere chocolates, each with radius 1·5 cm.

Calculate the height of Lily's original cylindrical block of chocolate.

Check-up 👍

1. Find the unknown length in each pair of similar shapes.

a)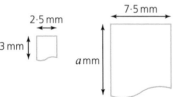
2·5 mm
7·5 mm
3 mm
a mm

b)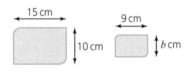
15 cm
9 cm
10 cm
b cm

c)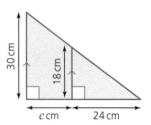
30 cm
18 cm
c cm
24 cm

2. A hotel has two mathematically similar pools with the dimensions shown.

   a) Calculate the linear enlargement scale factor.

   The base of the smaller pool has area 1 m².

   b) Calculate the area of the base of the larger pool.

   The smaller pool holds 320 litres of water.

   c) Calculate the volume of water in the larger pool.

0·6 m
1·5 m

3. Copy and complete these tables with the formulae you have learned in this chapter.

Circle	
Circumference	
Area	

Area of quadrilaterals	
Parallelogram	
Rhombus/Kite	
Trapezium	

Volume	
Prism	
Cylinder	
Pyramid	
Cone	
Sphere	

4. Calculate the circumference of each circle or circular object below. Give your answers to 2 d.p.

a)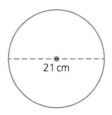
21 cm

b)
1·8 m

c)
10 cm

5. Calculate the area of each circle in question 4. Give your answers correct to 2 d.p.

6. A school library wants to buy new study desks that have an area of at least 2 m². Three options are shown. Which options would be suitable? Show working to justify your answer.

a)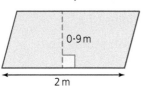
0·9 m
2 m

b)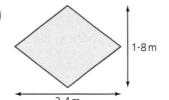
1·8 m
2·4 m

c)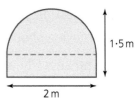
1·5 m
2 m

7 The dimensions of two boxes are shown. Every surface of the boxes is mirrored. Calculate the total surface area of mirror on each box. Give your answers correct to 3 significant figures.

a)

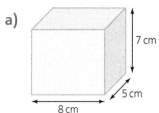

7 cm
5 cm
8 cm

b)

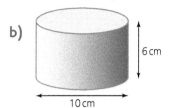

6 cm
10 cm

8 A chocolate factory makes a special mathematician's box of chocolates. The flavours are listed beside the volume of the chocolate.

Find out what flavour each chocolate is.

Flavour	Volume (to 3 s.f.) cm³
praline	2·79
truffle	2·83
caramel	3·25
toffee	1·5

a)

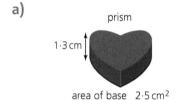

prism
1·3 cm
area of base 2·5 cm²

b)

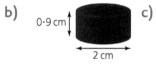

0·9 cm
2 cm

c)

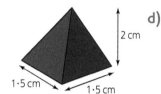

2 cm
1·5 cm   1·5 cm

d)

1·1 cm

9 The diagram shows a component which is a solid iron rhombus-shaped prism with a cylindrical hole drilled all the way through. Calculate the volume of iron in the component. Give your answer correct to 3 significant figures.

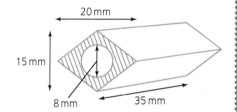

20 mm
15 mm
8 mm   35 mm

10 The diagram shows the dimensions of a plastic 3D model of an ice cream from outside a café.

a) Find the height of the conical part of the model.

b) Calculate the total volume of the model. Give your answer correct to 2 significant figures.

35 cm
95 cm

11 The dimensions of a water tank are shown. The water is 25 cm deep when a large solid sphere with radius 10 cm is dropped in. Will the water overflow? Show working to explain your answer.

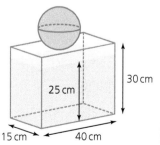

30 cm
25 cm
15 cm   40 cm

5 Algebra

▶ Expanding single brackets

Recall that $3x$ means '3 lots of x'. In the same way, $4(x + 2)$ means '4 lots of $(x + 2)$'. We could write this as $x + 2 + x + 2 + x + 2 + x + 2 = 4x + 8$. Note that we could have arrived at this answer simply by multiplying each term inside the bracket by the term outside. We call this process expanding brackets.

To expand a bracketed expression:

- Multiply each term inside the bracket by the term outside.

- When multiplying a pair of terms that involve both numbers and letters, multiply the numbers first and then the letters, taking care with indices.

- Take care with the signs of terms you are multiplying. Remember that multiplying a term by a negative changes its sign.

Worked examples

1 Expand:

a) $4(x+5)$

$= 4 \times x + 4 \times 5$

$= 4x + 20$

b) $9(x-6)$

$= 9 \times x + 9 \times (-6)$

$= 9x - 54$

c) $-3(4x+9)$

$= (-3) \times 4x + (-3) \times 9$

$= -12x - 27$

d) $-8(a-11)$

$= (-8) \times a + (-8) \times (-11)$

$= -8a + 88$

With practice, you should be able to go straight to the final answer.

e) $6(2x+7y)$

$= 12x + 42y$

f) $-5(9-7y)$

$= -45 + 35y$

g) $x(3+9x)$

$= 3x + 9x^2$

h) $2x(3x-2y)$

$= 6x^2 - 4xy$

2 Expand:

a) $2(x+2y+7)$

$= 2 \times x + 2 \times 2y + 2 \times 7$

$= 2x + 4y + 14$

b) $4x(9-3x+5y)$

$= 36x - 12x^2 + 20xy$

c) $3x(7x^2+3y-9)$

$= 21x^3 + 9xy - 27x$

Exercise 5A

1 Copy and complete to expand the following:

a) $3(x + 7)$

$= \square \times x + \square \times 7$

$= \square + \square$

b) $8(x - 4)$

$= \square \times x + \square \times \square$

$= \square - \square$

c) $7(2x + 5)$

$= 7 \times \square + \square \times \square$

$= \square + \square$

d) $8(3x - 11)$

$= \square \times \square + \square \times \square$

$= \square - \square$

2 Expand:

a) $2(x + 3)$

b) $4(x + 5)$

c) $7(x + 6)$

d) $9(2x + 5)$

e) $11(7x + 4)$

f) $6(x - 5)$

g) $8(x - 13)$

h) $12(3x - 2)$

i) $7(4x - 7)$

j) $19(4x - 3)$

→

   k) $8(3+x)$ l) $14(2-5x)$ m) $5(7+4y)$ n) $9(4-8x)$ o) $25(7+5x)$

   p) $3(2x+3y)$ q) $5(3x+4y)$ r) $7(x-2y)$ s) $9(9a+8b)$ t) $12(4a-7b)$

3 Expand:

   a) $-3(x+5)$ b) $-7(x+2)$ c) $-5(x+1)$ d) $-9(2x+13)$ e) $-10(5x+6)$

   f) $-4(x-2)$ g) $-8(x-9)$ h) $-3(4x-15)$ i) $-8(3x-16)$ j) $-13(3x-6)$

   k) $-2(2x+3y)$ l) $-6(2x+3y)$ m) $-2(x-2y)$ n) $-5(3a+7b)$ o) $-4(11a-9b)$

   p) $-13(5+x)$ q) $-9(4-3x)$ r) $-21(3+8y)$ s) $-17(5-6x)$ t) $-19(8+7x)$

4 Expand:

   a) $x(x+5)$ b) $x(2x+3)$ c) $x(4x+7)$ d) $x(8-3x)$ e) $x(12-7x)$

   f) $2x(3x+8)$ g) $3x(x-2)$ h) $4x(3x+5)$ i) $9x(3x+1)$ j) $4x(13x-8)$

   k) $5x(x+2y)$ l) $x(6x-y)$ m) $15x(2x+5y)$ n) $4a(8a-3b)$ o) $6a(2b+3a)$

   p) $17p(2p-5q)$ q) $8s(12t+5s)$ r) $21y(3y-7z)$ s) $2x(3x^2+14y)$ t) $5x(7x+2xy)$

5 Expand:

   a) $x(2x+y+3)$ b) $x(5x+4y+2)$ c) $x(4x+9y-11)$ d) $x(7-5x-2y)$

   e) $5x(2x+3y-5)$ f) $9x(7x+y-11)$ g) $12x(8-3x+12y)$ h) $17x(6+y-4x)$

   i) $10x(7x+6y+9)$ j) $3x(13x-7y-8)$ k) $x^2(3x+2y+6)$ l) $3x^2(4x^2-3xy+8y)$

If asked to simplify an expression involving brackets, we expand the brackets then collect any like terms. Note: Be careful when brackets have negative coefficients!

Worked example

Expand and simplify:

a)
$$4(2x+3)-7$$
$$=8x+12-7$$
$$=8x+5$$

b)
$$9x-3(6x-5)$$
$$=9x-[18x-15]$$
$$=9x-18x+15$$
$$=-9x+15$$

c)
$$2x(7x-4)-(4x^2+2)$$
$$=14x^2-8x-(4x^2+2)$$
$$=14x^2-8x-4x^2-2$$
$$=10x^2-8x-2$$

6 Expand and simplify:

a) $2(x+8)+3$ 　　　b) $5(x+3)+9$ 　　　c) $6(x+9)-8$ 　　　d) $4(x+7)-11$

e) $4(x^2+3x)-15$ 　f) $6x(x+7)+8x$ 　g) $3(2x+7)-6$ 　　h) $8(4x-3)+10$

i) $7(5x+2)+19$ 　j) $5(5x+1)-8$ 　　k) $9x(3x-2)-7x^2$ 　l) $10(3x+4)-28$

m) $-3(x+2)+9$ 　n) $-4(x-4)-1$ 　o) $-7(2x+3)+15$ 　p) $-9(4x-5)-12$

7 Expand and simplify:

a) $5+2(3x+8)$ 　　b) $3x+4(2x+5)$ 　　c) $12+3(7x-6)$ 　d) $7x+8(2x-11)$

e) $7-2(2x+3)$ 　　f) $8x^2-4x(x-7)$ 　g) $12-3(4x+5)$ 　h) $11-4(2x-1)$

i) $5-6(2x-9)$ 　　j) $14x+7(3-5x)$ 　k) $19-7(3x-1)$ 　l) $14x^2-3x(9-4x)$

m) $6x^2+3x(4x+1)-5$ n) $11x(2x-7)-13x+5$ o) $8x-3-4(5x-9)$ p) $-8(8x+3)-8x-3$

8 Expand and simplify:

a) $2(x+1)+4(x+7)$ 　　b) $7(x+3)+9(x+8)$ 　　c) $4(x+6)+5(x-3)$

d) $9(x+7)+3(x-5)$ 　　e) $2x(x-7)+4(x^2+8)$ 　f) $3(4x^2+1)+2x(3x-2)$

g) $6(2x-3)+4(7x-2)$ 　h) $5(x+3)-4(x+1)$ 　　i) $12(x-2)-8(x-5)$

j) $9(3x^2-2)-5x(6x+7)$ 　k) $14(5x-3)+4(8-3x)$ 　l) $7(9x+12)-3(6-5x)$

9 Match the cards that have the same expression when expanded and simplified. Write down the matching card letters.

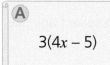

A

$3(4x-5)$

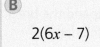

B

$2(6x-7)$

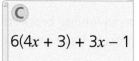

C

$6(4x+3)+3x-1$

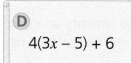

D

$4(3x-5)+6$

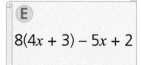

E

$8(4x+3)-5x+2$

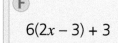

F

$6(2x-3)+3$

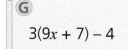

G

$3(9x+7)-4$

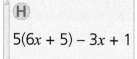

H

$5(6x+5)-3x+1$

10 Expand and simplify where possible:

a) $3x(x+2y)+y(5y+4x)$ 　b) $4x(3x+7y)+3y(4y+7x)$ 　c) $2x(x-5y)+2y(y-6x)$

d) $4x(x-5y)-3y(2y+4x)$ 　e) $6x(7x-3y)-5y(5y-8x)$ 　f) $9x(2x+7y)-4y(3x-2y)$

g) $x^2(3x+y)+y(2y^2-7x)$ 　h) $2x(x^3+2xy)-y(y^2+3x)$ 　i) $y^3(2x-y^2)-x^2(2y-3x)$

▶ Solving equations – recap

Recall that to solve an equation, we use the balancing technique.

- Perform balancing steps until you have a value for the variable.
- Check your solution by substituting the value back into the original expression.

Worked example

Solve

a)
$$4x + 3 = 39$$
$$-3 \quad -3$$
$$4x = 36$$
$$\div 4 \quad \div 4$$
$$x = 9$$

b)
$$6x - 7 = x + 8$$
$$-x \qquad -x$$
$$5x - 7 = 8$$
$$+7 \quad +7$$
$$5x = 15$$
$$\div 5 \quad \div 5$$
$$x = 3$$

c)
$$8x + 15 = 2x + 3$$
$$-2x \qquad -2x$$
$$6x + 15 = 3$$
$$-15 \quad -15$$
$$6x = -12$$
$$\div 6 \quad \div 6$$
$$x = -2$$

Exercise 5B

1   Solve:

a) $2x + 5 = 19$ b) $5x + 8 = 53$ c) $8x + 2 = 34$ d) $6x + 13 = 49$

e) $3x - 4 = 23$ f) $9x - 1 = 98$ g) $6x - 8 = 22$ h) $12x - 7 = 65$

i) $3x + 23 = 68$ j) $8x - 19 = 77$ k) $13x - 14 = 77$ l) $25x + 42 = 167$

m) $4x + 19 = 3$ n) $8x + 33 = 9$ o) $16x + 33 = 1$ p) $4x + 29 = 1$

2   Solve:

a) $4x + 5 = x + 11$ b) $6x - 1 = 2x + 23$ c) $9x + 3 = 2x + 52$ d) $8x - 7 = 3x + 33$

e) $12x + 23 = 4x + 39$ f) $7x - 5 = 4x + 13$ g) $3x + 7 = x - 1$ h) $9x + 6 = 2x - 22$

i) $4x - 9 = 2x - 3$ j) $8x + 17 = 5x + 2$ k) $12x + 7 = 3x - 38$ l) $11x - 19 = 5x - 73$

m) $3x + 1 = 7x - 3$ n) $2x - 3 = 7x + 17$ o) $3x + 12 = 11x - 52$ p) $10x - 13 = 17x + 71$

3   Solve the following equations, giving your answers as fractions in their simplest form.

a) $4x + 3 = 5$ b) $6x + 9 = 13$ c) $8x - 3 = 1$ d) $5x + 7 = 9$

e) $9x + 4 = x + 6$ f) $12x - 7 = 3x - 1$ g) $9x + 3 = 2x - 1$ h) $7x + 5 = x + 2$

i) $13x - 16 = 5x - 10$ j) $2x + 3 = 5x + 1$ k) $3x - 1 = 7x - 3$ l) $9x + 15 = 15x + 12$

m) $7x - 9 = 17x - 5$ n) $7x + 4 = 13x + 2$ o) $9x + 17 = 23x + 7$ p) $17x - 4 = 4x + 2$

4   Look at the solutions to the equations below. In each case, a mistake has been made. Describe the mistake in words and then solve the equation correctly.

a)
$$2x + 7 = 11$$
$$+7 \qquad +7$$
$$2x = 18$$
$$\div 2 \quad \div 2$$
$$x = 9 \quad \text{✗}$$

b)
$$6x - 3 = 2x - 11$$
$$-2x \qquad -2x$$
$$4x - 3 = 11$$
$$+3 \quad +3$$
$$4x = 14$$
$$\div 4 \quad \div 4$$
$$x = 3{\cdot}5 \quad \text{✗}$$

c)
$$2x + 13 = 9x - 29$$
$$-9x \qquad -9x$$
$$7x + 13 = -29$$
$$-13 \quad -13$$
$$7x = -42$$
$$\div 7 \quad \div 7$$
$$x = -6 \quad \text{✗}$$

▶ Solving equations

When solving an equation, we are trying to find the value that makes it 'balance'. To solve an equation:

- Expand brackets.
- Gather like terms.
- Take balancing steps.

Worked example

Solve:

a)
$$3(7x-2)+13=-77$$
$$21x-6+13=-77$$
$$21x+7=-77$$
$$-7\quad-7$$
$$21x=-84$$
$$\div 21\quad\div 21$$
$$x=-4$$

b)
$$3(2x+7)=2x+29$$
$$6x+21=2x+29$$
$$-2x\qquad-2x$$
$$4x+21=29$$
$$-21\;-21$$
$$4x=8$$
$$\div 4\quad\div 4$$
$$x=2$$

c)
$$4(3x+4)=7(x-2)$$
$$12x+16=7x-14$$
$$-7x\qquad-7x$$
$$5x+16=-14$$
$$-16\;-16$$
$$5x=-30$$
$$\div 5\quad\div 5$$
$$x=-6$$

Exercise 5C

1 Solve:

a) $4(x+3)+2=18$ b) $7(x-1)+4=39$ c) $8(x-4)-3=45$ d) $3(x-1)-16=44$

e) $5(x+3)-7=53$ f) $9(x+2)+6=42$ g) $3(x-2)+4=-17$ h) $6(x+9)-2=-32$

2 Solve:

a) $3(4x+1)=10x+15$ b) $9(6x-7)=24x-3$ c) $8(2x+1)=18x-6$

d) $3(3x-17)=4x+19$ e) $12(4x-9)=11x+3$ f) $8(5x+4)=5x-3$

g) $16(1-2x)=9x+16$ h) $5(3-8x)=4x-7$ i) $2(1-x)=2x+34$

3 Solve:

a) $3(2x-5)=7(x-3)$ b) $4(8x+1)=2(5x+24)$ c) $6(2x-13)=2(3x-9)$

d) $9(2x+3)=15(4x-1)$ e) $4(3x-7)=13(x-3)$ f) $4(5-2x)=-3(x+5)$

g) $3(10-x)=13(x+6)$ h) $7(5x+6)=-4(3-2x)$ i) $9(8+3x)=2(5x+2)$

4 Solve:

a) $7(3x-8)=3(2x+5)+4$ b) $4(4x-7)=8(3x-10)-20$ c) $5(6x+2)+7=7(2x-3)-10$

d) $9(6x+7)=7(2+3x)+16$ e) $5(4x+3)=11(2x+1)+4$ f) $4(2x-3)-7=9(8-x)-6$

g) $3(4x-3)=10(x-2)+35$ h) $8(x-2)-17=5(2x-11)+2$ i) $5-2(3x+1)=7(3-2x)-50$

When solving an equation involving one or more fractions:

- Multiply every term in the equation by the lowest common multiple (LCM) of the denominators.
- Simplify fractions.
- Expand brackets if necessary, then balance and solve as normal.

Worked example Solve:

a) $\dfrac{2x}{3} + 4 = 10$

$\times 3 \ \times 3 \ \times 3$

$2x + 12 = 30$

$-12 \ \ -12$

$2x = 18$

$\div 2 \ \div 2$

$x = 9$

Note that $\dfrac{2x}{3}$ multiplied by

3 is simply $2x$, since

$\dfrac{2x}{3} \times 3 = \dfrac{2x}{3} \times \dfrac{\cancel{3}}{1} = \dfrac{2x}{1} = 2x.$

b) $\dfrac{3x+5}{2} = \dfrac{4x+9}{5}$

$\dfrac{10(3x+5)}{2} = \dfrac{10(4x+9)}{5}$

$5(3x+5) = 2(4x+9)$

$15x + 25 = 8x + 18$

$-8x \ \ \ \ -8x$

$7x + 25 = 18$

$-25-25$

$7x = -7$

$\div 7 \ \div 7$

$x = -1$

LCM(2, 5) = 10 so we multiply both fractions by 10.

Alternatively, we could solve (b) by 'cross-multiplying'.

$\dfrac{3x+5}{2} \diagdown\!\!\!\!\diagup \dfrac{4x+9}{5}$

$5(3x+5) = 2(4x+9)$

$15x + 25 = 8x + 18$

$-8x \ \ \ \ -8x$

$7x + 25 = 18$

$-25 \ -25$

$7x = -7$

$\div 7 \ \div 7$

$x = -1$

Exercise 5D

1 Solve:

a) $\dfrac{x}{3} + 2 = 7$

b) $\dfrac{x}{2} - 4 = 5$

c) $\dfrac{x}{4} + 5 = 11$

d) $\dfrac{x}{8} - 3 = 1$

e) $\dfrac{x}{6} + 8 = 17$

f) $\dfrac{x}{5} - 6 = -4$

g) $\dfrac{x}{4} - 7 = -3$

h) $\dfrac{x}{2} - 19 = -13$

2 Solve:

a) $\dfrac{2x+1}{3} - 2 = 5$

b) $\dfrac{4x-3}{11} + 7 = 10$

c) $\dfrac{3x+1}{5} - 4 = 1$

d) $\dfrac{8x+6}{3} - 2 = 8$

e) $\dfrac{7x-9}{2} + 7 = 6$

f) $\dfrac{11x-8}{3} - 5 = 7$

g) $\dfrac{6x-3}{25} - 7 = -4$

h) $\dfrac{4x-7}{5} + 8 = 17$

3 Solve:

a) $\dfrac{3x-1}{2} = \dfrac{4x+2}{3}$

b) $\dfrac{3x+8}{4} = \dfrac{6x+1}{5}$

c) $\dfrac{7x-4}{5} = \dfrac{9x+2}{10}$

d) $\dfrac{5x+3}{4} = \dfrac{3x-1}{2}$

e) $\dfrac{2x+3}{9} = \dfrac{4x-3}{15}$

f) $\dfrac{4x+13}{3} = \dfrac{5-x}{2}$

g) $\dfrac{2x-3}{5} = \dfrac{3x-3}{7}$

h) $\dfrac{5x-7}{2} = \dfrac{9x-5}{4}$

4 Solve:

a) $\dfrac{3x+1}{2} + \dfrac{x}{3} = \dfrac{7x+3}{4}$

b) $\dfrac{4x+1}{3} + \dfrac{x}{2} = \dfrac{5x-1}{3}$

c) $\dfrac{5x+1}{3} - \dfrac{3x-1}{10} = \dfrac{7x+11}{6}$

▶ Evaluating expressions

When asked to 'evaluate' an expression, we substitute given values into the variables in the expression.

- Substitute the given values into the expression.
- Use BIDMAS to evaluate the expression in the correct order.

Worked example

Given $a = 4, b = 3, c = -5$ and $d = -1$, evaluate:

a) $5(7b+2c)$
$= 5(7 \times 3 + 2 \times (-5))$
$= 5(21 - 10)$
$= 5 \times 11$
$= 55$

b) $(a+8d)^3$
$= (4 + 8 \times (-1))^3$
$= (4-8)^3$
$= (-4)^3$
$= -64$

c) $3d - 2(5 - 6b)$
$= 3 \times (-1) - 2(5 - 6 \times 3)$
$= -3 - 2(5 - 18)$
$= -3 - 2(-13)$
$= -3 + 26$
$= 23$

d) $\dfrac{4-c}{3d}$
$= \dfrac{4-(-5)}{3 \times (-1)}$
$= \dfrac{9}{-3}$
$= -3$

Exercise 5E

1   Given $a = 9, b = 5$ and $c = 2$, evaluate:

   a) $5a - 2b$
   b) $8b - 7a$
   c) $7c - 4b$
   d) $ab + 3c$

   e) $c^3 - 4a$
   f) $a^2 + 6b$
   g) $\dfrac{12b}{a+6}$
   h) $\dfrac{20a}{3c^2}$

2   Given $p = 2, q = -1$ and $r = 3$, evaluate:

   a) $4(p-r)$
   b) $5(2q+r)$
   c) $7q(r + 5)$
   d) $4r(3-2q)$

   e) $3(6p-7q)$
   f) $p(q-r)$
   g) $3q(2p-9)$
   h) $5p(3r-4)$

3   Given $a = 5, b = -3$ and $c = 4$, evaluate:

   a) $5b - c^2$
   b) $3ab^2$
   c) $(c-b)^2$
   d) $(2b)^2 - 4a$

   e) $8(a+b)$
   f) $(5c-b)^2$
   g) $(2a-13)^3$
   h) $(4a+5b)^3$

4   Given $x = 8, y = -3$ and $z = -2$, evaluate:

   a) $\sqrt{2x}$
   b) $\sqrt{4x-7}$
   c) $\sqrt{yz+3}$
   d) $\sqrt{z^2 + 4x}$

   e) $\sqrt{13-4y}$
   f) $\sqrt{1-z^3}$
   g) $5x - \sqrt{1-y}$
   h) $\sqrt{40-3y}+3z$

5   Given $w = 5, x = -4, y = -2$ and $z = 7$, evaluate:

   a) $\dfrac{2w}{y}$
   b) $\dfrac{yz-1}{-3}$
   c) $\dfrac{6(z-1)}{16-y}$
   d) $\dfrac{21}{z+x}$

   e) $\dfrac{8z-16}{5y^2}$
   f) $\dfrac{-10}{3z+8y}$
   g) $\dfrac{9}{3w-2z}$
   h) $\dfrac{4(2w+3x)}{y^3}$

➜

6   In these number pyramids, the value in each box is equal to the sum of the values in the two boxes directly underneath. Evaluate each expression and complete the pyramids.

   a) $a = 14, b = -8, c = -3$

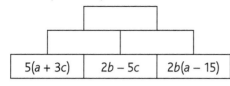

| $5(a + 3c)$ | $2b - 5c$ | $2b(a - 15)$ |

   b) $a = -4, b = 5, c = -2$

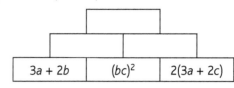

| $3a + 2b$ | $(bc)^2$ | $2(3a + 2c)$ |

   c) $d = 7, e = 3, f = -9$

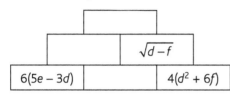

   d) $d = 12, e = -3, f = -2$

7   In this question, $a = 6, b = -1, c = 3$ and $d = -4$.

   Evaluate each expression and then use the table below to convert your answer to a letter. You will then unscramble your letters to make a maths word.

A	B	C	D	E	F	G	H	I	J	K	L	M	N	O	P	Q	R	S	T	U	V	W	X	Y	Z
1	2	3	4	5	6	7	8	9	10	11	12	13	14	15	16	17	18	19	20	21	22	23	24	25	26

   a) $\dfrac{18}{b+c}$ $\dfrac{a-d}{2}$ c^3-9 $3bd$ $-d+b$ $\dfrac{2a+3}{3-2b}$

   b) b^4 $c(10+d)+b$ $\sqrt{c^2+d^2}$ $\dfrac{9d}{a-8}+c$ $d(b-6)-(c^2+1)$ d^2+c

   c) $\dfrac{7}{a-b}$ $\sqrt{\dfrac{(a-d)^2}{4}}$ $3b(4d+10)$ $\sqrt{10-3c}$ $\dfrac{-14}{a+2d}$ $\sqrt[3]{\dfrac{48}{a}}$ $\dfrac{6(c^2-b)}{5}$

8   In this question $p = 4, q = -3, r = -1$ and $s = 5$.

   Every row and column in the square below must contain a $-2, -1, 0, 1$ and 2 once and only once. Draw an empty 5×5 square in your jotter. Evaluate the clues given and enter your answer in the corresponding square of your grid to get you started. Complete the empty squares with the numbers remaining.

	$\dfrac{-2}{s+q}$		$\sqrt{25+2pq}$	
q^2-2p		$q-r$		$\sqrt{6p+8q}$
$\dfrac{q^3}{30+q}$			$3p+5(q-r)$	$\dfrac{50}{3q-4p}$
		$3s-p^2$	q^2p-36	
$s^2-p^2-q^2$				

▶ Inequalities

An equation relates two expressions that are equal to each other. We can relate pairs of expressions that are not equal by using inequalities. The inequality symbols and their definitions are as follows:

< means 'less than', e.g. $x < 7$ means 'x is less than 7'.

> means 'greater than', e.g. $x > -2$ means 'x is greater than –2'.

$\leq$ means 'less than or equal to', e.g. $a \leq -5$ means 'a is less than or equal to –5'.

$\geq$ means 'greater than or equal to', e.g. $b \geq 12$ means 'b is greater than or equal to 12'.

An easy way to remember how the symbols work is that the larger end (the 'open' end) points to the larger value.

Worked example

Place the symbol < or > between the following pairs of numbers to produce a valid inequality.

a) 3 ☐ 5	3 is less than 5 so the appropriate symbol is <	3 < 5
b) 8 ☐ –2	8 is greater than –2 so the appropriate symbol is >	8 > –2
c) –7 ☐ 10	–7 is less than 10 so the appropriate symbol is <	–7 < 10
d) –12 ☐ –30	Be careful! –12 is greater than –30 (it is further up the number line) so the appropriate symbol is >	–12 > –30

Exercise 5F

1   Place the symbol < or > between each pair of numbers to produce a valid inequality.

a) 5 ☐ 8   b) 13 ☐ 9   c) 1 ☐ 12   d) 14 ☐ 13   e) –6 ☐ 4   f) 3 ☐ –3

g) –10 ☐ –7   h) –25 ☐ –30   i) 0 ☐ –1   j) –17 ☐ –15   k) 0·25 ☐ 0·5   l) –1·8 ☐ –1·81

2   Write each of the following statements as an inequality.

a) Five is less than ten
b) Six is greater than one
c) Negative four is less than zero
d) One half is greater than one quarter
e) x is less than or equal to two
f) x is greater than negative nine
g) Eight is greater than or equal to x
h) Forty-six is less than x
i) $2x$ is less than or equal to x plus one
j) $10x$ is greater than negative two fifths

3   The equivalent statement to '3 is less than 5' is ' 5 is greater than 3'. That is, 3 < 5 is equivalent to 5 > 3.
Copy and complete the pairs of equivalent statements.

a) 3 < 5   b) 4 > 1   c) 12 < 20   d) –5 ☐ 3   e) –1 ☐ –7
    5 > ☐   1 ☐ 4   ☐ > 12   3 ☐ –5   –7 ☐ –1

f) –3 < 1   g) –10 ☐ –13   h) x < 10   i) 4 ≥ x j) 8 ≤ y
    1 ☐ –3   –13 ☐ –10   ☐ > ☐   ☐ ≤ 4 y ☐ 8

4   Which values from the set {–23, –19, –5, 0, 4, 5, 7, 13} satisfy each of the following inequalities?

a) $x > 3$ b) $x \leq 0$ c) $x > 7$ d) $-5 \leq x \leq 5$ e) $-30 \leq x \leq -15$

▶ Solving inequalities

To solve inequalities, we follow the same procedures as we would to solve equations, namely balancing terms. Remember that we are not solving an equation, so we must not replace the inequality symbol by an 'equals'!

We must take care when multiplying or dividing by a negative number as this changes the direction of the inequality symbol. For example, consider the inequality 3 < 8. If we multiply both sides by −1 we have −3 and −8. But since −3 is greater than −8 we must write −3 > −8.

Worked example Solve the following inequalities:

a)
$$4x + 5 < 29$$
$$-5 \quad -5$$
$$4x < 24$$
$$\div 4 \quad \div 4$$
$$x < 6$$

b)
$$3x + 11 \le 6x + 35$$
$$-3x \qquad -3x$$
$$11 \le 3x + 35$$
$$-35 \quad -35$$
$$-24 \le 3x$$
$$3x \ge -24$$
$$x \ge -8$$

Note that when we 'flip' the inequality, the sign must change direction.

c)
$$3(2x + 9) \le 7(x - 2)$$
$$6x + 27 \le 7x - 14$$
$$-7x \qquad -7x$$
$$-x + 27 \le -14$$
$$-27 \quad -27$$
$$-x \le -41$$
$$\div -1 \quad \div -1$$
$$x \ge 41$$

Exercise 5G

1 Solve:

a) $x + 3 < 5$
b) $x + 19 > 40$
c) $x - 4 \le 6$
d) $x - 7 > 9$
e) $x + 2 \le 1$
f) $x + 13 \ge 10$
g) $x - 8 < -6$
h) $x + 4 < -2$
i) $7 + x \ge 0$
j) $12 + x < 3$
k) $4x < 16$
l) $3x > 24$
m) $7x \le 21$
n) $9x < 63$
o) $12x \ge 60$
p) $7x < 91$
q) $2x < -8$
r) $-3x \ge -9$
s) $-5x > 45$
t) $-14x \le -56$

2 Solve:

a) $3x - 4 < 17$
b) $2x + 8 > 26$
c) $7x - 3 \le 53$
d) $8x + 11 < 27$
e) $5 - x > 7$
f) $7 - 2x > -3$
g) $15 \le 2x - 3$
h) $12 \ge 7 - 5x$

3 Solve:

a) $2x - 7 \ge x + 11$
b) $6x + 4 > 2x + 32$
c) $9x + 5 < 2x - 16$
d) $12x - 18 \ge x - 7$
e) $6x + 5 > 8x - 1$
f) $9 - 3x \le x - 7$
g) $4 - 5x > 2x + 25$
h) $4 - 2x > 3x + 29$

4 Solve:

a) $8(x + 3) \ge 6x + 38$
b) $4(6x + 9) > -12$
c) $-5(2x + 3) \ge 3(7 - 3x)$
d) $7(2x - 7) - 12 \ge 5(x + 4)$
e) $2(8 - 3x) < 7(x - 8) - 5x$
f) $3(4x + 7) \le 2(2x - 8) - 3$

5 Solve:

a) $\dfrac{8(2x - 7)}{9} \le 2x + 6$
b) $\dfrac{3x + 8}{4} < \dfrac{6x + 1}{5}$
c) $\dfrac{7(x + 3)}{3} < \dfrac{5x - 12}{6}$
d) $\dfrac{2x - 9}{5} > \dfrac{4x - 19}{11}$
e) $\dfrac{6(4x - 5)}{11} < \dfrac{3(5x - 12)}{4}$
f) $\dfrac{5(3x + 8)}{8} \ge \dfrac{2(x - 7)}{3}$

▶ Forming equations and inequalities

Being able to take a problem in context and transform it into an equation or inequality is a crucial skill in Mathematics. Once a problem has been framed in this way we can use all our algebraic skills to solve it. Forming and solving equations is fundamental in professions as diverse as finance, engineering and weather forecasting.

To form an equation or inequality from a given problem:

- Identify the variables in the problem (the things that can change or that we are trying to find out).
- Identify the operations that are applied to the variables (multiplication, addition etc.).
- Write these down as an equation or inequality and solve to find the unknown(s).

Worked examples

1 Isma thinks of a number. She subtracts 6 from it then multiplies the answer by 8 to achieve a final answer of 72. Write down and solve an equation to find Isma's starting number.

Let x be Isma's starting number.	x	$8(x-6)=72$
She subtracts 6 from it.	$x-6$	$8x-48=72$
Then multiplies the answer by 8.	$8(x-6)$	$8x=120$
Her answer is 72.	$8(x-6)=72$	$x=15$
		Isma started with 15.

2 James has some marbles. Craig has three fewer marbles than James does. Carys has half as many marbles as James. Altogether, they have 32 marbles. How many marbles do they each have?

Let x be the number of marbles James has.	x	$x+x-3+\dfrac{x}{2}=32$	James: x James has 14 marbles
Craig has three fewer.	$x-3$	$2x-3+\dfrac{x}{2}=32$	Craig: $x-3$
Carys has half as many.	$\dfrac{x}{2}$	$4x-6+x=64$	Craig has 11 marbles.
Altogether they have 32 marbles.	$x+x-3+\dfrac{x}{2}=32$	$5x-6=64$	Carys: $\dfrac{x}{2}$
		$5x=70$	Carys has 7 marbles.
		$x=14$	

3 A stall in a food market costs £100 for the day. Irene is selling home baking for £3 per item and it costs her £4 in petrol to get to the market. Form and solve an inequality to find the minimum number of items Irene must sell to make a profit.

Let x be the number of items Irene sells.	x	$3x-4>100$
Each costs £3.	$3x$	$3x>104$
She pays £4 for petrol.	$3x-4$	$x>\dfrac{104}{3}$
For a profit, this must be greater than £100.	$3x-4>100$	$x>34\dfrac{2}{3}$
		She must sell 35 items to make a profit.

Exercise 5H

1   Sean has some marbles. Rowan has 7 fewer marbles than Sean. Let x be the number of marbles Sean has.

   a)   Write down an expression for the number of marbles Rowan has.

   b)   Together they have 35 marbles. Use this information to form an equation.

   c)   Solve your equation to find the number of marbles each has.

2   Jenna, Hughie and Pauline buy some raffle tickets at the school fayre. Hughie buys 3 more tickets than Jenna. Pauline buys twice as many tickets as Hughie. Let j be the number of tickets Jenna buys.

   a)   Write down an expression for the number of tickets Hughie has.

   b)   Write down an expression for the number of tickets Pauline has.

   c)   Altogether they have 37 tickets. Form an equation and solve it to find the number of tickets each person has.

3   A theme park sells child and adult tickets. An adult ticket costs £5 more than a child ticket. Let c be the cost of a child ticket.

   a)   Write down an expression for the cost of an adult ticket.

   b)   Two adult tickets and three child tickets come to a total cost of £45. Write down an equation and solve it to find the cost of a child ticket and the cost of an adult ticket.

4   Diane is organising a charity fundraiser. She sells tickets for £4 each and pays £30 for a band to play. The cost of hiring the community centre is £150. Form and solve an inequality to find the minimum number of tickets Diane must sell to make a profit.

5   Jenny has a budget of £300 to decorate her conservatory. Potted plants cost £15 each and a local decorator will paint the room for £200. Form and solve an inequality to find the maximum number of potted plants Jenny can buy if she also has the room painted.

Worked example   A taxi firm charges a flat rate of £4 for a hire plus £1·20 per mile travelled.

   a)   Write down a formula for the cost in pounds C of a journey of m miles.

   We call the quantity we are trying to find the subject of the formula, and write it on the left-hand side. The cost per mile is £1.20, so we will multiply the number of miles by 1·2.

   The flat rate is fixed, so it is added on. $C = 1 \cdot 2m + 4$

   b)   Use your formula to calculate the cost of a 13-mile journey.

   The number of miles is 13, so we $C = 1 \cdot 2m + 4$
   substitute 13 in place of m in our $= 1 \cdot 2 \times 13 + 4$
   formula and evaluate. $= 15 \cdot 6 + 4$

   We must finally give our answer in context. $= 19 \cdot 6$ The cost of the journey is £19·60.

   c)   A particular journey cost £25·60. Use your formula to determine the length of the journey.

   The cost is £25·60 so we must $C = 1 \cdot 2m + 4$
   substitute 25·6 in place of C and solve $25 \cdot 6 = 1 \cdot 2m + 4$
   the equation for m. $1 \cdot 2m = 21 \cdot 6$
                                          $m = 21 \cdot 6 \div 1 \cdot 2$
                                          $= 18$ The journey was 18 miles.

6 A courier service charges a base rate of £3 for a delivery plus 15p for every mile that the parcel is transported.

   a) Write down a formula for the cost C pounds of a delivery of m miles.

   b) Use your formula to calculate the cost of delivering a parcel to a location 20 miles away.

   c) One particular delivery costs £8·40. Use your formula to find the distance that the parcel was transported.

7 A gym barbell weighs 20 kg and can be fitted with pairs of 20 kg plates to increase the weight.

   a) Write down a formula for the weight W kg of a barbell fitted with p pairs of 20 kg plates.

   b) Use your formula to calculate the weight of a barbell fitted with 7 pairs of 20 kg plates.

   c) One weightlifter is exercising with a 500 kg barbell. Use your formula to determine the number of pairs of 20 kg plates fitted.

8 A street vendor sells mixed fruit and nuts in 50 g scoops. There is a base cost of 50p plus 30p per scoop.

   a) Write down a formula for the cost C pence for s scoops of fruit and nuts.

   b) Use your formula to calculate the cost of a bag containing 9 scoops. Give your answer in pounds and pence.

   c) A particular bag cost £6·20. Use your formula to determine the weight of fruit and nuts purchased.

9 A plumber charges a call-out fee of £30 and an hourly rate of £12. They also charge for any parts used.

   a) Write down a formula for the earnings in pounds (E) of a plumber who is called out for a job that lasts h hours and requires P pounds worth of parts.

   b) Use your formula to calculate the cost of a job that lasts 4 hours and requires £56 in parts.

   c) A particular job comes to a total cost of £398 and requires £260 in parts. Use your formula to calculate the number of hours the job takes.

   d) Another job comes to a total cost of £196·50 and takes 7 hours to complete. Use your formula to determine the cost of the parts used.

10 Andrew uses an online system to mark the homework for his classes. It takes him 10 minutes to log on to the system. Each junior homework takes 8 minutes to mark and each senior homework takes 11 minutes.

   a) Write down a formula to find the time, t minutes, it will take to log on and mark j junior homework exercises and s senior homework exercises.

   b) Use your formula to calculate the time it would take to log on and mark 32 junior homework exercises and 15 senior homework exercises.

   c) One night, Andrew takes 4 hours and 4 minutes to log on and complete his marking. This includes 10 junior homework exercises. Use your formula to determine how many senior homework exercises he marked.

   d) On another night the marking takes a total of 4 hours and 55 minutes. Andrew marked the same number of junior and senior homework exercises. Use your formula to determine how many homework exercises he marked in total.

▶ *nth* **term formulae**

To find the general term for a linear pattern we take the following steps.

- Find the common difference.
- Multiply the term numbers by the common difference.
- Look for the adjustment required to match the *n*th term.
- Write down the general term.

Worked examples

1 Complete the tables and find a general term for each sequence

a)
n	1	2	3	4	5	6
*n*th term	5	12	19	26	33	40

Common difference = 7

Multiples of 7: 7, 14, 21, 28...

Adjustment required: −2

General term is 7*n* − 2

b)
n	1	2	3	4	5	6
*n*th term	8	5	2	−1	−4	−7

Common difference = −3

Multiples of −3: −3, −6, −9, −12...

Adjustment required: +11

General term is 3*n* + 11

2 A given sequence has general term 8*n* + 7. Use the general term to calculate:

a) the 10th term

We substitute *n* = 10 into the general term and evaluate:

$8n+7$
$=8\times10+7$
$=87$

b) the 50th term

$8n+7$
$=8\times50+7$
$=407$

Exercise 5I

1 For each of the following sequences, copy and complete the table and find the general term.

a)
n	1	2	3	4	5	6
*n*th term	8	14	20			

b)
n	1	2	3	4	5	6
*n*th term	12	23	34			

c)
n	1	2	3	4	5	6
*n*th term	3	7	11			

d)
n	1	2	3	4	5	6
*n*th term	3	11	19			

e)
n	1	2	3	4	5	6
*n*th term	9	7	5			

f)
n	1	2	3	4	5	6
*n*th term	5	1	−3			

g)
n	1	2	3	4	5	6
*n*th term	23	31	39			

h)
n	1	2	3	4	5	6
*n*th term	−6	−10	−14			

i)
n	1	2	3	4	5	6
*n*th term	−2	5	12			

j)
n	1	2	3	4	5	6
*n*th term	4·5	5	5·5			

2 For each of the sequences in question 1, use the general term to calculate:
 i) the 10th term
 ii) the 40th term

Worked example   Consider the following matchstick pattern.

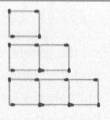

a) Complete the table below.

No. squares (s)	1	2	3	4	5	6	...	10
No. matchsticks (m)	4	7	10	13	16	19	...	31

b) Find a formula for the number of matchsticks m required for s squares.

Find the general term in terms of s: $3s + 1$

Write your formula for the number of matchsticks using the letters given:

$m = 3s + 1$

c) Use your formula to determine the number of matchsticks needed for 26 squares.

$s = 26$:

$m = 3s + 1$

$\quad = 3 \times 26 + 1$

$\quad = 79$

79 matchsticks are needed.

d) A particular pattern uses 97 matchsticks. How many squares are there?

$m = 97$:

$m = 3s + 1$

$97 = 3s + 1$

$3s = 96$

$s = 32$

There are 32 squares in the pattern.

---

3   Copy the matchstick pattern shown opposite.

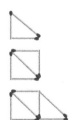

   a)   Draw the next matchstick pattern in the sequence.

   b)   Copy and complete the following table.

Pattern number (n)	1	2	3	4	5	6	...	10
Number of matchsticks (m)	5	8	11					

   c)   Find a formula for the number of matchsticks required for pattern n.

   d)   Use your formula to find the number of matchsticks required for pattern 30.

   e)   One particular pattern uses 128 matchsticks. What is the pattern number?

4   Copy the matchstick pattern shown opposite.

   a)   Draw the next matchstick pattern in the sequence.

   b)   Copy and complete the following table.

Pattern number (n)	1	2	3	4	5	6	...	10
Number of matchsticks (m)	3	5	7					

   c)   Find a formula for the number of matchsticks required for pattern n.

   d)   Use your formula to find the number of matchsticks required for pattern 18.

   e)   One particular pattern uses 97 matchsticks. What is the pattern number?

5   Copy the matchstick pattern shown opposite.

   a)   Draw the next matchstick pattern in the sequence.

   b)   Copy and complete the following table.

Pattern number (n)	1	2	3	4	5	6	...	10
Number of matchsticks (m)	4	8	12					

   c) Find a formula for the number of matchsticks required for pattern *n*.

   d) Use your formula to find the number of matchsticks required for pattern 60.

   e) What is the largest pattern that could be created with 131 matchsticks?

6 Copy the matchstick pattern shown opposite.

   a) Draw the next matchstick pattern in the sequence.

   b) Copy and complete the following table.

Pattern number (*n*)	1	2	3	4	5	6	...	10
Number of matchsticks (*m*)	5	11	17					

   c) Find a formula for the number of matchsticks required for pattern *n*.

   d) Use your formula to find the number of matchsticks required for pattern 18.

   e) What is the largest pattern that could be created with 210 matchsticks?

7 Look at the table below, converting temperatures in degrees Celsius to degrees Fahrenheit.

Degrees Celsius (C)	1	2	3	4	5	6	...	10
Degrees Fahrenheit (F)	33·8	35·6	37·4					

   a) Copy and complete the table.

   b) Use the table to find a formula to convert temperatures in degrees Celsius to degrees Fahrenheit.

   c) Convert 35 degrees Celsius to Fahrenheit.

   d) Convert 68 degrees Fahrenheit to degrees Celsius.

   e) There is one temperature that has the same numerical value in degrees Fahrenheit and degrees Celsius. Using your formula, form an equation and solve it to find this temperature.

8 Look at the table below.

n	1	2	3	4	5	6
*n*th term	4	7	12	19	28	39

   a) Consider the 'gaps' between the terms. Describe in words how the gap changes.

   b) Write down the first six square numbers. Compare them with the second row in that table. What do you notice?

   c) Copy and complete the *n*th term formula for the sequence: $n^2 + \square$

9 In the same way, find an *n*th term formula for each of the following sequences.

a)

n	1	2	3	4	5	6
*n*th term	6	9	14	21	30	41

b)

n	1	2	3	4	5	6
*n*th term	11	14	19	26	35	46

c)

n	1	2	3	4	5	6
*n*th term	0	3	8	15	24	35

d)

n	1	2	3	4	5	6
*n*th term	−8	−5	0	7	16	27

e)

n	1	2	3	4	5	6
*n*th term	−1	−4	−9	−16	−25	−36

f)

n	1	2	3	4	5	6
*n*th term	1	−2	−7	−14	−23	−34

▶ Subject of the formula

It can be useful to rearrange a given formula to calculate an unknown in it. For example, in Chapter 4 you used $C = \pi D$ to find the circumference of a circle and then $D = \dfrac{C}{\pi}$ to calculate the diameter of the circle.

The subject of the formula is the variable you want to find when rearranged: In $C = \pi D$, C is the subject of the formula and in $D = \dfrac{C}{\pi}$, D is the subject of the formula. To rearrange the subject of the formula:

- Use balancing steps as you would when balancing an equation.

Worked examples

Rearrange to make the subject of the formula the variable asked for.

1 $A = bh$ to b Rewrite
   $\div h \ \div h$ to have
   $\dfrac{A}{h} = b$ the
              subject
   $b = \dfrac{A}{h}$ on the
              LHS.

2 $y = 3x + c$ to x
   $-c \quad -c$
   $y - c = 3x$
   $\div 3 \quad \div 3$
   $x = \dfrac{y - c}{3}$

3 $4a = bl - d$ to l
   $+d \quad +d$
   $4a + d = bl$
   $\div b \quad \div b$
   $l = \dfrac{4a + d}{b}$

4    $6p = rk - m$ to m
     $-rk \ -rk$
     $6p - rk = -m$
     $\div(-1) \ \div(-1)$
     $m = \dfrac{6p - rk}{-1}$
     $= -6p + rk$

In Worked example 1 we balanced and then rewrote the answer with the subject of the formula on the LHS. We can do this in one step as we did in Worked examples 2–4.

Exercise 5J

1   Rearrange these well-known formulae to make the subject of the formula the variable asked for:

   a) $A = lb$ to l b) $Q = IT$ to T c) $D = ST$ to S d) $F = ma$ to a

2   The weight of an object W measured in Newtons (N), can be calculated using $W = mg$, where m kg is the mass of the object and g N/kg is the gravitational acceleration.

   a)   Rearrange to make mass m the subject of the formula.

   b)   Use your answer to a) to calculate the mass of a person on the Moon if their weight is 120 N and the gravitational acceleration is 1·6 N/kg.

3   Rearrange to make the subject of the formula the variable asked for:

   a) $y = c + 5x$ to c b) $p = q + 6r$ to q c) $m = 5k + 3$ to k d) $6r = 7s - 1$ to s

   e) $a = 4b + d$ to b f) $f = 9g - h$ to g g) $n = 10p - q$ to p h) $3v = y + 8w$ to w

   i) $d = ef + g$ to f j) $3k = lm + n$ to m k) $2a = bc - d$ to c l) $4e = fg - h$ to g

4   When accelerating, the final velocity of an object v m/s can be found using $v = u + at$, where u m/s is the initial velocity, a m/s^2 is the acceleration and t is time in seconds.

   a)   Rearrange to make t the subject of the formula.

   b)   Find the time it takes an object to travel between two points if $v = 20$ m/s, $u = 2$ m/s and $a = 6$ m/s^2.

   ➜

5 Rearrange to make the subject of the formula the variable asked for:

a) $y = 2 - x$ to x b) $r = 5 - s$ to s c) $p = 1 - q$ to q d) $h = 4 - g$ to g

e) $2p = 3s - t$ to t f) $5q = 6n - p$ to p g) $7k = mn - 2l$ to l h) $xy = vw - 3z$ to z

When a formula contains a fraction, we balance by multiplying through by the denominator. Brackets should be introduced when multiplying more than one term.

Worked examples

Rearrange to make the variable asked for the subject of each formula.

1 $2r = \dfrac{s + 5t}{7}$ to t

$\times 7 \quad \times 7$

$14r = s + 5t$

$-s \quad -s$

$14r - s = 5t$

$\div 5 \quad \div 5$

$t = \dfrac{14r - s}{5}$

2 $3w = \dfrac{u}{v} + 7$ to u

$\times v \quad \times v \quad \times v$

$3vw = u + 7v$

$-7v \quad -7v$

$u = 3vw - 7v$

Alternatively, we could have subtracted first.

Make sure to multiply every term in the formula when balancing a fraction.

$3w = \dfrac{u}{v} + 7$

$-7 \quad -7$

$3w - 7 = \dfrac{u}{v}$

$\times v \quad \times v$

$u = v(3w - 7)$

$u = 3vw - 7v$

6 Rearrange to make the subject of the formula the variable asked for:

a) $p = \dfrac{q + 3}{8}$ to q b) $m = \dfrac{n - 4}{9}$ to n c) $2t = \dfrac{a - 5}{b}$ to a d) $6k = \dfrac{2l - b}{h}$ to l

e) $a = \dfrac{b}{2} + 3$ to b f) $6c = \dfrac{d}{5} - e$ to d g) $p = \dfrac{q}{r} + s$ to q h) $3g = \dfrac{2h}{m} - t$ to h

i) $7a = \dfrac{2b}{3} + c$ to b j) $11f = \dfrac{6g}{h} - 5$ to g k) $3p = \dfrac{6q}{7} + 9$ to q l) $5x = 4 - \dfrac{3y}{2}$ to y

7 To convert a temperature from degrees Farenheit, F, to degrees Celsius, C, we can use the formula

$$C = \dfrac{5F - 160}{9}.$$

a) Rearrange to make F the subject of the formula.

b) If $C = 30$ degrees Celsius, find the corresponding temperature in degrees Farenheit.

8 The displacement of a particle, s metres, from an origin is given by $s = ut + \dfrac{1}{2}at^2$ where u m/s is the initial velocity, a m/s^2 is the acceleration and t is time in seconds.

a) Rearrange to make a the subject of the formula.

b) Find the acceleration of the particle after 10 seconds if $s = 80$ m and $u = 5$ m/s.

Worked example

Rearrange to make x the subject of the formula.

$$\frac{6}{x-3}=5y$$
$$\times(x-3)\ \times(x-3)$$
$$6=5y(x-3)$$
$$\div5y\ \div5y$$
$$\frac{6}{5y}=x-3$$
$$+3\quad+3$$
$$x=\frac{6}{5y}+3$$

When we have more than a single term on a numerator or denominator, we must put brackets around them as we are multiplying by every term.

Alternatively, if the formula can be written with a single fraction either side of the equals sign, we can cross multiply. Any expression can be written as a fraction by writing it over 1.

$$\frac{6}{x-3}\ \times\ \frac{5y}{1}\ \text{to }x$$
$$6=5y(x-3)$$
$$\div5y\ \div5y$$
$$\frac{6}{5y}=x-3$$
$$+3\quad+3$$
$$x=\frac{6}{5y}+3$$

9 Rearrange to make the subject of the formula the variable asked for:

a) $\frac{10}{x+4}=y$ to x

b) $\frac{31}{k-5}=2m$ to k

c) $\frac{P}{q-3}=8r$ to q

d) $\frac{w}{v-x}=5h$ to v

e) $\frac{2}{3s+1}=t$ to s

f) $\frac{7}{2a-1}=b$ to a

g) $\frac{12}{3p+5}=q$ to p

h) $\frac{5}{4x+1}=3y$ to x

When rearranging formulae where the subject is inside brackets, raised to a power or inside a root, balance from 'the outside in'. That is, balance any terms outside the brackets, powers or roots before dealing with them.

Worked examples

Rearrange these formulae to make the variable asked for the subject.

1 $A=\frac{1}{2}h(a+b)$ to b
$$\times2\quad\times2$$
$$2A=h(a+b)$$
$$\div h\quad\div h$$
$$\frac{2A}{h}=a+b$$
$$-a\quad-a$$
$$b=\frac{2A}{h}-a$$

2 $V=\frac{4}{3}\pi r^3$ to r
$$\times3\quad\times3$$
$$3V=4\pi r^3$$
$$\div4\pi\quad\div4\pi$$
$$r^3=\frac{3V}{4\pi}\quad\text{To balance the cube, take the cube root.}$$
$$r=\sqrt[3]{\frac{3V}{4\pi}}$$

3 $A=3p\sqrt{\frac{q}{r}}$ to q
$$\div3p\quad\div3p$$
$$\frac{A}{3p}=\sqrt{\frac{q}{r}}\quad\text{Square both sides}$$
$$\left(\frac{A}{3p}\right)^2=\frac{q}{r}$$
$$\times r\quad\times r$$
$$q=r\left(\frac{A}{3p}\right)^2$$

10 Rearrange to make the subject of the formula the variable asked for:

a) $b = \dfrac{1}{3}c(x+y)$ to x b) $m = \dfrac{1}{4}n(p-q)$ to p c) $k = 5r(2s+t)$ to t d) $g = fk(h-6)$ to h

e) $a = 7b(c-d)$ to c f) $x = \dfrac{y}{3}(z+10)$ to z g) $p = 4q(2r-s)$ to s h) $p = \dfrac{2}{3}q(r-5s)$ to r

11 In this question, rearrange to make the subject of the formula the variable asked for. Recall from Chapter 1 that there may be more than one possible answer when finding a root. We only expect the positive root where more than one may exist.

a) $K = m^2 - 1$ to m b) $w = 8v^2$ to v c) $A = \pi r^2$ to r d) $E = \dfrac{1}{2}mv^2$ to v

e) $y = x^3 + 4$ to x f) $r = 6t^3$ to t g) $a = 2f^3 - 1$ to f h) $c = 4d^3 - g$ to d

12 The energy in a body, E measured in Joules (J), is linked to the mass of the body, m kg, and the speed of light, c m/s, by the formula $E = mc^2$. Rearrange to make the speed of light the subject of the formula.

13 Rearrange to make the variable asked for the subject of the formula:

a) $k = \sqrt{x}$ to x b) $m = \dfrac{1}{5}\sqrt{y}$ to y c) $p = \sqrt{q+1}$ to q d) $c = \pi\sqrt{\dfrac{f}{g}}$ to f

e) $a = \sqrt{2b}$ to b f) $x = \sqrt{3y-1}$ to y g) $p = 5\sqrt{2q+r}$ to q h) $f = \dfrac{3}{5}\sqrt{4g-h}$ to g

14 The nose of a model rocket is in the shape of a paraboloid, as shown.

The volume, V cm^3, of the paraboloid is given by $V = \dfrac{1}{2}\pi r^2 h$, where r cm is the base radius and h cm is the height.

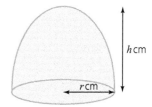

h cm

r cm

a) Rearrange the formula to make r the subject of the formula.

b) Using $\pi = 3\cdot14$, find the radius of a paraboloid with volume $62\cdot8$ cm^3 and height 10 cm.

15 A circle is defined by $x^2 + (y-2)^2 = 10$, where (x, y) is a point on the circumference of the circle.

a) Rearrange to make x the subject of the formula.

b) When $y = 5$, there are two possible values for x on the circumference of the circle. What are they?

c) Explain why there are two possible values for x in this question but in Q14 there was only one value for the base radius of the paraboloid. A sketch may help you.

16 An ellipse is defined by $\dfrac{x^2}{5} + \dfrac{y^2}{20} = 1$ where (x, y) is a point on the perimeter of the ellipse.

a) Rearrange to make x the subject of the formula.

b) What are the two possible values for x on the perimeter of the ellipse when $y = 2$?

17 The period of a simple pendulum is approximated by the formula $T = 2\pi\sqrt{\dfrac{L}{g}}$ where T is the period (the time taken to complete a cycle), L is the length of the pendulum and g is the local acceleration of gravity.

a) Rearrange to make L the subject of the formula.

b) Use your formula to calculate the length in metres of a pendulum with period 10 seconds and local acceleration of gravity $9\cdot8$ m/s^2. Give your answer correct to three significant figures.

Check-up

1 Expand:

   a) $7(x-8)$ b) $3(x+9)$ c) $5(2x-11)$ d) $-7(4x-9)$

   e) $2x(5x+6)$ f) $-9x(3-9x)$ g) $12(4x^2+5x-1)$ h) $-7(2-6x+3x^2)$

2 Expand and simplify:

   a) $3(4x+5)-7$ b) $8(2x-3)-11$ c) $7(1-4x)+23$ d) $9(5x-7)+19$

   e) $8x+2(9x-4)$ f) $17-3(4x+1)$ g) $5x^2-2x(7-3x)$ h) $23+7(8-11x)$

3 Expand and simplify:

   a) $5(x+2)+3(x-1)$ b) $4(2x+9)+5(x+7)$ c) $3(8x-1)+6(2x+9)$

   d) $7(x-10)-3(2x+1)$ e) $12(5-6x)-3(2x+9)$ f) $11(10x-3)-5(8-3x)$

   g) $x(5x+4)-2(x-9)$ h) $7x(6x-7)+4(8-x)$ i) $5x(x-8)+3x(5-6x)$

4 Given $a = 5$, $b = -2$, $c = 7$ and $d = -3$, evaluate:

   a) b^2c+ad b) d^3-3ab c) $\sqrt{ac-7b}$ d) $\sqrt{a+2c+bd}$

5 Given $p = 8$, $q = -4$, $r = 9$ and $s = -5$, evaluate:

   a) $s(r+q)$ b) $2p(q-4s)$ c) $ps(3r+7q)$ d) $4(q-r)+11(p+s)$

6 Given $u = 3$, $v = -6$, $w = 4$ and $x = -2$, evaluate:

   a) $\dfrac{4uv}{x}$ b) $\dfrac{3w-2v}{ux}$ c) $\dfrac{5(v^2+3x)}{u-2v}$ d) $\dfrac{2u(w-x)}{\sqrt{5u+3x}}$

7 Expand the brackets and solve:

   a) $7(x-4)+9=30$ b) $6(3x+1)-14=64$ c) $5(9x+13)-8=12$

   d) $4(7-3x)+51=7$ e) $5(2x-11)+3x=36$ f) $12(7-3x)+3=-57$

   g) $8(5x-7)=15x+19$ h) $6(3x-10)=7x+6$ i) $4(13-3x)+9=3x-14$

8 Solve:

   a) $9(3x-10)=8(2x-3)$ b) $5(2x+9)=3(2-x)$ c) $4(5x-1)=3(7x-2)$

   d) $2(4x+9)=3(2x+3)-1$ e) $4(3x-5)=5(2x+1)-11$ f) $6(4x+1)=2(11x-3)-6$

9 Solve:

   a) $\dfrac{4x}{3}=12$ b) $\dfrac{x+8}{5}=7$ c) $\dfrac{2x-11}{9}=7$ d) $\dfrac{3(4x-5)}{7}=15$

   e) $\dfrac{x-15}{3}=\dfrac{2x}{11}$ f) $\dfrac{5x+11}{8}=\dfrac{3x+1}{4}$ g) $\dfrac{3x-7}{2}+9=13$ h) $\dfrac{2x-3}{5}-\dfrac{x+4}{2}=\dfrac{x}{3}$

10 Write the following statements as inequalities:

   a) five is less than thirteen   b) negative one is greater than negative four

   c) negative seven is less than five   d) zero is greater than negative eleven

11 Solve:

a) $8x - 9 > 15$ 　　b) $6x + 7 < 31$ 　　c) $12x - 17 \le 31$ 　　d) $9x + 14 \ge -4$

e) $3x - 10 > -16$ 　f) $5 - 7x \le 40$ 　　g) $4 - 2x \ge x - 5$ 　　h) $9 - 7x > 21 - 4x$

12 Solve:

a) $7(4x - 7) \le 5(2x + 1)$ 　　b) $2(5x - 1) \ge 7(3x - 5)$ 　　c) $8(4x - 5) > 9(3x + 10)$

d) $2(1 - x) < 3(3x - 25)$ 　　e) $4(3x - 2) \le 5(12 - x)$ 　　f) $4(4x + 11) < 3(5x + 13)$

13 Adam, Logan and Irene spend time in the summer holidays picking strawberries. Let a be the number of strawberries Adam picks.

a) Logan picks three times as many strawberries as Adam. Write down an expression for the number of strawberries Logan picks.

b) Irene picks nine fewer strawberries than Logan. Write down an expression for the number of strawberries Irene picks.

c) Altogether, they pick 89 strawberries. Form an equation and solve it to find the number of strawberries that Adam picked.

14 A food delivery company sells boxes of fresh fruit. They charge £8 for the basic box with the option to add portions of fruit at an extra cost of £3 per portion.

a) Write down a formula for the cost C pounds of a box with p additional portions.

b) Use your formula to calculate the cost of a box with 7 additional portions of fruit.

c) One particular box costs £44. Use your formula to calculate the number of additional portions in this box.

15 For each sequence, copy and complete the table and find the general term.

a)

n	1	2	3	4	5	6
nth term	6	13	20			

b)

n	1	2	3	4	5	6
nth term	21	17	13			

16 Rearrange to make the subject of the formula the variable asked for:

a) $V = Ah$ to A 　　b) $y = 4x + w$ to w 　　c) $p = 5q - r$ to q 　　d) $2m = g - 3n$ to n

e) $a = \dfrac{b}{7} + c$ to b 　　f) $\dfrac{8}{y - 4} = 3x$ to y 　　g) $v = 3w^2$ to w 　　h) $c = \sqrt[3]{d - 5}$ to d

17 As you climb a mountain, the temperature decreases. A mountaineering group approximates the temperature, $T°C$, as you climb, using the formula:

$$T = B - 2h$$

where B is the temperature at the base of the mountain, in degrees Celsius, and h is how far you have climbed up the mountain, in thousands of feet.

a) Rearrange to make h the subject of the formula.

b) What height, in feet, have the group climbed if the temperature where they stand is 18·8°C and the temperature at the base of the mountain is 22°C?

18 The density of a cube, $\rho\,\text{g/cm}^3$, can be calculated by $\rho = \dfrac{m}{l^3}$, where m is the mass of the cube in grams and l is the length of a side of the cube in centimetres. The more dense a cube, the heavier it will feel for its size.

a) Rearrange to make l the subject of the formula.

b) A cube of steel with mass 1000 g has a density of 8 g/cm³. Find the length of the cube.

6 Coordinates and symmetry

▶ Rotational symmetry

A shape has **rotational symmetry** if after being turned around a **focal point** it looks the same. The number of ways the object can be rotated into the same configuration is called the **order** of rotational symmetry. Every shape looks the same after a rotation of 360°. If this is the only way in which a shape can be rotated into the starting position, its order of rotational symmetry is 1.

Worked example   State the order of rotational symmetry for each of these shapes.

a)

If we rotate this shape on its centre by 180° or 360° it will fit back in place perfectly.

This shape has rotational symmetry of order 2.

b)

This shape will look the same after a rotation of 90°, 180°, 270° or 360°.

This shape has rotational symmetry of order 4.

c)

This shape has rotational symmetry of order 3 (not 6, as the coloured triangles must match up).

Exercise 6A

1 State the order of rotational symmetry of each shape.

a)

b)

c)

d)

e)

f)

g)

h)

2 State the order of rotational symmetry and number of lines of symmetry of each design.

a)

b)

c)

d)

e)

f)

g)

h)

We can draw rotated copies of a shape to create pictures with a given order of rotational symmetry.

Worked examples

1   Complete the shape so that it has rotational symmetry of order 2 about the given focal point.

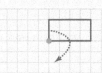

We think of 'spinning' the shape 180° about the focal point and draw the image it creates.

2   Complete the shape so that it has rotational symmetry of order 4 about the given focal point.

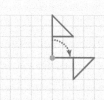   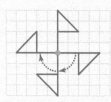

First, rotate through 90°   Complete all rotations

3   Copy the following shapes and complete them so that they have rotational symmetry of order 2.

a)   b)   c)   d)

e)   f)   g)   h)

4   Copy the following shapes and complete them so that they have rotational symmetry of order 4.

a)   b)   c)   d)

e)   f)   g)   h)

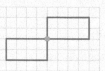

▶ Coordinates revisited

Recall the following about plotting and reading points on a coordinate grid.

When drawing a coordinate grid:

- Use a ruler.
- Label the axes.
- Space out numbers equally along the axes.
- Write numbers on the lines, not in the boxes.

When writing coordinates:

- Use brackets.
- Write the x-coordinate followed by the y-coordinate.
- Separate the coordinates with a comma.

Worked example

Look at the coordinate grid.

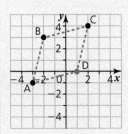

a) Write down the coordinates of points A, B and C.

A has an x-coordinate of –3 and a y-coordinate of –1, so we write A(–3, –1).

Similarly, we have B(–2, 3) and C(2, 4).

b) Plot the point D such that ABCD forms a rhombus, and write down the coordinates of point D.

D(1, 0)

Exercise 6B

1   Look at the coordinate grid opposite.

  a) Write down the coordinates of each point.

  b) Which four pairs of points have the same x-coordinate?

  c) Which two pairs of points have the same y-coordinate?

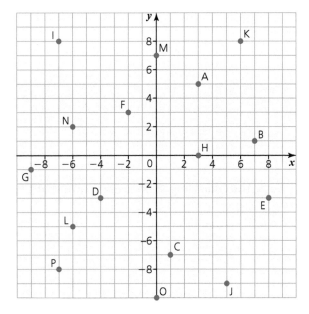

2   Draw a coordinate grid like the one in question 1 and plot the following points.

  A(4, 7)   B(–2, 3)   C(–5, –8)   D(1, –9)   E(0, 4)   F(–8, 0)

  G(–9, 1)   H(7, –10)   I(0, –2)   J(–4, 5)   K(7, 3)   L(–1, 8)

3 Draw a coordinate grid with the *x*- and *y*-axes going from −5 to 5.

   a) Plot the points A(−4, 2), B(4, 2) and C(4, −4).

   b) Plot the point D such that ABCD is a rectangle and write down the coordinates of point D.

   c) Draw the two diagonals of the rectangle ABCD.

   d) Write down the coordinates of the point where the diagonals intersect.

4 Draw a coordinate grid with the *x*- and *y*-axes going from −5 to 5.

   a) Plot the points A(0, −3), B(−3, 2) and C(0, 4).

   b) Plot the point D such that ABCD is a kite and write down the coordinates of point D.

   c) Draw the two diagonals of the kite ABCD.

   d) Write down the coordinates of the point where the diagonals intersect.

5 Look at the coordinate grid below and use it to decipher the following words.

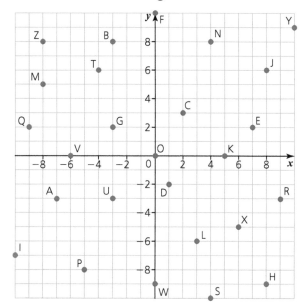

   a) (−5, −8), (7, 2), (4, 8), (2, 3), (−10, −7), (3, −6)

   b) (2, 3), (0, 0), (−8, 5), (−5, −8), (−7, −3), (4, −10), (4, −10)

   c) (2, 3), (10, 9), (3, −6), (−10, −7), (4, 8), (1, −2), (7, 2), (9, −3)

   d) (−4, 6), (0, 0), (9, −3), (−3, −3), (4, −10)

6 Now use the grid to encode the names of the following mathematicians.

   a) Hannah Fry

   b) Cedric Villani

   c) Terence Tao

   d) Katherine Johnson

   e) Maryam Mirzakhani

   f) Srinivasa Ramanujan

   g) Mary Cartwright

   h) Persi Diaconis

7 Draw a coordinate grid with *x*- and *y*-axes going from −10 to 10. Plot the following pairs of points, join each pair up with a straight line and write down the coordinates of the points of intersection.

   a) A(−2, 3) and B(6, 3)
       C(−2, 5) and D(4, 2)

   b) E(−7, −7) and F(−1, −1)
       G(−8, −3) and H(4, −6)

   c) I(−7, 9) and J(−5, 3)
       K(−2, 7) and L(−10, 5)

   d) M(0, 7) and N(7, 0)
       O(8, 10) and P(0, −2)

▶ Transformations

When we apply a **translation** to a shape on the coordinate grid, we move its position. It maintains its size, shape and orientation. The new shape is called the **image**. Image points are written with a dash.

Worked examples

1   Translate ABCD 4 units right and 3 units down. Write down the coordinates of the image points.

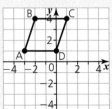 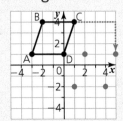 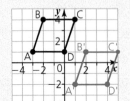

A'(1, −2)

B'(2, 1)

C'(5, 1)

D'(4, −2)

4 boxes right, 3 boxes down   Join image points and complete.

2   Translate EFGH 5 units left and 2 units up. Write down the coordinates of the image points

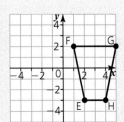 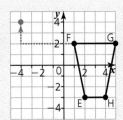 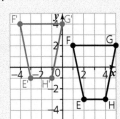

E'(−3, −1)

F'(−4, 4)

G'(0, 4)

H'(−1, −1)

5 boxes left and 2 boxes up   Join image points and complete.

Exercise 6C

1   Copy each shape, complete the translation and write down the coordinates of the image points.

a)
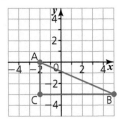

3 units left and 4 units up

b)
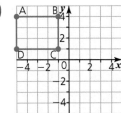

6 units down and 4 units right

c)
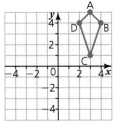

7 units left and 1 unit down

d)
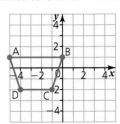

3 units right and 2 units up

e)
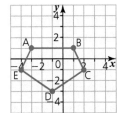

2 units right and 2 units down

f)
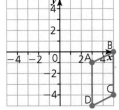

7 units left and 4 units up

2   Name each shape in Question **1**.

3   Look at the image points in your answers to Question **1**. What is the connection between the original points, the translation and the image points?

When we **reflect** a shape on the coordinate grid, we draw its mirror image relative to a given line of symmetry. The size of the shape is unaffected by reflection.

Worked examples

1   Reflect triangle ABC in the x-axis and then reflect its image in the y-axis.
    At each stage, write down the coordinates of the image points.

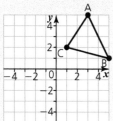 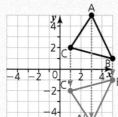

A′(3, −5)
B′(5, −1)
C′(1, −2)

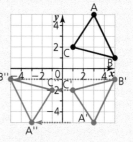

A″(−3, −5)
B″(−5, −1)
C″(−1, −2)

Reflect in x-axis   Reflect in y-axis

2   Reflect the kite DEFG in the line $y = x$ and write down the coordinates of the image points.

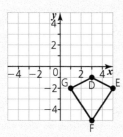 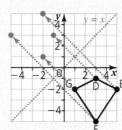 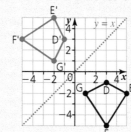

D′(−1, 3)
E′(−2, 5)
F′(−5, 3)
G′(−2, 1)

4   Reflect the following shapes in the x-axis and write down the coordinates of the image points.

a) 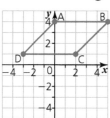   b) 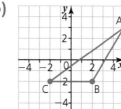   c)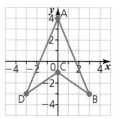

5   Reflect the following shapes in the y-axis and write down the coordinates of the image points.

a) 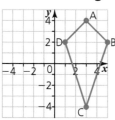   b) 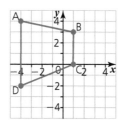   c)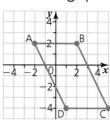

6   Reflect the following shapes in the line $y = x$ and write down the coordinates of the image points.

a) 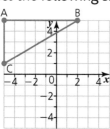   b) 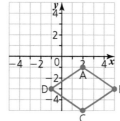   c)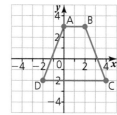

To **rotate** a shape on a coordinate grid, we must know the centre of rotation, the angle through which we wish to rotate and the direction.

Worked example

Write down the coordinates of the image points when the shape is rotated clockwise about the origin by a) 90°   b) 180°.

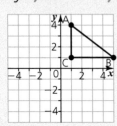
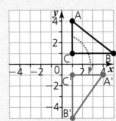

A'(4, −1)

B'(1, −5)

C'(1, −1)

90° clockwise

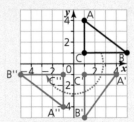

A"(−1, −4)

B"(−5, −1)

C"(−1, −1)

180° clockwise

7   Rotate the given shapes 90° clockwise about the origin and write down the coordinates of the image points.

a)
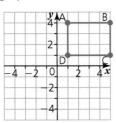

b)
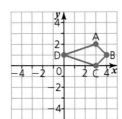

c)
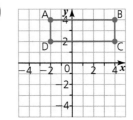

8   Rotate the given shapes 180° clockwise about the origin and write down the coordinates of the image points.

a)
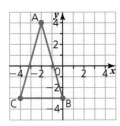

b)
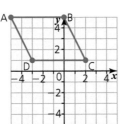

c)
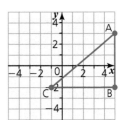

9   Rotate the given shapes 270° clockwise about the origin and write down the coordinates of the image points.

a)
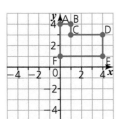

b)
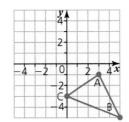

c)
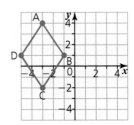

To **enlarge** or **reduce** a given shape we must know the **scale factor** and the **centre of enlargement** or **reduction**. We draw lines from the centre of enlargement to the vertices of the given shape and extend them to the required length (calculated using the scale factor).

Worked examples

1   Enlarge the given shape by a factor of 3 using the origin as the centre of enlargement and write down the coordinates of the image points.

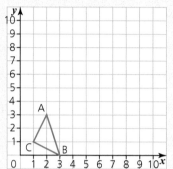

Draw dashed lines from the centre of enlargement to each vertex and then continue to three times the length (since the scale factor is 3).

It can help to 'count steps', for instance C is 1 unit right and 1 up from the origin, so C′ will be 3 units right and 3 up.

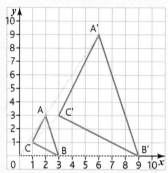

A′(6, 9) B′(9, 0) C′(3, 3).

2   Reduce the given shape by a factor of $\frac{1}{2}$ using the point (4,1) as the centre of reduction.

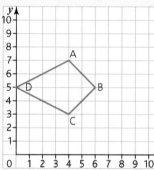

Draw dashed lines from the centre of reduction to each vertex and mark off each image vertex half-way along the lines.

The image points are A′(4, 4), B′(5, 3), C′(4, 2) and D′(2, 3).

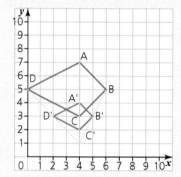

10   Using the given information, enlarge or reduce each of the following shapes and write down the coordinates of the image points.

a)   Centre (0, 0), scale factor 3

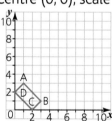

b)   Centre (0, 0), scale factor 4

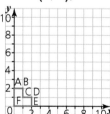

c)   Centre (0, 0), scale factor 2

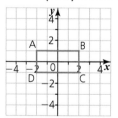

d)   Centre (5, 9), scale factor 3

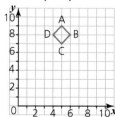

e)   Centre (1, 0), scale factor 3

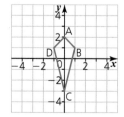

f)   Centre (0, 0), scale factor: $\frac{1}{3}$

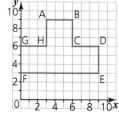

Check-up

1   State the order of rotational symmetry of each shape.

a)    b)   c)   d)

e)   f)    g)   h)

2   Copy and complete each of the following shapes so that they have rotational symmetry of order 2 about the given point.

a)   b) 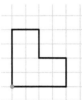   c) 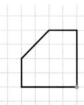   d)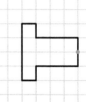

3   Copy and complete each of the following shapes so that they have rotational symmetry of order 4 about the given point.

a)   b)   c)   d)

4   Look at the coordinate grid below and write down the coordinates of each point.

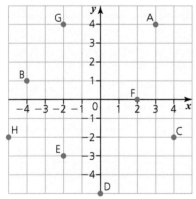

5   Draw a coordinate grid with x- and y-axes going from −5 to 5 and plot the following points.
A(3, −1)   B(0, 4)   C(−2, 5)   D(−4, 1)   E(−5, 0)   F(5, −4)

6   Draw a coordinate grid with x- and y-axes going from −5 to 5.
   a)   Plot the points A(−2, 3), B(3, −1) and C(0, −3).
   b)   Plot the point D such that ABCD is a parallelogram and write down the coordinates of D.
   c)   Draw AC and BD, the diagonals of parallelogram ABCD.
   d)   Write down the coordinates of the point of intersection of AC and BD.

7 Copy each shape, perform the specified transformation and write down the coordinates of the image points.

a) Translate 2 units right and 5 units down

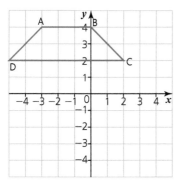

b) Reflect in the *x*-axis

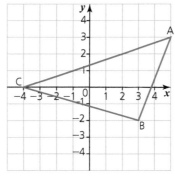

c) Reflect in the line *y* = *x*

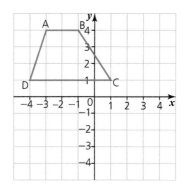

d) Rotate 180° clockwise about the origin

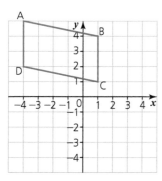

e) Rotate 90° clockwise about the origin

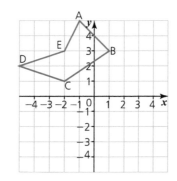

f) Reduce by a factor of $\frac{1}{2}$ using the origin as the centre of reduction

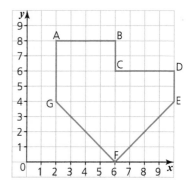

8 Copy the diagram.

a) Rotate parallelogram ABCD 270° clockwise about the origin and write down the coordinates of the image points.

b) Reflect the **image** in the line *y* = *x* and write down the coordinates of the new image points.

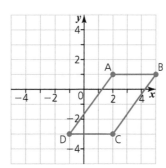

9 The following diagram shows triangle ABC as well as its image after three different transformations. Describe the transformation of the original shape that has resulted in:

a) triangle A′B′C′

b) triangle A″B″C″

c) triangle A‴B‴C‴.

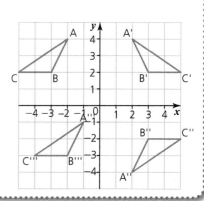

7 Money

▶ Foreign currency

A table of **exchange rates**, from the £ to different currencies, is shown.

- To convert £s to foreign currency:

$$£ \xrightarrow{\times \text{ by rate}} \text{foreign currency}$$

Currency	Symbol	Rate	Round to nearest...
US dollar	$	1·25	cent 2 d.p.
Euro	€	1·10	cent 2 d.p.
Japanese yen	¥	133·84	yen
Australian dollar	AU$	1·80	cent 2 d.p.
Swiss franc	SFr	1·20	cent 2 d.p.
UAE dirham	ﻼﺩ	4·59	fils 2 d.p.
South Korean won	₩	1490·84	won
Czech koruna	Kč	29·41	koruna

> **Worked example**   Convert each amount to the currency shown:
>
> **a)** £100 (US Dollar)
> £100 = 100 × 1·25
>   = $125
>
> **b)** £250 (Swiss franc)
> £250 = 250 × 1·2
>   = 300SFr
>
> **c)** £29·99 (Japanese yen)
> £29·99 = 29·99 × 133·84
>   = 4013·8...
>   = 4014¥

Exercise 7A

1 Convert £100 to each currency shown:

  a) £100 (euro)   **b)** £100 (Swiss francs)   **c)** £100 (UAE dirham)   **d)** £100 (Australian dollars)

2 Convert each amount into the currency shown:

  a) £40 (US dollars)   **b)** £800 (South Korean won)   **c)** £3000 (euro)

  d) £600 (Czech koruna)   **e)** £265 (Swiss franc)   **f)** £1280 (UAE dirham)

  g) £751·50 (Japanese yen)   **h)** £89·95 (Australian dollars)   **i)** £504·85 (Czech koruna)

3 Toby has £650 for a trip to Seoul. Convert his money into South Korean won.

4 Eve designs an app and sells it to a Swiss company for £45 000. Calculate the price in Swiss francs.

5 Mary sells £500 000 worth of Harris Tweed to a Japanese distributer. Calculate the price in yen.

- To convert foreign currency to £s: $£ \xleftarrow[\div \text{ by rate}]{} \text{foreign currency}$

> **Worked example**   Convert each amount into pounds sterling. Round your answers to the nearest penny.
>
> **a)** 500 SFr Swiss francs
> 500 = 500 ÷ 1·2
>   = 416·666...
>   = £416·67
>
> **b)** 640 000 ₩ South Korean won
> 640 000 = 640 000 ÷ 1490·84
>   = 429·288...
>   = £429·29
>
> **c)** 985·50 ﻼﺩ UAE dirham
> 985·50 = 985·50 ÷ 4·59
>   = 214·705...
>   = £214·71

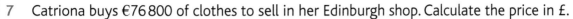

6 Convert each amount into £ sterling:

    a) US $1000   **b)** AU$1000   **c)** €1000   **d)** 6000 Kč

    e) 950 000 ¥   **f)** 150 000 ₩   **g)** 2200 ﺩ.ﺇ   **h)** AU$548·50

7 Catriona buys €76 800 of clothes to sell in her Edinburgh shop. Calculate the price in £.

8 Ruari imports 600 000 ﺩ.ﺇ UAE dirhams of paint from Dubai to sell in his UK online shop. Calculate the price in £.

9 Jody saves up £500 for a trip to Prerov in the Czech Republic. She spends 11 764 Kč and changes her remaining Koruna back to £s. How many £s will she receive?

10 Diana exchanges £650 for Japanese yen. She spends 65 940 ¥ in Yokohama and changes her remaining ¥ back to £s. How many £s will she receive?

11 Sally exchanges £3000 for Swiss Francs. She goes to study in Bern and spends 2750 SFr. She changes her remaining SFr back to £s. How many £s will she receive?

12 Desmond travels to Australia for six months. He takes £6000 worth of Australian dollars with him and spends AU$9000 when he is there. He changes his remaining AU$ back to £s. How many £s will he receive?

Worked example   Ciara's aunt gives her AU$1400. She exchanges it for Swiss francs. How many will she have?

convert AU$ → £

  AU$1400 = 1400 ÷ 1·8

          = 777·777...

          = £777·78

convert £ → Swiss francs

  £777·78 = 777·78 × 1·2

          = 933·336

          = 933·34 SFr

As we do not have the exchange rate for Australian dollars to Swiss francs our strategy must be AU$ → £

£ → Swiss Francs

---

13 Dominic buys 1800 Swiss francs but his trip falls through and he is going to Italy instead. Calculate how many euros his 1800 SFr is worth.

14 Tina buys $2700 US dollars for a trip to Texas. She cancels the trip and goes to Australia instead. Calculate how many Australian dollars her US dollars are worth.

15 Amy buys 298 168 000 ₩ for a business deal. The deal falls through and she has to exchange the money for Japanese yen instead. Find the value of the deal in Japanese yen.

16 Hannah wins a design competition. The prize is €500. Hannah buys UAE dirham with her winnings. Hannah hopes she can buy at least 2000 ﺩ.ﺇ. Can she?

17 Carla saves £2100 for a trip to Australia and the United Arab Emirates. She spends AU$2457 in Adelaide and will exchange the rest for UAE dirham. How many UAE dirham will she receive?

18 A company sells the same model of television at different prices around the world.

    £2499·99   400 000 ¥   16 500 ﺩ.ﺇ   US$2999·99

    In which currency is the television least expensive?

19 Pat buys 16 000 koruna of textiles from Prague. She sells the entire amount for a profit of 20% to a buyer in Japan. Calculate:

    a) the amount Pat paid in pounds   **b)** how many Japanese yen Pat received.

- When buying foreign currency, a higher rate is better; each £ is worth more.

20 Anas has £1000 to change into euros. He checks the exchange rate for the euro and finds these prices.

Supermarket	Bank	Airport
€1·18	€1·15	€1·11

   a) Which outlet offers the best exchange rate?

   b) How many more euros will Anas receive if he goes to the bank rather than the airport?

21 Liam has £300 to change into Indian rupees. He finds these prices.

   a) Which outlet offers the best exchange rate?

   b) How many more rupees will Liam receive if he goes with the best rather than worst deal?

Supermarket	Bank	Online
₹92·73	₹88·15	₹93·25

22 Roisin has £1500 to change into dollars. She finds these deals. The online deal charges commission: this is a fee which is deducted.

Online	Bank
$1·23 4% Commission (Fee)	$1·19 No fee

   a) How many dollars will Roisin receive with the online deal?

   b) Which deal is better for Roisin and by how much?

23 Salma has £1000 to exchange into Swiss francs. She finds these options.

   Which deal is better for Salma and by how much?

Bank	Airport
1·27 SFr 3% Commission	1·13 SFr No fee

Banks and shops that sell currency buy it back at a less favourable rate.

- £ → Foreign currency, multiply by the 'we sell' rate.
- Foreign currency → £, divide by the 'we buy' rate.

24 a) Bia plans a trip to Chicago and exchanges £900. She does not go and changes her money back to pounds sterling. How much money did she lose by doing this?

	We sell	We buy
Euro	1·14	1·36
Dollar	1·27	1·40

   b) Leanne converts £30 000 to euros for a business deal. The deal falls through and she changes the money back. How much money did she lose by doing this?

25 Darren, Marc and Bob each have €800 to exchange for pounds.

   The exchange rates they use are shown.

We buy	Darren	Bob	Marc
Euro	€1·21	€1·57	€1·44

   a) How many £s do each of them receive?

   b) Bob thinks that when he exchanges foreign currency for £s a higher rate is better. Is he correct?

26 Holly is returning to Glasgow after working in Poland, she has 8500 zł to exchange for £ sterling.

   She finds the options shown for selling her Polish zloty.

We buy	Bank	Supermarket	Airport
Zloty	5·21zł	4·87zł	6·04zł

   a) Which option should Holly choose?

   b) How much more £ sterling will Holly receive choosing the best over the worst option?

▶ Payslips

Income Tax and National Insurance (NI) are collected by the government to pay for public services like the NHS. Employees contribute different amounts depending on their salary.

Gross pay is the sum of all earnings before deductions while **net pay** is your 'take home' pay after tax, NI etc. have been taken away.

● Net pay = Gross pay − Total deductions

Worked example Dara works in construction. Complete her monthly payslip.

Basic pay	Overtime	Bonus	Gross pay
£1850	£187	–	£2037
Income Tax	National Insurance	Pension	Total deductions
£207	£149	£102	£458
		Net pay	£1579

Gross pay = Basic pay + Overtime
= 1850 + 187 = £2037
Total deductions = Income Tax + NI + Pension
= 207 + 149 + 102 = £458
Net pay = Gross pay − Total deductions
= 2037 − 458
= £1579

Exercise 7B Complete the missing amounts in each of these monthly payslips.

1

Basic pay	Overtime	Bonus	Gross pay
£1350	–	£150	
Income Tax	National Insurance	Pension	Total deductions
£75	£85	£75	
		Net pay	

2

Basic pay	Overtime	Bonus	Gross pay
£1400	£142	–	
Income Tax	National Insurance	Pension	Total deductions
£83	£90	£77	
		Net pay	

3

Basic pay	Overtime	Bonus		Gross pay
£2383	–	–		
Income Tax	National Insurance	Pension	Student loan	Total deductions
£244	£191	£119	£69	
			Net pay	

4

Basic pay	Overtime	Bonus		Gross pay
£1401	£400	£50		
Income Tax	National Insurance	Pension	Student loan	Total deductions
£130	£127		£21	£426
			Net pay	

5

Basic pay	Overtime	Bonus		Gross pay
£1647		–		£1667
Income Tax	National Insurance	Pension	Student loan	Total deductions
	£105	£83	£5	£299
			Net pay	

6

Basic pay	Overtime	Bonus		Gross pay
£2945		–		£3083
Income Tax	National Insurance	Pension	Student loan	Total deductions
£384		£154	£132	
			Net pay	£2138

▶ Budgeting

We should all **budget**; that is plan our spending carefully. If possible, we should **save** for emergencies and bigger essential purchases.

Worked example   After housing costs, a father has £561 a month to spend.

a) Complete his budget for a **4-week month**.

b) How much money will he have left?

561 – 524 = £37

	Weekly	Monthly
Food	£65	£260
Non-food	£30	£120
Clothes		£50
Travel pass	£15	£60
Swimming lessons		£24
Mobile		£10
Total monthly spending		£524

Exercise 7C

1 After paying her mortgage Pearl has £800 a month to spend.

a) Complete Pearl's budget for a 4-week month.

b) How much money will be left?

Pearl	Weekly	Monthly
Food	£70	£
Non-food	£30	£
Gas + electricity		£67
Council tax		£143
Broadband/phone		£34
Socialising	£30	£
Total monthly spending		£

2 After housing costs and bills Tyler has £720 a month to spend. Here is her budget for a 4-week month.

a) Complete Tyler's budget for a 4-week month.

b) Tyler is planning to spend more than she has. How much more?

c) Set out a new spending plan for Tyler. List how much she should spend on each item. Include an amount for savings. Remember the total amount Tyler has available is £720.

d) If Tyler saved the amount you have planned every month for a year, how much would she have saved at the end of the year?

Tyler	Weekly	Monthly
Food	£120	£
Household/ toiletries	£20	£
Clothes		£75
Socialising	£50	£
Gym membership		£45
TV subscription		£9·99
Savings		£0
Total monthly spending		£

3 a) What are the benefits of having some money saved up?

b) What might be the consequences of not having some money saved up?   →

4   After housing costs and bills, Adi has £621 a month to spend. Here is his budget for a 4-week month.

a) Complete Adi's budget for a 4-week month.

b) Using this plan what is the maximum amount Adi could save?

c) After 9-months of saving this amount each month, Adi needs a new washing machine which is £289. Will Adi have enough savings?

Adi	Weekly	Monthly
Food	£85	£
Household/ toiletries	£22	£
Clothes		£30
Insurance	£5	£
Savings		£0
Days out		£100
Total monthly spending		£

5   After housing costs and bills, Tony has £950 a month to spend. Here is his budget plan for a 4-week month. Money that is owed is called **debt**. Tony spends £380 a month paying back debt.

a) Complete Tony's budget.

b) Tony is planning to spend more than he has. How much more?

c) Tony is thinking about taking out a loan to pay for a luxury holiday. The repayments would be £205 a month every month for 3 years. Do you think this is a good idea? Give a reason for your answer.

d) What percentage of Tony's available money is allocated to repaying debt?

e) Set out a new spending plan for Tony. How much should he spend on each item? Tony can't reduce the debt repayments but he could pay more to clear the debt faster.

Tony	Weekly	Monthly
Food	£135	£
Non-food	£95	£
Socialising	£70	£
Hair and beauty		£150
Debt repayments		£380
Total monthly spending		£

6   Emily is a franchise manager. After paying her mortgage, she has £1450 a month to spend.

a) Complete Emily's budget plan for a 4-week month. Her **fixed** costs have already been filled in.

b) Emily is planning a holiday to Florida with her two children. She wants to save up £2500. How many months would Emily have to save up for if she follows your budget plan?

Emily	Weekly	Monthly
Food	£	£
Household	£	£
Gas + electricity		£90
Council tax		£224
Broadband/phone		£52
Socialising	£	£
Clothes	£	£
Savings		£
Electric car		£279
Total monthly spending		£

7   After housing costs, Kate has £125 a week to spend as she likes. She manages to save £19 a week.

a) How much will she save in a year?

b) Kate puts 1 year of these savings in an account paying 2·5% p.a. Calculate her balance if she makes no withdrawals for 3 years.

8   Some costs are **fixed**, this means they are the same each month, e.g. council tax and broadband, usually gas and electricity bills are fixed amounts too.

a) In Q1 Pearl's budget plan is for a 4-week month. Make a new spending plan for Pearl for a 5-week month.

b) In Q6 Emily's budget is for a 4-week month. Make a new spending plan for Emily based on a 5-week month.

▶ National Insurance and Income Tax

Employees pay National Insurance 'at source'. This means it is deducted directly from their wages before it reaches their account. To calculate how much to pay:

Weekly earnings bands	Rate
£0–£183	0%
£183–£962	12%
over £962	2%

* Identify which **bands** and **rates** the earnings fall into.
* Calculate each percentage and add the total due.

Worked example Calculate the weekly amount of National Insurance due on the salaries shown.

a) Hannah: £380 a week

£380 is in the £183–£962 band. Split £380 into 2 bands: 183 @ 0%
380 − 183 = 197 @12%
Check: 197 + 183 = £380

Earnings	Rate	NI due
183	0%	0
197	12%	0·12 × 197 = 23·64
Total NI due		£23·64

b) Elisabeth: £1010 a week

£1010 is in the over £962 band. Split £1010 into 3 bands: 183 @ 0%
962 − 183 = 779 @ 12%
1010 − 962 = 48 @ 2%
Check: 183 + 779 + 48 = £1010

Earnings	Rate	NI due
183	0%	0
779	12%	0·12 × 779 = £93·48
48	2%	0·02 × 48 = 0·96
Total NI due		£94·44

Exercise 7D

1 Ian earns £160 a week. How much National Insurance will he pay?

2
Earnings	Rate
183	0%
100	12%

Joe and Anna are trying to calculate National Insurance for someone who earns £283 a week. Joe thinks they will pay 12% of the full £283. Anna thinks their money will be split as shown and they will only pay 12% of £100.

a) Who is correct?   b) Calculate the amount of National Insurance due.

3 Copy and complete to calculate the amount of National Insurance due on each weekly salary

a) Ben: £180

Earnings	Rate	NI due
180	0%	
Total NI due		

b) Dan: £383

Earnings	Rate	NI due
183	0%	
200	12%	
Total NI due		

c) Leanne: £1062

Earnings	Rate	NI due
183	0%	
779	12%	
100	2%	
Total NI due		

4 Calculate the amount of National Insurance due on each weekly salary shown.

a) Alba: £550   b) Esther: £810   c) Fiona: £950   d) Darci: £1300

➡

Employees also pay Income Tax. A table of rates and bands like this is used to calculate how much each person pays.

Annual salary bands	Rate
£0 – £12 500	0%
£12 500 – £14 585	19%
£14 585 – £25 158	20%
£25 158 – £43 430	21%
£43 430 – £150 000	41%

- The first £12 500 earned is tax free. This is called the Personal Allowance.
- Identify which bands and rates the earnings fall into.
- Calculate each percentage and add the total due.

Worked example

Calculate the annual tax due for the annual salary shown:

Helen is a marketing manager earning £25 000

£25 000 is in the £14 585–£25 158 band. Split over 3 bands:

12 500 @ 0%

14 585 − 12 500 = 2085 @ 19%

25 000 − 14 585 = 10 415 @ 20%

Check: 12 500 + 2085 + 10 415 = £25 000

Salary	Rate	Tax due
12 500	0%	0
2085	19%	0·19 × 2085 = 396·15
10 415	20%	0·2 × 10 415 = 2083
Total due		£2479·15

Exercise 7E

1   Danni earns £12 400 a year. Does she have to pay any Income Tax?

2   Arjan earns £16 000 a year. He thinks he has to pay 19% of the full £16 000. His mum says he has a Personal Allowance of £12 500 which is tax free and his money splits as shown in the table.

Salary	Rate
12 500	0%
3500	19%

   a)   Who is correct, Arjan or his mum?

   b)   Calculate the Income Tax due.

3   Farah earns £20 000. Her salary splits into different bands like this.
Calculate the total amount Farah will have to pay in Income Tax.

Salary	Rate	Tax Due
12 500	0%	
2085	19%	
5415	20%	

4   Paul earns £24 800.

   a)   How much of Paul's income falls into the 20% tax band?

   b)   Calculate how much Paul will have to pay in Income Tax.

Salary	Rate	Tax Due
12 500	0%	
2085	19%	
a)	20%	

5   Carmel earns £100 000.

   a)   How much of Carmel's income falls in the 21% tax band?

   b)   How much of Carmel's income falls in the 41% tax band?

   c)   Calculate the total amount Carmel will have to pay in Income Tax.

Salary	Rate
12 500	0%
2085	19%
10 573	20%
a)	21%
b)	41%

6   Joe earns £43 430. He is offered a new job with a salary of £48 000.
How much more Income Tax will Joe pay if he takes the new job?

▶ Loans and finance agreements

A **loan** is a sum of money borrowed that will be paid back with **interest**. This could stand alone or be part of a **finance agreement** for a product like a car or furniture. People who take out loans and finance agreements should shop around to find the lowest **APR** available to them. The annual percentage rate (APR) is the yearly cost of borrowing the money. Loans and finance agreements will cost extra and tie up your money for a long time after you have spent the money. Debts like these can eat up income and become a problem.

Worked example Calculate the cost of each loan or finance agreement:

a) Romey borrows £3000 at APR 8·5%.
She repays £94·26 a month for 3 years.

total repaid = 94·26 × 12 × 3   12 months a year, 3 years
= £3393·36
cost of loan = total repaid − loan amount
= 3393·36 − 3000
= £393·36

b) Scott buys a new sofa.
The cash price is £895.

He agrees to finance at 13·5% APR.

He pays £20·24 a month for 5 years.

total repaid = 20·24 × 12 × 5   12 months a year, 5 years
= £1214·40
cost of finance = total repaid − cash price
= 1214·40 − 895
= £319·40

Exercise 7F

1 Calculate the total cost of each loan

Loan amount	Repayments
a) £1000	£43·18 a month for 2 years
b) £1000	£49·69 a month for 2 years
c) £7500	£157·09 a month for 5 years
d) £7500	£128·66 a month for 8 years

2 Calculate the cost of each finance agreement.

Item		Cash Price	Repayments
Used car	a)	£5300	£169·49 a month for 3 years
Washer/dryer	b)	£650	£29·89 a month for 2 years
Computer	c)	£1800	£37·06 a month for 5 years
Electric guitar	d)	£300	£10·67 a month for 3 years

3 List these APRs from lowest to highest charges: 21·8%, 316·5%, 4·9%, 1500%, 6·8%, 17·9%

4 Scott is borrowing £1000. He finds these deals.

Easy Loans APR 6·8%	Wow Pay APR 6·8%
£44·59 a month, 2 years	£13·43 a month, 8 years

a) The APR for both loans is the same. Why will Scott pay more interest overall on the Wow Pay deal?

b) How much more interest does Wow Pay charge?

5 Zara borrows £2000 to pay for home repairs and finds these options.

a) Which loan has the lowest APR?

b) Calculate the cost of each loan.

c) How much interest could Zara save by choosing the best over the worst deal?

Alba Bank APR 8·4%	Saving Society APR 3·8%	Pearl Credit APR 20·2%
£90·54 a month, 2 years	£86·61 a month, 2 years	£100·37 a month, 2 years

▶ Credit cards

Debit cards are used to spend your own money straight from your bank account. When you make a payment or withdraw cash using a credit card you are borrowing money. Unless you quickly pay in full credit card interest is calculated daily and added to the amount you owe.

Credit card debt can grow quickly and take a very long time to repay. If you borrowed £1000 on your 21st birthday using a card with a 29% APR and only made the minimum repayments, you would be 40 years old before you fully repaid it. To find the new balance (debt owed to the card company):

● Add up debt, interest, new spending, fees and charges and subtract the payment (credit CR) made.

Worked example

a) Calculate the new balance owed.

Money owed = 215·42 + 4·47 = £219·89

New balance = 219·89 − 5·00

= £214·89

b) Why has the balance (amount owed) only reduced by a very small amount?

Most of the £5 payment went towards paying the interest on the debt.

Date of transaction	Description	Amount £
	Balance from previous statement	215·42
18 June	Interest	4·47
25 June	Minimum payment received	5·00 CR
	New balance	£214·89

Exercise 7G

1. a) Calculate the new balance owed.
   b) Why did the payment made not reduce the balance owed?

Date of transaction	Description	Amount £
	Balance from previous statement	1825·32
18 Aug	Interest	38·07
25 Aug	Minimum payment received	27·38 CR
	New balance	

2. a) Find the new balance owed.
   b) Withdrawing cash from a credit card can affect your ability to get other loans.

   Can you think of another reason it is a bad idea?

Date of transaction	Description	Amount £
	Balance from previous statement	175·32
03 June	Cash withdrawal	200·00
03 June	Cash advance fee	6·00
20 June	Interest	8·93
25 June	Fixed payment received	30·00 CR
	New balance	

3. a) Calculate the new balance owed.
   b) This person is paying more than the minimum amount to try to pay off the debt. What other advice would you give them?

Date of transaction	Description	Amount £
	Balance from previous statement	510·20
03 Sept	Spa Treats Co	84·10
10 Sept	In game purchases.com	195·00
21 Sept	Interest	21·05
25 Sept	Payment received	80·00 CR
	New balance	

▶ Making financial decisions

Draw a timeline from the age of 14 to 30. Draw arrows for how old you think you might be when you make these decisions.

- Starting to pay rent or buying a house
- Setting up a direct debit
- Borrowing some money
- Buying a car

- Paying for a holiday
- Choosing a job to apply for
- Buying some insurance
- Opening a savings account

In this exercise each question asks you to make a financial decision.

- Calculate the cost of any contracts, loans or financial products.
- Explain the pros and cons of your decision.

Worked example

| new car | Car Finance APR 9·8% |
| Pay £2500 and |
| £249·63 a month for 5 years. |

| preowned car | Personal Loan APR 3·4% |
| Pay £250 a month for 3 years. |

Theresa needs a car. She finds these options. Which option would you recommend and why?

new car

$249{\cdot}63 \times 12 \times 5 = 14\,977{\cdot}80$ total monthly payments

$+ \quad 2500{\cdot}00$ first payment

£17 477·80 total cost

preowned car

$250 \times 12 \times 3$

$= £9000$ total payment

The new car costs nearly £8500 more. The APR on the new car finance is higher than the loan. I would recommend the used car as long as it was safe and reliable.

Exercise 7H

1 Isma runs 3 times a week and needs to buy new trainers. She finds these options. Which option would you recommend and why?

| Fashion shop £6 | Basic athletic shoe £20·99 | Big brand £125 |

2 Sarah wants a new computer game that is coming out. She finds these options.

| Brand new disc copy £49·99 | Wait 6 months and buy disc copy second hand £20 | Digital copy, can be updated with new content £89·99 |

Which option would you recommend and why?

3 Jonny is in S5. He has two younger sisters in P7 and S1. His dad gives him a choice for lunches. Which option would you choose and why?

| £2 a day for the school canteen. | A £30 weekly budget to shop and make lunches for him and his sisters. He gets to keep any savings each week. |

4 Gavin is 18 and planning a holiday with friends. The total cost is £415. He is deciding whether to save up for 6 months or to pay for the holiday using his credit card which has an APR of 29·8%.

a) If he saved the same each month, how much will that be?

b) What do you recommend and why?

5   Pam is deciding whether to move out and pay rent when she goes to college. She makes a list of pros and cons. What would you recommend and why?

Pros: independence, live closer to college, fewer travelling expenses.	Cons: cost of rent and bills £550 a month. Can't save up to buy own house.

6   Anne plays guitar in a local band. Her guitar was stolen and she needs to buy a replacement. She has these options.

Brand new	Second hand
Cash price £359	£60
Finance APR 12·5%	Save up £20 a week
£11·82 a month for 3 years	

   a) Calculate the cost of the finance for the new guitar.

   b) What would you recommend and why?

Anne finds insurance cover for damage or theft of her guitar. It costs £1·50 a month for an annual policy. It has an excess of £50. This is the amount you have to pay if a claim is made.

   c) Would you buy the insurance policy? Explain your decision.

7   Christopher wants to buy a car. He finds these options.

New car	Used 3 years old	Used 8 years old
Pay £1450 and	Finance APR 6·9%	Cash price £3000
Finance APR 9·9%	Pay £157·25 a month, 5 years	Save up, 1 year
£241·39 a month, 5 years	Mileage 18 105	Mileage 71 340
Free servicing	Safety checked	Safety checked

   a) Calculate the total cost of the new car.

   b) How much less is the 3-year-old used car?

   c) Which option would you choose and why?

8   Basic car insurance that covers damage caused to other drivers (third party) and fire and theft is a legal requirement. Christopher finds these options for 1-year policies:

Quicksure	Total Cover Co	Oxford Insurance
£28·60 a month	Annual payment £209	One-off annual payment of £362 or
Third party fire and theft plus loss of keys and windscreen cover	Third party fire and theft only	12 monthly payments of £31·67
	£300 excess	Fully comprehensive, all damage
£150 excess		Road-side assistance
		Excess £0

   a) Find the total cost of the Quicksure policy.

   b) What are the differences between the Quicksure and Total Cover Co policies?

   c) Does Oxford Insurance charge extra to pay their policy in monthly instalments?

   d) Which policy would you choose and why?

9   Miriam is 16. She spends £20 a month on a music streaming subscription service. She finds these options.

Free Stream (with adverts)	Harmony (1 advert per hour)
£0 a month	£5 a month

   a) How much could Miriam save in a year using each option? Which would you recommend?

   b) Miriam has a savings account paying 4% p.a. After 1 year she puts the money she saves in the account. Calculate the balance she would have in her account after 5 years if she followed your recommendation from part a).

Check-up

1   Sian works as a social media manager. Complete her payslip.

Basic pay	Overtime	Bonus		Gross pay
£1858	–	£100		
Income Tax	National Insurance	Pension	Student loan	Total deductions
£162	£140	£98	£31	
			Net pay	

2   After housing costs Ivy has £810 a month to spend.
    Her spending plan for a 4-week month is shown.

a) Complete Ivy's monthly budget.

b) Ivy is planning on spending more than she has.
    How much more?

c) Set out a new spending plan for Ivy. List
    how much she should spend on each item.

    Include an amount for savings. Remember
    the total amount she has available is £810.

Ivy	Weekly	Monthly
Food	£95	£
Travel		£80
Subscriptions		£39
Clothes	£50	£
Socialising	£100	£
Debt repayments		£75
Total spending		£

3   Ryan earns £500 a week as a graduate accountant.

The bands and rates of National Insurance
contributions are shown.

Calculate how much National Insurance
Ryan will pay in a week.

Weekly earnings bands	Rate
£0–£183	0%
£183–£962	12%
over £962	2%

4   Rose borrows £6000 at APR 7·3% for home improvements.
    She repays £119 a month for 5 years.
    Calculate the cost of the loan.

5   Margaret earns £40 000.

a) How much of Margaret's income falls in the 21% tax band?

b) Calculate the total Margaret will have to pay in Income Tax.

Salary	Rate
12 500	0%
2085	19%
10 573	20%
a)	21%

6   Gill buys white goods for her new home. The total cost is £810 and she agrees to a finance deal
    with APR 8·2%. Gill pays £25·35 a month for 3 years. Calculate the total cost of the finance.

7   A couple are planning a wedding. They have these options.

Luxury	Simple
Cost of wedding £12 000	Cost of wedding £1750
Loan APR 12·4%	Pay with savings
Pay £193·37 a month for 8 years	

a) Calculate the total cost of the loan.

b) Which wedding would you recommend and why?

8 John is borrowing £2500 to buy a second-hand car. He finds these options.

| Trinity Credit APR 20·9%
£51·04 a month for 8 years | Fillan Finance APR 20%
£73·92 a month for 4 years | Brendan Bank APR 6·1%
£75·98 a month for 3 years |

  a) John is keen on the low payment from Trinity Credit. Would you recommend this deal?

  b) Calculate the total cost of each offer.

  c) Which would you recommend and why?

9 A credit card statement is shown.

  a) Calculate the new balance owed.

  b) What are the pros and cons of withdrawing travel money this way?

Date of transaction	Description	Amount £
	Balance from previous statement	472·83
01 Jun	Foreign cash withdrawal $500	458·72
01 Jun	Foreign payment charge 2·5%	11·47
18 Jun	Interest	19·22
21 Feb	Minimum Payment Received	12·80 CR
	New balance	

10 A table of exchange rates from the £ to different currencies is shown.

Calculate the value of each amount in the currency shown:

Currency	Symbol	Rate	Rounding
US dollar	$	1·25	Nearest cent 2 d.p.
South African rand	R	21·1	Nearest cent 2 d.p.
Hungarian forint	Ft	389·47	Nearest forint
Brazilian real	R$	6·68	Nearest centavo 2 d.p.

  a) £500 (US dollars)   **b)** £500 (South African rand)   **c)** £500 (Brazilian real)

  d) £169·59 (US dollars)   **e)** £1480·50 (Hungarian forint)   **f)** £200 000 (Brazilian real)

  g) 40 000 Ft (£)   **h)** R$ 8010·42 (£)   **i)** R 4 000 000 (£)

11 George plans a trip to Boston. He buys US $1500. The trip falls through and he is going to Brazil instead. Find the value of his US dollars in Brazilian real.

12 Louise imports 47 816 400 Ft of cast iron cookware from Hungary. She sells it to a distributer in the USA. Calculate the value of her import in US dollars.

13 Peggy has £900 and wants to buy euros. She finds these rates:

Online Forex Store	Online Next Day	Supermarket
1·13	1·18	1·1

  a) Which outlet offers the best exchange rate for Peggy?

  b) How many more euros will Peggy receive if she goes with the best rather than the worst deal?

14 Exchange rates for the euro in two different outlets are shown. Convert each amount choosing the best deal you can.

Bank		Supermarket	
We sell	We buy	We sell	We buy
€1·15	€1·25	€1·10	€1·41

  a) £7000 (€)   **b)** €9000 (£)

Convert each amount choosing the worst deal you can.

  c) £7000 (€)   **d)** €9000 (£)

8 Straight line

▶ Gradient of a straight line

The **gradient** of a slope is a measure of how steep it is. The greater the gradient, the steeper the slope. It is important for people to know the gradient of ramps or roads, for example.

To calculate the gradient of a slope:

● Use the formula gradient = $\dfrac{\text{vertical height}}{\text{horizontal distance}}$ or gradient = $\dfrac{V}{H}$ for short.

● A line has a positive gradient if it slopes upwards from left to right and a negative gradient if it slopes downwards from left to right.

● Remember always to simplify a fraction if possible.

Worked example Calculate the gradients of the following slopes. The direction of travel is left to right.

a)
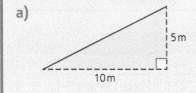
5 m
10 m

Substitute into the formula

gradient = $\dfrac{V}{H}$

= $\dfrac{5}{10}$

= $\dfrac{1}{2}$

b)
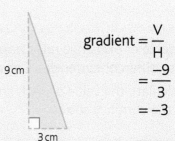
9 cm
3 cm

gradient = $\dfrac{V}{H}$

= $\dfrac{-9}{3}$

= -3

Exercise 8A

1   Calculate the gradients of the following slopes from left to right.

a)
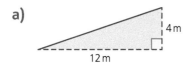
4 m
12 m

b)
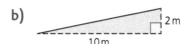
2 m
10 m

c)
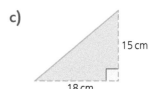
15 cm
18 cm

d)
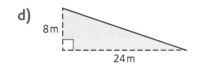
8 m
24 m

e)
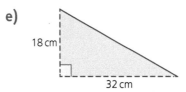
18 cm
32 cm

f)
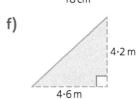
4·2 m
4·6 m

2   A particular slope has a gradient of 3. Its horizontal distance is 2 m. What is its vertical height?

3   A particular slope has a gradient of 4. Its vertical height is 32 m. What is its horizontal distance?

4   Slope A has a vertical change of 15 cm and a horizontal change of 36 cm. Slope B has a vertical change of 10 m and a horizontal change of 30 m. Which slope is steeper? Give a reason for your answer.

5   Slope A has a vertical change of 40 cm and a horizontal change of 100 cm. Slope B has a vertical change of 34 cm and a horizontal change of 85 cm. Which slope is steeper? Give a reason for your answer.

We can calculate the gradient of a line by using any two points on it.

The vertical change is the difference in the y-coordinates $(y_2 - y_1)$.
The horizontal change is the difference in the x-coordinates $(x_2 - x_1)$.

This results in the gradient formula $m = \dfrac{y_2 - y_1}{x_2 - x_1}$.

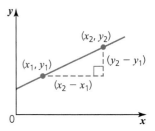

Worked examples

1   Calculate the gradient of the line shown.

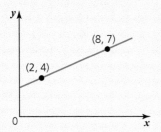

Label the points. $(2, 4)\ (8, 7)$

Substitute into $(x_1, y_1)\ (x_2, y_2)$
the formula and
evaluate.

$$m = \frac{y_2 - y_1}{x_2 - x_1}$$

$$= \frac{7 - 4}{8 - 2}$$

$$= \frac{3}{6}$$

$$= \frac{1}{2}$$

2   Calculate the gradient of the line passing
through the points $(-1, 10)$ and $(4, -5)$.

$(-1, 10)\quad (4, -5)$

$(x_1, y_1)\quad (x_2, y_2)$

$$m = \frac{y_2 - y_1}{x_2 - x_1}$$

$$= \frac{-5 - 10}{4 - (-1)}$$

$$= \frac{-15}{5}$$

$$= -3$$

Exercise 8B

1   Calculate the gradient of each of the following lines.

a) 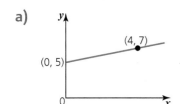   b) 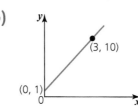   c) 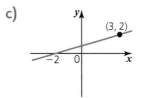   d)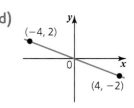

2   Calculate the gradient of the line that passes through each pair of points.

a) $(1, 6)$ and $(3, 10)$ b) $(0, 7)$ and $(2, -3)$ c) $(-2, -9)$ and $(3, 11)$
d) $(0, -3)$ and $(2, 21)$ e) $(-5, -35)$ and $(3, 1)$ f) $(0, 10)$ and $(2, -4)$
g) $(-5, -35)$ and $(1, 1)$ h) $(3, 4)$ and $(7, 4)$ i) $(-4, 1)$ and $(6, 6)$
j) $(8, -3)$ and $(20, 0)$ k) $(-9, 1)$ and $(18, 19)$ l) $(-10, 4)$ and $(5, 1)$
m) $(-8, 6)$ and $(4, -3)$ n) $(-16, -17)$ and $(0, -3)$ o) $(-4, -9)$ and $(2, 6)$

▶ Problem solving with gradient

There are many contexts in which gradient is important. When checking whether a slope has a suitable gradient, ensure that you use an inequality statement to compare it to the critical value.

Worked examples

1 To pass safety regulations, a ramp must have a gradient no greater than 0·8. Does the ramp shown below meet safety regulations?

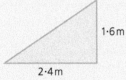

1·6 m

2·4 m

$$\text{gradient} = \frac{V}{H}$$

$$= \frac{1 \cdot 6}{2 \cdot 4}$$

$$= \frac{16}{24}$$

$$= \frac{2}{3}$$

$$= 0 \cdot \dot{6}$$

Since $0 \cdot \dot{6} \leq 0 \cdot 8$, the ramp meets safety regulations.

2 A line has gradient 4 and passes through the points $(3, -5)$ and $(7, t)$. Find the value of t.

We can substitute the coordinates and the value 4 into the gradient formula, giving:

$$m = \frac{y_2 - y_1}{x_2 - x_1}$$

$$4 = \frac{t - (-5)}{7 - 3}$$

$$4 = \frac{t + 5}{4}$$

Multiplying both sides by 4, we have

$$16 = t + 5$$

$$t = 11$$

Exercise 8C

1 A theme park claim their new rollercoaster, 'The Algebrageddon', has a stretch in which the gradient exceeds 5. A diagram of the steepest stretch is shown. Are their claims justified? Give a reason for your answer.

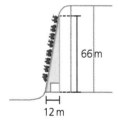

66 m

12 m

2 The maximum allowable decline for a certain category of zip line is $\frac{3}{100}$. Does the zip line shown meet this requirement? Give a reason for your answer.

5 m

150 m

3 A company manufactures accessibility ramps in different sizes. A particular ramp must have a maximum gradient of $\frac{1}{12}$ to be safe.

  a) A ramp is installed so that it rises 20 cm across a length of 2·2 metres. Is the ramp safe? Give a reason for your answer.

  b) A ramp is to be installed that rises 35 cm. What is the minimum horizontal length the ramp must cover to be safe?

4 A straight line with gradient 7 passes through the points $(2, -15)$ and $(8, t)$. Find the value of t.

5 A straight line with gradient -5 passes through the points $(-1, 18)$ and $(6, t)$, Find the value of t.

6 A straight line with gradient $\frac{3}{2}$ passes through the points $(-3, 7)$ and $(9, t)$. Find the value of t.

7 A straight line with gradient $-\frac{3}{5}$ passes through the points $(-17, -2)$ and $(t, -14)$. Find the value of t.

▶ Vertical and horizontal lines

- Horizontal lines have a gradient of zero. Their equations take the form $y = $ a number.
- Vertical lines have a constant x-coordinate. Their equations take the form $x = $ a number.

Worked examples

1 Sketch each of the given lines.

a) $x = 3$ b) $y = -2$

$x = 3$ represents a vertical line passing through all points which have an x-coordinate of 3.

$y = -2$ represents a horizontal line passing through all points which have a y-coordinate of -2.

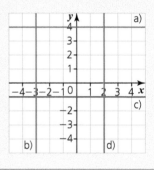

2 Write down the equation of each of the given lines.

a) The line is horizontal. Every point on it has y-coordinate 4.
Its equation is $y = 4$.

b) The line is vertical. Every point on it has x-coordinate -3.
Its equation is $x = -3$.

c) $y = -1$

d) $x = 2$

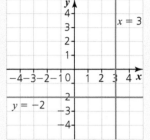

Exercise 8D

1 Write down the equation of each of the given lines.

2 Draw a coordinate grid with x- and y-axes going from -10 to 10 and sketch each of the following lines.

a) $y = 4$ b) $y = -7$ c) $x = 8$ d) $x = -5$

e) $y = -9$ f) $x = 2$ g) $x = -10$ h) $y = -1$

i) $x = 6$ j) $y = 0.5$

3 Write down the equation of the **horizontal** line that passes through each of the given points.

a) $(1, 3)$ b) $(5, 2)$ c) $(0, -7)$ d) $(-8, 11)$ e) $(16, -34)$

4 Write down the equation of the **vertical** line that passes through each of the given points.

a) $(5, 1)$ b) $(8, -2)$ c) $(-7, 0)$ d) $(-10, 11)$ e) $(0, 45)$

5 For each given pair of points, write down the equation of the line which passes through **both** points. Drawing a sketch may help you.

a) $(2, 8)$ and $(7, 8)$ b) $(5, -2)$ and $(5, 10)$ c) $(0, 6)$ and $(12, 6)$

d) $(-3, -3)$ and $(-3, 4)$ e) $(-10, -2)$ and $(10, -2)$ f) $(0, -3)$ and $(0, 7)$

6 What is the equation of the x-axis?

7 What is the equation of the y-axis?

▶ Drawing straight lines

We can sketch a straight line by finding two or more points that lie on it. To do so:

- Construct a table of x and y values.
- Select some sensible values of x (choose numbers that are not too large or too far apart).
- Substitute each value of x into the equation and evaluate to find the corresponding value of y.
- Take each x and y value as a coordinate pair and plot on a coordinate grid.
- Join up the points with a straight line (extending it in both directions).

Worked examples

1 Given the equation $y = 2x + 1$:

a) Complete the table of values.

x	0	1	2
y	1	3	5

We have the points (0, 1), (1, 3) and (2, 5).

b) Sketch the line.

Mark on the points calculated in the table of values then join with a ruler and label.

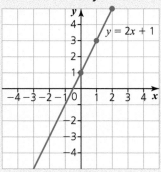

We substitute each value of x into the equation, evaluate and fill in the table.

$x = 0$:
$y = 2x + 1$
$= 2 \times 0 + 1$
$= 0 + 1$
$= 1$

$x = 1$:
$y = 2x + 1$
$= 2 \times 1 + 1$
$= 2 + 1$
$= 3$

$x = 2$:
$y = 2x + 1$
$= 2 \times 2 + 1$
$= 4 + 1$
$= 5$

c) Write down the coordinates of the y-intercept.

The y-intercept is the point where the line crosses the y-axis.

y-intercept: (0, 1)

2 Given the equation $y = -\dfrac{1}{2}x - 3$:

a) Sketch the line.

This time we are not given a table of values, so we must choose sensible values for x. Since we must halve the x-coordinate, it is practical to choose even numbers.

x	0	2	4
y	−3	−4	−5

Substituting each value of x into the equation, we have:

$x = 0$:
$y = -\dfrac{1}{2}x - 3$
$= -\dfrac{1}{2} \times 0 - 3$
$= 0 - 3$
$= -3$

$x = 2$:
$y = -\dfrac{1}{2}x - 3$
$= -\dfrac{1}{2} \times 2 - 3$
$= -1 - 3$
$= -4$

$x = 4$:
$y = -\dfrac{1}{2}x - 3$
$= -\dfrac{1}{2} \times 4 - 3$
$= -2 - 3$
$= -5$

Mark on the points.

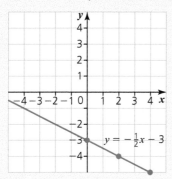

Now use a ruler to join them and label the line.

b) Write down the coordinates of the *y*-intercept.

y-intercept: (0, −3)

Exercise 8E

1 Given the equation $y = 3x$:

x	0	1	2
y	0		

    a) Copy and complete the following table of values.

    b) Draw a coordinate grid with *x*- and *y*-axes going from −10 to 10 and sketch the line.

    c) Write down the coordinates of the *y*-intercept.

2 Given the equation $y = x + 4$:

x	0	1	2
y	4		

    a) Copy and complete the following table of values.

    b) Draw a coordinate grid with *x*- and *y*-axes going from −10 to 10 and sketch the line.

    c) Write down the coordinates of the *y*-intercept.

3 Given the equation $y = -x$:

x	0	1	2
y		−1	

    a) Copy and complete the following table of values.

    b) Draw a coordinate grid with *x*- and *y*-axes going from −10 to 10 and sketch the line.

    c) Write down the coordinates of the *y*-intercept.

4 Given the equation $y = -2x + 3$:

x	0	1	2
y			

    a) Copy and complete the following table of values.

    b) Draw a coordinate grid with *x*- and *y*-axes going from −10 to 10 and sketch the line.

    c) Write down the coordinates of the *y*-intercept.

5 For each of the following:

    i) Draw and complete a suitable table of values.

    ii) Draw a coordinate grid with *x*- and *y*-axes going from −10 to 10 and sketch the line.

    iii) Write down the coordinates of the *y*-intercept.

    a) $y = x - 3$ **b)** $y = 2x + 4$ **c)** $y = 3x + 5$ **d)** $y = 2x - 4$

    e) $y = -2x - 1$ **f)** $y = -3x + 7$ **g)** $y = -x + 6$ **h)** $y = 6x - 1$

    i) $y = \dfrac{1}{2}x$ **j)** $y = \dfrac{1}{4}x + 3$ **k)** $y = -\dfrac{1}{5}x + 2$ **l)** $y = \dfrac{3}{2}x + 4$

6 For each of the lines in question **5**, use two of your coordinate points to calculate the gradient.

7 What connections can you see between the equation of a line, the gradient and the *y*-intercept?

► Equation of a straight line: $y = mx + c$

To find the equation of a straight line:

- Calculate the gradient (m) and the y-intercept $(0, c)$.
- Substitute into the formula $y = mx + c$.

Worked examples

1 Write down the gradient and y-intercept of the following lines.

   a) $y = 3x + 2$

   This line is in the format $y = mx + c$ so we simply read off the values.

   Gradient: 3, y-intercept: $(0, 2)$

   b) $y = -\dfrac{1}{3}x - 1$

   Gradient: $-\dfrac{1}{3}$, y-intercept: $(0, -1)$.

   c) $4x + 2y = 5$

   Be careful! The equation is not in the form $y = mx + c$ so we must first rearrange.

   $4x + 2y = 5$

   $\underset{-4x}{} \qquad \underset{-4x}{}$

   $2y = -4x + 5$

   $\div 2 \quad \div 2 \quad \div 2$

   $y = -2x + \dfrac{5}{2}$

   Gradient: -2, y-intercept: $(0, \dfrac{5}{2})$

2 Write down the equation of the line with gradient -4 and y-intercept $(0, 7)$.

   We know $m = -4$ and $c = 7$, so we substitute directly into $y = mx + c : y = -4x + 7$

Exercise 8F

1 Write down the gradient and y-intercept of the following lines.

   a) $y = 2x + 5$ b) $y = 6x + 1$ c) $y = 9x - 2$ d) $y = -3x + 12$

   e) $y = -7x - 1$ f) $y = x - 6$ g) $y = -11x - 4$ h) $y = -x$

   i) $y = \dfrac{1}{2}x + 1$ j) $y = \dfrac{3}{5}x - \dfrac{1}{5}$ k) $y = -\dfrac{1}{4}x + \dfrac{2}{5}$ l) $y = -\dfrac{5}{2}x - \dfrac{3}{2}$

2 Write down the equations of the straight lines with the following gradients and y-intercepts.

   a) gradient: 4, y-intercept: $(0, 6)$ b) gradient: 9, y-intercept: $(0, -5)$

   c) gradient: -2, y-intercept: $(0, 0)$ d) gradient: 1, y-intercept: $(0, 11)$

   e) gradient: -5, y-intercept: $(0, 3)$ f) gradient: -1, y-intercept: $(0, -7)$

3 By first rearranging to the form $y = mx + c$, write down the gradient and y-intercept of the following lines.

   a) $3y = 6x - 9$ b) $4y = 2x + 1$ c) $y + 5x = 3$ d) $y - x + 7 = 0$

   e) $8y + 6x = 4$ f) $8y - 4 = 14x$ g) $\dfrac{y}{3} = 2x - 3$ h) $\dfrac{y}{11} + x = 8$

4 What is the gradient and y-intercept of the line with equation $y = \dfrac{4}{5}$?

If you are not given the gradient explicitly, you can calculate it using any two points that lie on the straight line by using the gradient formula.

Worked examples

1 Find the equation of the straight line passing through the points (0, 4) and (3, 13).

   The y-intercept is (0, 4) so $c = 4$.

   The equation of the line is $y = 3x + 4$.

   $$m = \frac{y_2 - y_1}{x_2 - x_1}$$
   $$= \frac{13 - 4}{3 - 0}$$
   $$= \frac{9}{3}$$
   $$= 3$$

2 Find the equation of the line shown.

   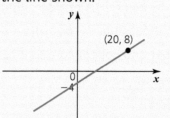

   $$m = \frac{y_2 - y_1}{x_2 - x_1}$$
   $$= \frac{8 - (-4)}{20 - 0}$$
   $$= \frac{12}{20}$$
   $$= \frac{3}{5}$$

   The y-intercept is (0, −4) so $c = −4$.

   The equation of the line is $y = \frac{3}{5}x - 4$.

3 Find the equation of the line with y-intercept (0, −5) parallel to $y = 7x + 2$.

   If two lines are parallel they have the same gradient, so we have $c = −5$ and $m = 7$.

   The equation of the line is $y = 7x − 5$.

---

5 Find the equation of the straight line that passes through the given pair of points.

   a) (0, 2) and (1, 6)
   b) (0, 12) and (4, 4)
   c) (0, −5) and (6, 7)
   d) (0, 13) and (4, 15)
   e) (−4, −10) and (0, 2)
   f) (0, 0) and (7, 21)

6 Find the equation of each straight line.

   a)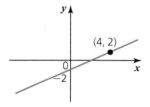

   b)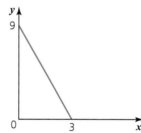

   c)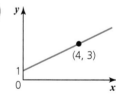

   d)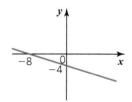

   e)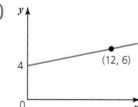

   f)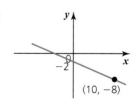

7 Find the equation of the line with y-intercept (0, 6) parallel to $y = 4x − 9$.

8 Find the equation of the line with y-intercept (0, −13) parallel to $y = \frac{1}{2}x - 1$.

9 Find the equation of the line with y-intercept (0, 5) parallel to $8y − 4x = 5$.

10 Find the equation of the line with y-intercept (0, −12) parallel to $7x + 3y − 5 = 0$.

▶ Equation of a straight line: $y - b = m(x - a)$

If we do not know the y-intercept, we can use the alternative form for the equation of a straight line, $y - b = m(x - a)$ where m is the gradient and (a, b) is any point on the line. To do so:

- Calculate the gradient.
- Select one point to be (a, b). Consider whether one will make the arithmetic easier.
- Substitute into the formula $y - b = m(x - a)$.
- Expand the bracket and collect constant terms.

Worked examples

1 Find the equation of the line passing through $(1, 5)$ and $(4, -1)$.

First, calculate the gradient.

$$m = \frac{y_2 - y_1}{x_2 - x_1}$$
$$= \frac{-1 - 5}{4 - 1}$$
$$= \frac{-6}{3}$$
$$= -2$$

We now choose one point to be (a, b) and substitute. Using $(1, 5)$ we have:

$$y - b = m(x - a)$$
$$y - 5 = -2(x - 1)$$
$$y - 5 = -2x + 2$$
$$y = -2x + 7$$

Note that we would obtain the same equation had we used the point $(4, -1)$.

2 Find the equation of the line passing through the points $(15, -18)$ and $(33, -12)$.

Calculating the gradient gives:

$$m = \frac{y_2 - y_1}{x_2 - x_1}$$
$$= \frac{-12 - (-18)}{33 - 15}$$
$$= \frac{6}{18}$$
$$= \frac{1}{3}$$

Using $(15, -18)$, we have:

$$y - b = m(x - a)$$
$$y - (-18) = \frac{1}{3}(x - 15)$$
$$y + 18 = \frac{1}{3}(x - 15)$$
$$3y + 54 = x - 15$$
$$3y = x - 69$$
$$y = \frac{1}{3}x - 23$$

Exercise 8G

1 Find the equations of the lines passing through the given points.

a) $(2, 5)$ and $(4, 9)$
b) $(5, 1)$ and $(8, 13)$
c) $(3, -2)$ and $(7, 2)$
d) $(-4, 6)$ and $(7, 39)$
e) $(7, 1)$ and $(9, -5)$
f) $(-1, -18)$ and $(4, 7)$
g) $(2, -6)$ and $(5, -27)$
h) $(11, 5)$ and $(6, 65)$
i) $(-2, -7)$ and $(3, 38)$
j) $(-3, 52)$ and $(2, -3)$
k) $(-8, -1)$ and $(6, 6)$
l) $(-8, -11)$ and $(20, -4)$
m) $(-9, -12)$ and $(18, 6)$
n) $(8, 12)$ and $(40, 24)$
o) $(-35, 9)$ and $(14, 2)$
p) $(-4, -3)$ and $(2, 5)$
q) $(-5, 43)$ and $(0, 8)$
r) $(4, -3)$ and $(20, -7)$

2 Find the equations of the lines shown.

a)
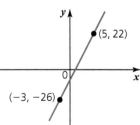

b)
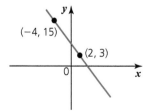

c)
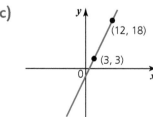

Worked example

The cost of a SuperGas bill depends on the number of units of gas used. The graph shows the cost in pounds (c) versus gas consumption (g).

a) Find an equation in terms of c and g for the cost incurred by a customer using g units.

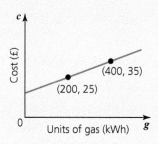

$$m = \frac{y_2 - y_1}{x_2 - x_1}$$

$$= \frac{35 - 25}{400 - 200}$$

$$= \frac{10}{200}$$

$$= \frac{1}{20}$$

Using (200, 25) we have:

$$y - b = m(x - a)$$

$$y - 25 = \frac{1}{20}(x - 200)$$

$$y - 25 = \frac{1}{20}x - 10$$

$$y = \frac{1}{20}x + 15$$

Note: it can be easier to expand the brackets on the right-hand side rather than multiply through by 20.

Rewrite the equation in terms of the letters given: $c = \frac{1}{20}g + 15$

b) What is the fixed charge?

The fixed charge corresponds to the y-intercept, so it is £15.

c) Use your equation to calculate the bill for a customer who uses 140 units of gas.

Substitute 140 for g and evaluate:

$$c = \frac{1}{20}g + 15$$

$$= \frac{1}{20} \times 140 + 15$$

$$= 7 + 15$$

$$= 22$$

The bill will be £22.

---

3 The charge from a pizza delivery company depends on the distance of the delivery. The graph shows the cost of the delivery versus the distance.

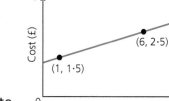

a) Find an equation for the cost, C pounds, to deliver a pizza d miles away.

b) Use your equation to calculate the cost of delivering a pizza to a house 15 miles away.

c) One particular delivery cost £3·30. Use your equation to calculate the distance it was delivered.

4 A taxi company base their charges on the distance driven. The graph shows the cost of a journey versus the distance driven.

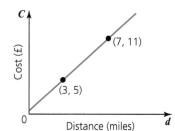

a) Find an equation for the cost, C pounds, of a taxi journey of d miles.

b) Use your equation to calculate the cost for a journey of 13 miles.

c) A particular journey costs £9·50. Use your equation to calculate the distance driven.

Check-up

1   Calculate the gradient of the following slopes.

a)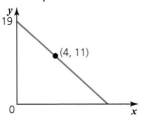
9 m
18 m

b)
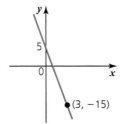
25 m
35 m

c)
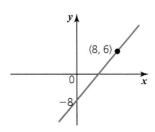
4 m
12 m

2   Calculate the gradient of the line that passes through each pair of points.

a) (2, 8) and (5, 14)   b) (7, −2) and (11, −4)   c) (−5, 3) and (2, 24)

3   In order to comply with building regulations, a particular ramp must
have a gradient no greater than $\frac{1}{15}$. Does the ramp shown comply
with regulations? Give a reason for your answer.

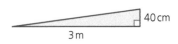
40 cm
3 m

4   A straight line goes through the points (6, 13) and (11, t) and has gradient −5. Find the value of t.

5   Draw a coordinate grid with x- and y-axes going from −5 to 5 and sketch the following lines.

a) $y = 3$ b) $x = −4$ c) $x = 1$ d) $y = −2$

6   Write down the equation of the **vertical** line that passes through the point (3, 7).

7   Write down the equation of the **horizontal** line that passes through the point (−6, −5).

8   Write down the equation of the line that passes through the points (−2, 8) and (7, 8).

9   Write down the equation of the line that passes through the points (4, −5) and (4, 12).

10 What is the equation of the y-axis?

11 Draw a coordinate grid with x- and y-axes going from −10 to 10. On the same grid, sketch the
following lines. Drawing a table of values for each line may help.

a) $y = 4x$ b) $y = 4x − 2$ c) $y = −3x + 5$ d) $y = \frac{1}{2}x$

12 Write down the equation of the straight line with the following gradient and y-intercept.

a) gradient: 4, y-intercept: (0, 9)   b) gradient: −5, y-intercept: (0, 7)

c) gradient: $\frac{1}{4}$, y-intercept: (0, −2)   d) gradient: $-\frac{5}{2}$, y-intercept: $\left(0, -\frac{3}{8}\right)$

13 Find the equation of the lines shown below.

a)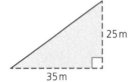
19
(4, 11)

b)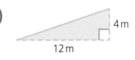
5
(3, −15)

c)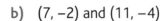
(8, 6)
−8

14 Find the equation of the straight lines passing through each given pair of points.

a) (0, 7) and (5, 27)   b) (0, 1) and (8, 49)   c) (0, −11) and (5, −21)

d) (−14, 8) and (0, 1)   e) (−10, 6) and (0, 0)   f) (0, 23) and (8, 3)

15 Find the equation of the line with y-intercept (0, 12) parallel to $y = −3x − 7$.

16 Find the equation of the line with y-intercept (0, −1) parallel to $y − 7x = 5$.

17 Find the equations of the lines shown below.

a)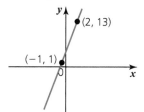

b)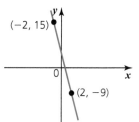

c)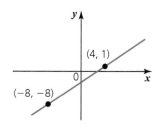

18 Find the equations of the lines passing through the following pairs of points.

a) $(3, 27)$ and $(8, 67)$

b) $(1, -2)$ and $(4, 37)$

c) $(-7, 8)$ and $(9, -8)$

d) $(7, 11)$ and $(13, 29)$

e) $(-3, 3)$ and $(4, -25)$

f) $(15, -3)$ and $(60, 6)$

19 Find the equation of the line through $(4, -5)$ parallel to $y = 7x + 12$.

20 Find the equation of the line through $(-6, 8)$ parallel to $4y + 2x = 12$.

21 Find the equation of the line through $(7, -3)$ parallel to $5y - 2x + 3 = 0$.

22 A delivery company calculates their packaging charge based on the number of items being packed. The graph shows the charge in pounds versus the number of items.

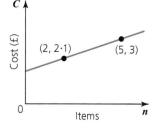

a) Form an equation for the charge, C pounds, to package n items.

b) Use your equation to calculate the charge to package 9 items.

c) A particular packaging charge comes to £5·70. Use your equation to find the number of items packaged.

23 A phone company calculate their customers' bills by charging a fixed fee plus an amount based on the duration of calls in minutes. One customer makes 100 minutes of calls and is charged £16. Another customer makes 150 minutes of calls and is charged £18.

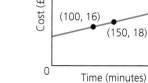

a) Form an equation for the cost, C pounds, to a customer who makes t minutes of calls.

b) What is the fixed fee?

c) What is the charge per minute?

d) Use your equation to calculate the cost to a customer who makes 70 minutes of calls.

24 The time Brian takes to mark a set of homework exercises depends on the number of jotters to be marked. Brian also spends a fixed amount of time reading the marking instructions. One day Brian marks 15 jotters and takes 1 hour 55 minutes. On another day, he takes 2 hours 58 minutes to mark 24 jotters.

a) Form an equation for the time, t minutes, it will take to mark j jotters.

b) Use your formula to calculate the time Brian would expect to take to mark 30 jotters.

25 An electrician charges customers a fixed callout fee plus an amount per hour worked. One job takes 2 hours and incurs a charge of £180. Another job takes 5 hours and incurs a charge of £390.

a) Find an equation for the cost, C pounds, of a job that lasts t hours.

b) How much is the callout fee?

c) Use your equation to calculate the cost of a job that lasts 8 hours.

9 Pythagoras' theorem and trigonometry

▶ Pythagoras' theorem

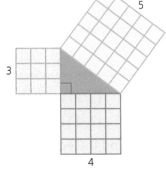

Pythagoras' theorem is one of the most important results in mathematics. It says that for any right-angled triangle, the square of the longest side (called the hypotenuse) is equal to the sum of the squares of the two shorter sides.

For example, in the triangle shown, the length of the hypotenuse is 5 and the shorter sides are 3 and 4.

Note that the hypotenuse is **always** the side opposite the right angle.

$5^2 = 25$ and $3^2 + 4^2 = 9 + 16$
$$= 25$$
In short, $5^2 = 3^2 + 4^2$

In general, we say that for a right-angled triangle ABC with a right angle at C, $c^2 = a^2 + b^2$.

❯❯ Calculating the hypotenuse

To calculate the length of the hypotenuse given the other two sides:

- Write down Pythagoras' theorem.
- Substitute the lengths of the known sides.
- Square and add (following the rules of BIDMAS).
- Take the positive square root to find the hypotenuse, writing your answer with appropriate units.

Worked example

Calculate the length of the unknown side in each triangle, giving your answers correct to three significant figures where necessary.

a)

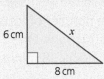

$c^2 = a^2 + b^2$ Write down the formula.
$x^2 = 6^2 + 8^2$ Substitute the known sides.
$\quad = 36 + 64$ Square and add.
$\quad = 100$
$x = \sqrt{100}$ Remember, this is x^2, so we must
$\quad = 10\,\text{cm}$ square root to find the value of x.

b)

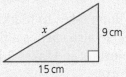

$c^2 = a^2 + b^2$
$x^2 = 15^2 + 9^2$
$\quad = 225 + 81$
$\quad = 306$
$x = \sqrt{306}$
$\quad = 17 \cdot 49\ldots$
$\quad = 17 \cdot 5\,\text{cm}$ to 3 s.f.

Note: If a right-angled triangle has integer sides (e.g. 6, 8, 10), we call them a 'Pythagorean triple'.

Exercise 9A

1 Copy and complete the working to calculate the hypotenuse in the triangle below.

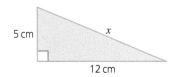

$$c^2 = a^2 + b^2$$

$$x^2 = \square^2 + \square^2$$

$$= \square + \square$$

$$= \square$$

$$x = \sqrt{\square}$$

$$= \square \text{ cm}$$

2 Calculate the length of the hypotenuse in each of the following triangles.

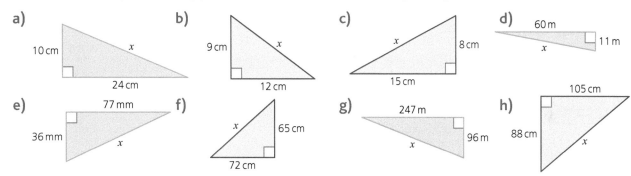

a)   10 cm x 24 cm

b)   9 cm x 12 cm

c) x 8 cm   15 cm

d)   60 m   11 m x

e)   77 mm   36 mm x

f) x 65 cm   72 cm

g)   247 m   96 m x

h)   105 cm   88 cm x

3 Calculate the length of the hypotenuse in each of the following triangles. Give your answers correct to three significant figures.

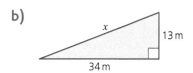 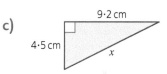

a)   7 cm x 18 cm

b) x 13 m   34 m

c)   9·2 cm   4·5 cm x

4 A football pitch is 120 m long and 90 m wide. What is the length of the diagonal of the pitch?

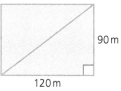

90 m

120 m

5 A rectangular television screen is 70 cm wide and 40 cm high. What is the diagonal length of the screen? Give your answer correct to three significant figures.

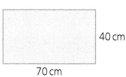

40 cm

70 cm

6 Sort the following sets of three numbers into those that are Pythagorean triples and those that are not.

5, 12, 13   33, 56, 64   12, 84, 85   36, 77, 85   57, 176, 183   60, 91, 109

>> Calculating a shorter side

If we are given the hypotenuse and one other side then we may use Pythagoras' theorem to find the remaining side. To do so:

- Write down Pythagoras' theorem.
- Substitute, taking care to ensure that c is the hypotenuse.
- Rearrange so that the unknown length is on the left-hand side of the formula.
- Solve.

Worked example   Calculate the length of the unknown side in each triangle.

a)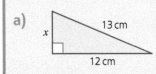

$$c^2 = a^2 + b^2$$
$$13^2 = x^2 + 12^2$$
$$\underline{-12^2 \qquad -12^2}$$
$$x^2 = 13^2 - 12^2$$
$$= 169 - 144$$
$$= 25$$
$$x = \sqrt{25}$$
$$= 5\,cm$$

b)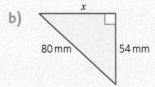

$$c^2 = a^2 + b^2$$
$$80^2 = x^2 + 54^2$$
$$\underline{-54^2 \qquad -54^2}$$
$$x^2 = 80^2 - 54^2$$
$$= 6400 - 2916$$
$$= 3484$$
$$x = \sqrt{3484}$$
$$= 59 \cdot 02\ldots$$
$$= 59 \cdot 0\,mm \text{ to } 3\,s.f.$$

Exercise 9B

1   Calculate the length of the unknown side in each of the following triangles.

a)   b) 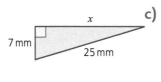   c) 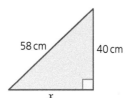   d)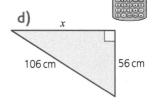

2   Calculate the length of the unknown side in each of the following triangles. Give your answers correct to three significant figures.

a) 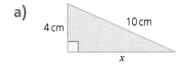   b) 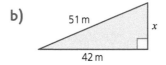   c)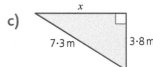

3   Helen measures her kitchen. It has a rectangular floor and is $3 \cdot 2$ metres long. She measures the distance from one corner to the opposite corner and finds that it is $7 \cdot 1$ metres.

   a)   Draw a floor plan of Helen's kitchen, marking on the lengths you know.

   b)   Use your diagram to help calculate the breadth of the kitchen, giving your answer correct to one decimal place.

›› Pythagoras – a mixture

When using Pythagoras' theorem:

- Identify a right-angled triangle and make a sketch, labelling all known sides.
- Take care to decide whether you are finding the hypotenuse or a short side.

Exercise 9C

Throughout this exercise, where necessary, give your answers correct to three significant figures.

1   Calculate the length of the unknown side in each triangle.

a)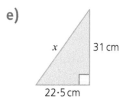

b)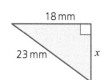

c)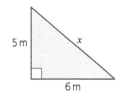

d)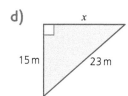

e)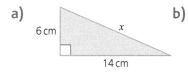

f)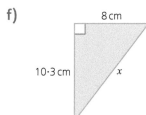

g)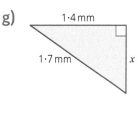

h)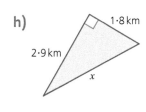

2   A club logo is in the shape of a right-angled triangle.

The length of the logo is 18 cm and the longest side is 27 cm.

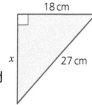

What is the minimum length of braiding, to the nearest centimetre, that is required to go the whole way around the outside of the logo?

3   The front of a storage unit is rectangular. The unit has length 3 m and breadth 1·2 m.

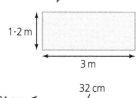

Find the length of the diagonal of the unit.

4   A rectangular book has a front cover with diagonal 32 cm and breadth 21 cm.

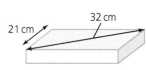

a)   Find the length of the book

b)   The book has a depth of 5 cm. Find the volume of the book.

5   The paper size A3 has width 297 mm and height 420 mm.

a)   Calculate the length of the diagonal of an A3 sheet of paper.

The paper size A4 has the same width as A3 but is half the height.

b)   Alison thinks the diagonal of A4 paper will be half the length of the diagonal of A3. Is Alison correct? Justify your answer.

6   Dougie walks in a straight line for 175 m. He then turns due west and walks for 62 m until he is due north of his starting point.

a)   How far north of his starting point is he?

b)   Dougie walks directly back to his starting point. It takes him 2 minutes and 5 seconds. What was Dougie's average speed walking back to the starting point?

▶ The converse of Pythagoras

Pythagoras' theorem states that in a triangle ABC with a right angle at C, $c^2 = a^2 + b^2$.

The **converse of Pythagoras** states that, given a triangle ABC, if $c^2 = a^2 + b^2$ then triangle ABC **must** be right-angled at C.

Worked example

Determine whether each of the following triangles is right-angled.

a)

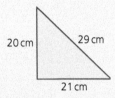

$29^2 = 841$

$20^2 + 21^2 = 400 + 441$

$\qquad = 841$

Since $29^2 = 20^2 + 21^2$, by the converse of Pythagoras, this triangle is right-angled.

Square the longest side.

Square the other two sides and add.

Write your conclusion.

b)

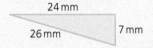

$26^2 = 676$

$24^2 + 7^2 = 576 + 49$

$\qquad = 625$

Since $26^2 \neq 24^2 + 7^2$, by the converse of Pythagoras, this triangle is **not** right-angled.

Exercise 9D

1 Determine whether each of the following triangles is right-angled.

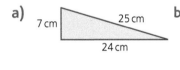

b)

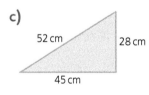

c)

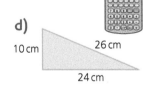

d)

2 A hockey pitch is marked out on a field. One side measures 100 yards and another measures 60 yards. A diagonal is measured as shown and is found to be 115 yards. Is the pitch right-angled? Explain your answer.

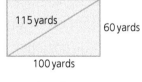

3 The central pole of a tent is 1·5 metres in length. One of the guy ropes extends from the top of the pole to a point 2 metres from the base of the pole. The guy rope is 2·5 metres long. Is the central pole vertical? Explain your answer.

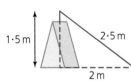

4 In his Design and Technology class, Gerry uses a set square to ensure things have right angles. He measures the edges of his set square and finds that two of them are 10 cm in length and the third is 14·5 cm. Is Gerry's set square fit for purpose? Explain your answer.

5 Nigel goes out sailing in his boat. He sails 8·5 km in one direction, turns his boat to the port side and sails 13·2 km before turning to head back to his starting point. The last leg of the journey was 15·7 km. Were any of Nigel's turns 90 degrees? Explain your answer.

▶ Problem solving with Pythagoras

When using Pythagoras' theorem:

● Identify a right-angled triangle and make a sketch, labelling all known sides.

● Take care to decide whether you are finding the hypotenuse or a short side.

Exercise 9E   Throughout this exercise, give your answers correct to three s.f.

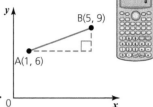

1 Look at the coordinate diagram shown.

　a) What is the length of the horizontal line?

　b) What is the length of the vertical line?

　c) Use Pythagoras' theorem to calculate the distance between points A and B.

2 In the same way as question 1, calculate the distance between each pair of points shown below.

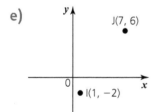
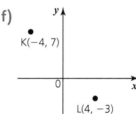
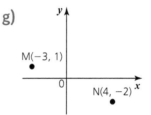
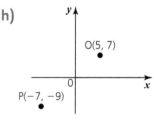

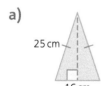
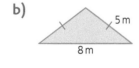

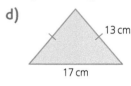

3 Calculate the height of each isosceles triangle by first splitting it into two right-angled triangles.

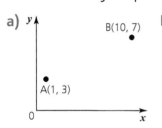
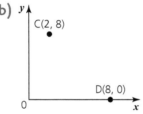

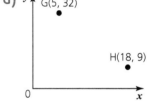

4 A Mathematics award is designed in the shape of a pencil. It is made up of an equilateral triangle joined to a rectangle as shown. What is the full length of the award?

5 A gift tag is in the shape of an isosceles triangle. It has two sides of length 8 cm and its height is 6 cm. How long is the base of the gift tag?

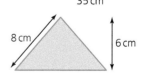

6 The flags on a crazy golf course are in the shape of equilateral triangles with sides of length 28 cm. What is the horizontal distance from the flagpole to the tip of the flag?

▶ Pythagoras in three dimensions

Pythagoras' theorem can be used to calculate lengths in three-dimensional objects.

Worked example

For the cuboid shown, calculate:

a) the length AC (shown in red on the diagram)

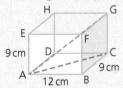

First, sketch the triangle ABC which contains AC as hypotenuse.

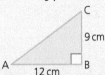

Now use Pythagoras' theorem to calculate AC.

$$c^2 = a^2 + b^2$$
$$AC^2 = 12^2 + 9^2$$
$$= 144 + 81$$
$$= 225$$
$$AC = \sqrt{225}$$
$$= 15 \text{ cm}$$

b) the space diagonal AG (shown in blue on the diagram).

Sketch the triangle ACG which contains AG as hypotenuse.

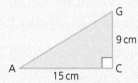

Pythagoras' theorem gives

$$c^2 = a^2 + b^2$$
$$AG^2 = 15^2 + 9^2$$
$$= 225 + 81$$
$$= 306$$
$$AG = \sqrt{306}$$
$$= 17 \cdot 49...$$
$$= 17 \cdot 5 \text{ cm to 3 s.f.}$$

Exercise 9F

Throughout this exercise, give your answers to three significant figures where necessary.

1   For each of the following cuboids, calculate:

　　i)　the length AC　　　　ii)　the space diagonal AG.

a)
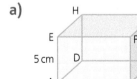

b)
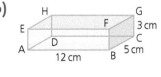

c)
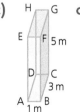

d)
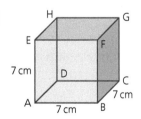

2   For the square-based pyramid shown, calculate:

　a)　the length AC

　b)　the length of the sloping edge AE.

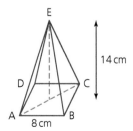

▶ Trigonometry: labelling your triangle

Trigonometry is the study of triangles and the relationships between their sides and their angles. At this level we only consider right-angled triangles. To study trigonometry effectively we must first learn to label triangles properly.

To label a right-angled triangle relative to a given angle:

- Find the right angle. The side directly across from this is called the **hypotenuse**. Label it 'H'.
- The side directly across from the marked angle is called the **opposite**. Label it 'O'.
- The remaining side (between the marked angle and the right angle) is called the **adjacent**. Label it 'A'.

Worked examples

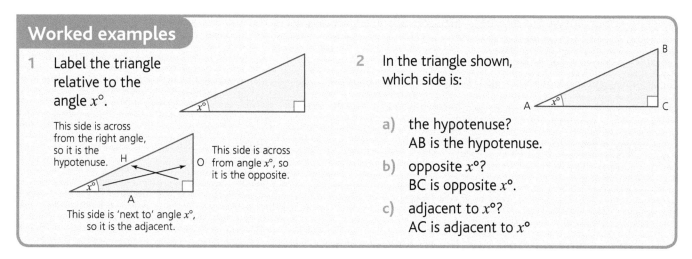

1　Label the triangle relative to the angle $x°$.

This side is across from the right angle, so it is the hypotenuse.

This side is across from angle $x°$, so it is the opposite.

This side is 'next to' angle $x°$, so it is the adjacent.

2　In the triangle shown, which side is:

a) the hypotenuse?
AB is the hypotenuse.

b) opposite $x°$?
BC is opposite $x°$.

c) adjacent to $x°$?
AC is adjacent to $x°$

Exercise 9G

1　Copy each of the following triangles and label the sides relative to the angle $x°$.

a)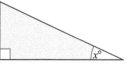

b)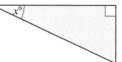

c)

2　For each triangle below, use two letters to name:

　　i)　the hypotenuse　　　　ii)　the opposite　　　　iii)　the adjacent.

a)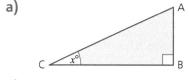

b)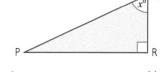

c)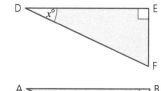

d)

e)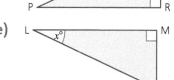

f)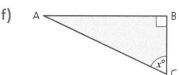

▶ Calculating an angle: sine

The sine ratio ('sin' for short) for a right-angled triangle is defined as

$$\sin x° = \frac{\text{opposite}}{\text{hypotenuse}} \quad \text{or} \quad \sin x° = \frac{O}{H} \text{ for short.}$$

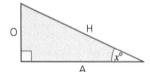

To calculate an angle using the sine ratio:

- Label the triangle.
- Write down the ratio and substitute in the lengths of the opposite and hypotenuse.
- Take the 'inverse sine' of this fraction. This is written as $\sin^{-1}$ and can be found using SHIFT and then sin on a scientific calculator.
- Give your answer in degrees with appropriate rounding.

Worked examples

Calculate the size of the marked angles, giving your answers correct to three significant figures.

1

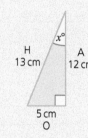

$$\sin x° = \frac{O}{H}$$
$$= \frac{7}{9}$$
$$x° = \sin^{-1}\left(\frac{7}{9}\right)$$
$$= 51·057…$$
$$= 51·1° \text{ to 3 s.f.}$$

2

$$\sin x° = \frac{O}{H}$$
$$= \frac{5}{13}$$
$$x° = \sin^{-1}\left(\frac{5}{13}\right)$$
$$= 22·619…$$
$$= 22·6° \text{ to 3 s.f.}$$

Exercise 9H

In this exercise, give your answers correct to three significant figures where necessary.

1 Copy and complete to find the size of angle x.

a) $\sin x° = \frac{O}{H}$

$$= \frac{7}{\square}$$

$$x° = \sin^{-1}\left(\frac{\square}{\square}\right)$$

$$= \square$$

$$= \square \text{ to 3 s.f.}$$

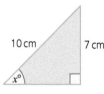

b) $\sin x° = \frac{O}{H}$

$$= \frac{\square}{15}$$

$$x° = \sin^{-1}\left(\frac{\square}{\square}\right)$$

$$= \square$$

$$= \square \text{ to 3 s.f.}$$

2 Use the sine ratio to calculate the size of the marked angles.

a)

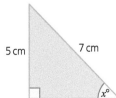

b)

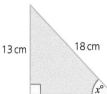

c)

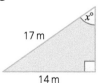

d)

e) 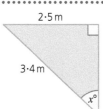 2·5 m 3·4 m $x°$

f) 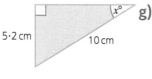 $x°$ 5·2 cm 10 cm

g) $x°$ 39 mm 54 mm

h) 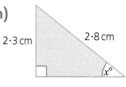 2·3 cm 2·8 cm $x°$

3 Look at the solutions to the questions below. In each case a mistake has been made. Describe the mistake in words and then calculate the correct value of angle $x°$.

a) 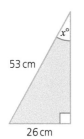 $x°$ 53 cm 26 cm

$$\sin x° = \frac{O}{H}$$
$$= \frac{53}{26}$$
$$x° = \sin^{-1}\left(\frac{53}{26}\right)$$

My calculator says 'math error'.   **X**

b) 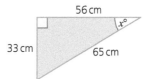 35 cm 12 cm $x°$ 37 cm

$$\sin x° = \frac{O}{H}$$
$$= \frac{12}{37}$$
$$x° = \sin^{-1}\left(\frac{12}{37}\right)$$
$$= 18·92...$$
$$= 18·9° \text{ to 3 s.f.}$$

X

4 Use the sine ratio to calculate the size of the marked angles, taking care to use the correct sides.

a)  37 cm 12 cm $x°$ 35 cm

b) 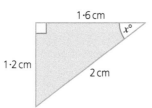 56 cm $x°$ 33 cm 65 cm

c) 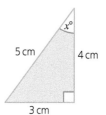 $x°$ 5 cm 4 cm 3 cm

d) 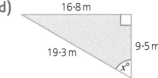 16·8 m 19·3 m 9·5 m $x°$

e) 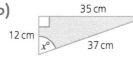 1·6 cm $x°$ 1·2 cm 2 cm

f) 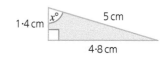 $x°$ 1·4 cm 5 cm 4·8 cm

5 A zip line extends from the top of a tree 15·4 metres tall to a point on the ground. The line itself is 600 metres long. At what angle does the zip line meet the ground?

 15·4 m 600 m

6 For each triangle below:

    i)   Calculate the missing side using Pythagoras' theorem

    ii) Calculate the size of angle $x°$.

a) 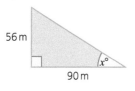 56 m $x°$ 90 m

b) 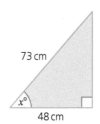 73 cm $x°$ 48 cm

c) 5·3 cm 2·8 cm $x°$

▶ Calculating an angle: cosine

The cosine ratio ('cos' for short) for a right-angled triangle is defined as

$$\cos x^\circ = \frac{\text{adjacent}}{\text{hypotenuse}} \quad \text{or} \quad \cos x^\circ = \frac{A}{H} \text{ for short.}$$

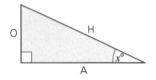

Worked examples

Calculate the size of the marked angles, giving your answers correct to three significant figures.

1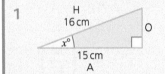

$$\cos x^\circ = \frac{A}{H}$$
$$= \frac{15}{16}$$
$$x^\circ = \cos^{-1}\left(\frac{15}{16}\right)$$
$$= 20 \cdot 364\ldots$$
$$= 20 \cdot 4^\circ \text{ to 3 s.f.}$$

2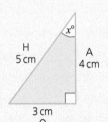

$$\cos x^\circ = \frac{A}{H}$$
$$= \frac{4}{5}$$
$$x^\circ = \cos^{-1}\left(\frac{4}{5}\right)$$
$$= 36 \cdot 869\ldots$$
$$= 36 \cdot 9^\circ \text{ to 3 s.f.}$$

Exercise 9I

Throughout this exercise, give your answers correct to three significant figures.

1 Use the cosine ratio to calculate the size of the marked angles.

a)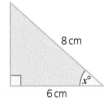

b)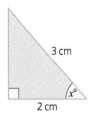

c)

d)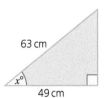

e)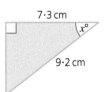

f)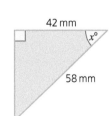

g)

h)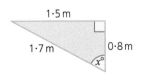

2 Use the cosine ratio to calculate the size of the marked angles.

a)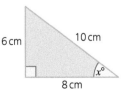

b)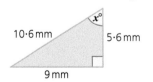

c) 1·5 m, 1·7 m, 0·8 m

3 The top of a slide in a children's playground is 140 cm high. The slide itself is 3·2 metres in length. What is the angle of depression a°, shown in the diagram?

140 cm, 3·2 m, a°

▶ Calculating an angle: tangent

The tangent ratio ('tan' for short) for a right-angled triangle is defined as

$$\tan x° = \frac{\text{opposite}}{\text{adjacent}} \text{ or } \tan x° = \frac{O}{A} \text{ for short.}$$

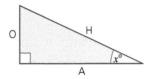

Worked examples

Calculate the size of the marked angles, giving your answers correct to three significant figures.

1

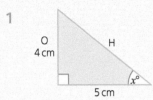

$$\tan x° = \frac{O}{A}$$
$$= \frac{4}{5}$$
$$x° = \tan^{-1}\left(\frac{4}{5}\right)$$
$$= 38 \cdot 659\ldots$$
$$= 38 \cdot 7° \text{ to 3 s.f.}$$

2

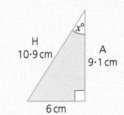

$$\tan x° = \frac{O}{A}$$
$$= \frac{6}{9 \cdot 1}$$
$$x° = \tan^{-1}\left(\frac{6}{9 \cdot 1}\right)$$
$$= 33 \cdot 398\ldots$$
$$= 33 \cdot 4° \text{ to 3 s.f.}$$

Exercise 9J

Throughout this exercise, give your answers correct to three significant figures.

1 Use the tangent ratio to calculate the size of the marked angles.

a)

6 cm
5 cm

b)

11 cm
11 cm

c)

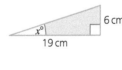
6 cm
19 cm

d)

1·2 m
0·8 m

e)

13·5 cm
6·7 cm

f)

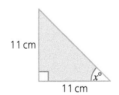
1 mm
3·2 mm

g)

3·9 cm
1·9 cm

h)

17 mm
18 mm

2 Use the tangent ratio to calculate the size of the marked angles.

a)

11·2 cm
21·2 cm
18 cm

b)

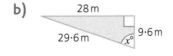
28 m
29·6 m
9·6 m

c)

37·3 cm
27·5 cm
25·2 cm

3 A ship sails 240 kilometres due north before turning and sailing 180 kilometres due west. What is the bearing of the ship from its starting position?

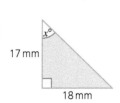
180 km   N
240 km

4 Susanne is standing 50 metres from the base of the Glasgow Tower, looking up at its peak. If the tower is 127 metres high and Susanne's eye level is 1·8 metres above the ground, what is the angle of elevation at which she is looking?

▶ Calculating a side: sine

We can also use trigonometric ratios to calculate missing sides in right-angled triangles.

To calculate a missing side using the sine ratio:

- Label the triangle.
- Write down the sine ratio and substitute in the angle and the given side. Do not confuse angle $x°$ in the ratio with an unknown side labelled x in the triangle!
- Balance the equation so that the missing side is on the left.
- Calculate, giving your answer with appropriate units and rounding.

Worked examples

Calculate the size of the marked sides, giving your answers correct to three significant figures.

1

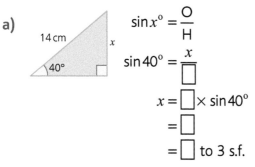

$$\sin x° = \frac{O}{H}$$

$$\sin 50° = \frac{x}{10}$$

$$\times 10 \quad \times 10$$

$$x = 10 \times \sin 50°$$
$$= 7 \cdot 660\ldots$$
$$= 7 \cdot 66 \text{ cm to 3 s.f.}$$

Note that when finding the opposite there is only one balancing step: multiplying. When finding the hypotenuse there are two balancing steps: multiplying and then dividing.

2

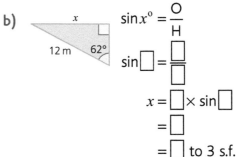

$$\sin x° = \frac{O}{H}$$

$$\sin 70° = \frac{4}{x}$$

$$\times x \quad \times x$$

$$x \sin 70° = 4$$

$$\div \sin 70° \quad \div \sin 70°$$

$$x = \frac{4}{\sin 70°}$$
$$= 4 \cdot 256\ldots$$
$$= 4 \cdot 26 \text{ cm to 3 s.f.}$$

Exercise 9K

Throughout this exercise, give your answers correct to three significant figures.

1 Copy and complete the working to calculate the marked sides.

a)

$$\sin x° = \frac{O}{H}$$

$$\sin 40° = \frac{x}{\square}$$

$$x = \square \times \sin 40°$$
$$= \square$$
$$= \square \text{ to 3 s.f.}$$

b)

$$\sin x° = \frac{O}{H}$$

$$\sin \square = \frac{\square}{\square}$$

$$x = \square \times \sin \square$$
$$= \square$$
$$= \square \text{ to 3 s.f.}$$

2 Use the sine ratio to calculate the lengths of the marked sides.

a)

b)

c)

d)

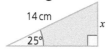

e)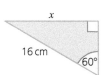

f)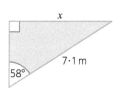

g)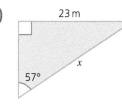

h)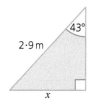

3 Copy and complete the working to calculate the lengths of the marked sides.

a)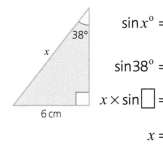

$$\sin x° = \frac{O}{H}$$

$$\sin 38° = \frac{\square}{x}$$

$$x \times \sin\square = \square$$

$$x = \frac{\square}{\sin\square}$$

$$= \square$$

$$= \square \text{ to 3 s.f.}$$

b)

$$\sin x° = \frac{O}{H}$$

$$\sin\square = \frac{\square}{x}$$

$$x \times \sin\square = \square$$

$$x = \frac{\square}{\sin\square}$$

$$= \square$$

$$= \square \text{ to 3 s.f.}$$

4 Use the sine ratio to calculate the lengths of the marked sides.

a)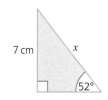

b)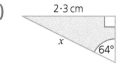

c)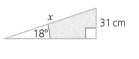

d)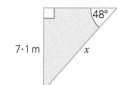

5 A mistake has been made in each of the two solutions below. Describe the mistake in words and then correctly calculate the length of side x in each case.

a)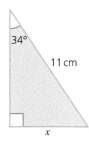

$$\sin x° = \frac{O}{H}$$

$$\sin 11° = \frac{x}{34}$$

$$x = 34 \times \sin 11°$$

$$= 6·487...$$

$$= 6·49 \text{ cm to 3 s.f.} \quad \text{X}$$

b)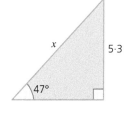

$$\sin x° = \frac{O}{H}$$

$$\sin 47° = \frac{5·3}{x}$$

$$x = 5·3 \times \sin 47°$$

$$= 3·876...$$

$$\text{X} \quad = 3·88 \text{ m to 3 s.f.}$$

6 A radio tower has a 32-metre cable stretched from its peak to a point on the ground. The cable meets the ground at an angle of 30°. How tall is the radio tower?

7 A ramp is being installed at a loading dock. It is 5 metres in length and safety regulations state that it must be at an angle of 15°. The ramp is intended to reach a point 1·3 metres higher than ground level. Is the ramp fit for purpose? Give a reason for your answer.

8 For the triangle shown

    a) Use the sine ratio to calculate the length of the side marked x.

    b) Use Pythagoras' theorem to calculate the length of the side marked y.

▶ Calculating a side: cosine

Calculate the size of the marked sides, giving your answers correct to three significant figures.

1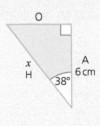

$$\cos x° = \frac{A}{H}$$

$$\cos 67° = \frac{x}{3.5}$$

$$\times 3.5 \quad \times 3.5$$

$$x = 3.5 \times \cos 67°$$
$$= 1.367...$$
$$= 1.37 \text{ m to 3 s.f.}$$

2

$$\cos x° = \frac{A}{H}$$

$$\cos 38° = \frac{6}{x}$$

$$\times x \quad \times x$$

$$x \cos 38° = 6$$

$$\div \cos 38° \quad \div \cos 38°$$

$$x = \frac{6}{\cos 38°}$$
$$= 7.614...$$
$$= 7.61 \text{ cm to 3 s.f.}$$

Exercise 9L

Throughout this exercise, give your answers correct to three significant figures.

1 Use the cosine ratio to calculate the lengths of the marked sides.

a)
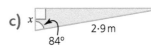

b)
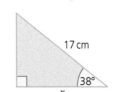

c)

d)

e)

f)
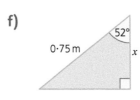

g)
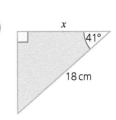

h)

2 Use the cosine ratio to calculate the lengths of the marked sides.

a)
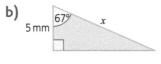

b)
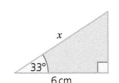

c)

d)
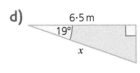

3 John walks up a road that has an incline of 24°. The length of the road is 36 metres. What is the horizontal distance that John has travelled?

4 An archway in a Museum of Mathematics is in the shape of an isosceles triangle as shown. What are the lengths of the sloping sides of the arch?

▶ Calculating a side: tangent

Calculating a side: tangent

Worked examples

Calculate the size of the marked sides, giving your answers correct to three significant figures.

1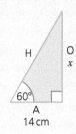

$$\tan x° = \frac{O}{A}$$

$$\tan 60° = \frac{x}{14}$$

$$\times 14 \qquad \times 14$$

$$x = 14 \times \tan 60°$$
$$= 24 \cdot 24 \ldots$$
$$= 24 \cdot 2 \text{ cm to 3 s.f.}$$

2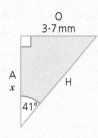

$$\tan x° = \frac{O}{A}$$

$$\tan 41° = \frac{3 \cdot 7}{x}$$

$$\times x \qquad \times x$$

$$x \tan 41° = 3 \cdot 7$$

$$\div \tan 41° \quad \div \tan 41°$$

$$x = \frac{3 \cdot 7}{\tan 41°}$$
$$= 4 \cdot 256 \ldots$$
$$= 4 \cdot 26 \text{ mm to 3 s.f.}$$

Exercise 9M

Throughout this exercise, give your answers correct to three significant figures.

1 Use the tangent ratio to calculate the length of the marked sides.

a) 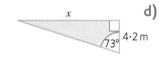   b)   c)   d)

e) 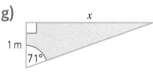   f)   g)   h)

2 Use the tangent ratio to calculate the lengths of the marked sides.

a) 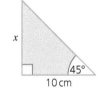   b) 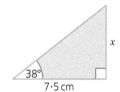   c) 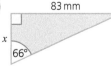   d)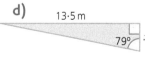

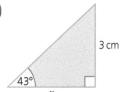  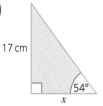 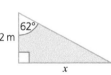 

3 A staircase in a house has an angle of elevation of 48°. The top of the staircase is 2·47 metres above the ground floor. What horizontal distance does the staircase cover?

▶ Which formula?

You will not always be told which formula to use. Sometimes you must decide based on the information you have about a given triangle. We can use the acronym "SOH CAH TOA" to help us decide which formula to use and to help memorise the formulae themselves. Each part is an acronym of a trig ratio:

SOH	CAH	TOA
$\sin x° = \dfrac{\text{opposite}}{\text{hypotenuse}}$	$\cos x° = \dfrac{\text{adjacent}}{\text{hypotenuse}}$	$\tan x° = \dfrac{\text{opposite}}{\text{adjacent}}$

To decide which formula to use:

● Write out the acronym SOH CAH TOA.

● Tick off the sides you know or want to find.

● The part of the acronym that has two letters ticked tells you which trig ratio to use.

Worked examples

In each triangle, calculate the size of the marked side or angle, giving your answers correct to three significant figures.

1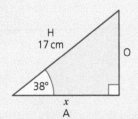

Write out the acronym.

We know the **hypotenuse** so we tick 'H'. We want the **adjacent** so we tick 'A'.

'CAH' has two letters ticked so we must use the cosine ratio.

S O H ✓ C A ✓ H ✓ T O A ✓

$\cos x° = \dfrac{A}{H}$

$\cos 38° = \dfrac{x}{17}$

$x = 17 \times \cos 38°$

$= 13\cdot 39…$

$= 13\cdot 4\,\text{cm to 3 s.f.}$

2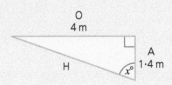

This time we are trying to find an angle. We know the **adjacent** and the **opposite** so we tick off 'A' and 'O'.

'TOA' has two letters ticked so we must use the tangent ratio.

S O ✓ H C A H T O ✓ A ✓

$\tan x° = \dfrac{O}{A}$

$= \dfrac{4}{1\cdot 4}$

$x° = \tan^{-1}\left(\dfrac{4}{1\cdot 4}\right)$

$= 70\cdot 70…$

$= 70\cdot 7°\text{ to 3 s.f.}$

Exercise 9N

1 Calculate the lengths of the marked sides, giving your answers correct to three significant figures.

a)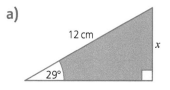

b)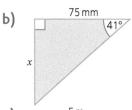

c)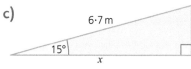

d)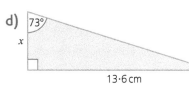

e)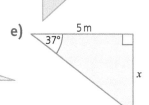

f)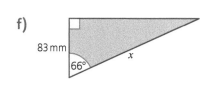

2 Calculate the sizes of the marked angles, giving your answers correct to three significant figures.

a)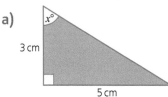

b)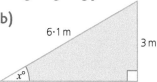

c)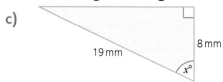

d)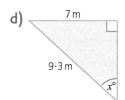

e)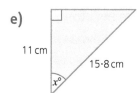

f)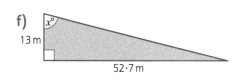

3 A chocolatier designs a chocolate slab in the shape of a right isosceles triangle. The longest edge of the slab is to be 15 cm. How long must the other two edges be?

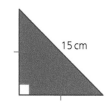

4 Callum is building a ramp for his remote-controlled car. He uses a piece of plywood 32 cm long and props it up on a 13 cm high paperweight as shown. What is the angle of elevation of Callum's ramp?

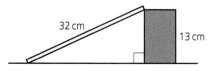

5 Calculate the length of the marked side at the top of the stack of triangles. All lengths are in centimetres. Work to two decimal places throughout.

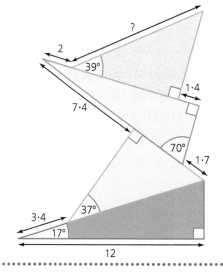

Check-up 👍

1 Use Pythagoras' theorem to calculate the length of the missing side in each of the following triangles, giving your answers correct to three significant figures.

a)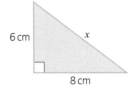
6 cm   x
8 cm

b)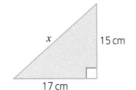
x   15 cm
17 cm

c)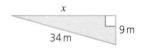
x
34 m   9 m

d)
4·9 mm
x   7·8 mm

e)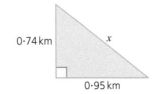
0·74 km   x
0·95 km

f)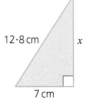
12·8 cm   x
7 cm

2 A computer monitor has a rectangular display with a length of 53·4 cm and a breadth of 42·8 cm. The manufacturers market the display as having a diagonal length of 68·5 cm. Are the manufacturers being honest? Give a reason for your answer.

42·8 cm
53·4 cm

3 From a starting point, a hillwalker walks 4·9 km on a given bearing. She then turns and walks 2·7 km due west until she is directly north of her starting point. What distance is she from her starting position?

N   2·7 km
4·9 km

4 On an orienteering course, a group of pupils follow a walking route in the shape of a right-angled triangle. They walk south for 1·53 km, then turn and walk for 1·85 km until they are directly east of their starting point.

a) Make a sketch of the route followed by the orienteers.

b) Use Pythagoras' theorem to calculate the **total distance** covered by the orienteers.

5 Determine whether each of the following triangles is right-angled.

a)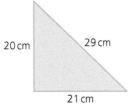
20 cm   29 cm
21 cm

b)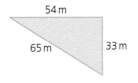
54 m
65 m   33 m

c)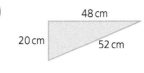
48 cm
20 cm   52 cm

6 A joiner is fitting a window frame. They connect two strips of wood measuring 135 cm and 52 cm to form an 'L' shape. The diagonal length between the two strips is 153 cm. Has the joiner successfully made a right angle? Give a reason for your answer.

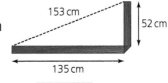
153 cm   52 cm
135 cm

7 A shopper orders a new metre stick online.

The delivery company's only available box is shown opposite. Will the metre stick fit in the box? Give a reason for your answer.

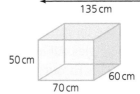
50 cm
60 cm
70 cm

8 Calculate the size of the marked angles, giving your answers correct to three significant figures.

a)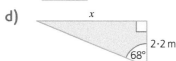

12 cm

10 cm

b)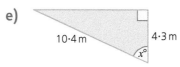

23 mm

24 mm

c)

14 m

17 m

d)

5·8 cm   7·1 cm

e)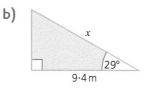

10·4 m   4·3 m

f)

0·43 mm

0·65 mm

9 Calculate the size of the marked sides, giving your answers correct to three significant figures.

a)

15 cm   *x*

63°

b)

x

29°

9·4 m

c)

x   3·14 mm

37°

d)

x

2·2 m

68°

e)

31·7 cm

x

81°

f)

83·9 mm

78°

x

10 A ship sets off on a bearing of 053° and sails for 18 km. How far east of its starting point is the ship? Give your answer correct to four significant figures.

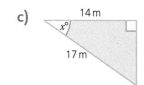

N

053°   18 km

11 A 20-foot ladder is leaning against a wall, forming an angle of 65° with the ground.

a) How high up the wall does the ladder reach?

b) The ladder slides backwards 3 feet. How high up the wall does the ladder now reach?

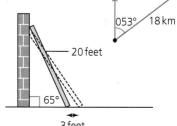

20 feet

65°

3 feet

12 In part of an assault course, participants have to descend on a zip-line from a treetop to the ground. The angle of depression from the horizontal is 1·7°. The zip-line starts at a point 16 metres above ground level.

a) What horizontal distance does the zip-line cover? Give your answer correct to three significant figures.

b) Use your answer to part **a)** to calculate the length of the zip-line itself.

13 During part of its flight, a plane climbs at an angle of 7°. It covers a horizontal distance of 20·7 km during this phase of flight.

a) How much altitude (height) does the plane gain during this phase of flight?

b) At the end of this phase, what distance has the plane flown?

c) The plane travels at a speed of 740 km/h. How long does this phase last? Give your answer to the nearest second.

10 Project

► Mathematics in the workplace

It is impossible to get through a day without using mathematics in one of its guises. Whether it be time management, using measurement, managing money or solving a problem, mathematics is everywhere!

1　Describe three occasions today when you have used mathematics.

2　Copy the table below and match the job titles with the descriptions of how mathematics is used.

Job title	Example of mathematics used
Sound engineer	Geometry is used to calculate angles and plan routes.
Meteorologist	Mathematical models are used to understand the complexities of a business cycle and the effects of inflation.
Doctor	Accurate measurements are taken to produce scale drawings and plans.
Architect	Mathematical models are used to track the progress of infectious diseases and the effectiveness of vaccines.
Economist	Numerical analysis and computer modelling techniques are used to make short- and long-term predictions about climate change.
Epidemiologist	The mathematical technique of Fourier analysis is used to manipulate pitch and sound.
Pilot	Building and structural safety features are calculated from known and measured data.
Estate agent	Dosages and treatment plans determined by numerical calculations and probabilities.

Studying mathematics will not only develop your subject expertise but will help you to develop many transferrable skills which are vital in a wide variety of careers. The ability to think clearly and logically allows mathematicians to break down often very complex problems into smaller, more manageable chunks. Mathematicians relish a problem and will show great determination and flexibility in investigating possible routes to a solution. Communicating all findings precisely and logically is also an important skill that mathematicians frequently demonstrate.

There is not enough room in this book to list all the possible careers you may enter as a mathematician but here are some broad areas.

Finance

Careers in finance revolve around money. This may be counting or managing money, considering the best products or services available on the market or analysing market trends. Scottish botanist Robert Brown first observed a trend in nature which demonstrated a now-famous technique used in mathematical finance called Brownian Motion. Career paths to consider within the finance sector include actuarial, accountancy, investments, insurance, stockbroking, banking and technology.

Statistics

Statisticians interpret and analyse the wealth of data which surrounds us. They often communicate their findings using graphs, charts and tables and identify trends in data. Florence Nightingale was a prolific mathematician who used statistics and statistical diagrams to improve nursing and hospital conditions after her experiences in the Crimean war. Fields of study statisticians focus on include market research, sport, environmental impacts, government policy, pharmaceuticals and medicine.

Science

You will already be aware of the intrinsic link between mathematics and the sciences. Mathematician Katherine Johnson's pioneering work facilitated flight paths to the Moon and logarithms, which were developed by John Napier, are widely used in chemistry calculations. The possibilities for mathematicians to work in science are plentiful; careers include research mathematicians, biomedical scientists, geologists, meteorologists and astronomists.

Engineering

The engineering industry uses mathematics and science to develop the world we live in. For example, in civil engineering, trigonometry is used to measure the suitability of land to be built on. Engineering is often a very practical and creative vocation where you invent, design, build and test new products. From James Watt's steam engine to Alexander Graham Bell's telephone, Scotland has a rich history of innovative engineers. Different engineering disciplines include chemical, mechanical, electrical, product design, sound, structural, biomedical, naval, aerodynamical and renewable energy.

Technology

It is hard to envisage everyday life without technology. IT and computers have an impact on all aspects of our lives. When mathematician Ada Lovelace first published details of a computer programme in the 1840s she had no way of knowing the developments which would follow. Employment opportunities within the technology industry include graphic designer, cryptographer, software developer, hardware developer, cloud computing architect, ethical hacker, computer game programmer, cyber security specialist and data analyst.

Other

Education, research, medicine, architecture, design, geophysics, genetics, climatology, oceanography, music, design, retail, farming, fishing, law, journalism – the list is endless...

Research and presentation task

Create a presentation on a career from one of the categories above. Your presentation can take the form of a poster, PowerPoint presentation or report, but it must include:

- a description of your chosen career
- examples of the mathematics used in it
- the skills required to be successful
- any examples of famous professionals in the field and/or notable discoveries
- subjects you could study at school in preparation for this career.

11 Further algebra

▶ Expanding double brackets

To expand double brackets, reduce the problem to the sum of single brackets.

- Split up the first bracket and then multiply everything in the second bracket by each term from the first bracket (see working in red in the Worked examples below).

- Expand the resulting single brackets and simplify by gathering like terms.

> **Worked examples**   Expand and simplify:
>
> 1 $(x + 3)(x + 4)$
> $= x(x + 4) + 3(x + 4)$
> $= x^2 + 4x + 3x + 12$
> $= x^2 + 7x + 12$
>
> 2 $(x + 5)(x - 1)$
> $= x(x - 1) + 5(x - 1)$
> $= x^2 - x + 5x - 5$
> $= x^2 + 4x - 5$
>
> 3 $(y - 7)(y + 7)$
> $= y(y + 7) - 7(y + 7)$
> $= y^2 + 7y - 7y - 49$
> $= y^2 - 49$
>
> 4 $(k - 10)^2$
> $= (k - 10)(k - 10)$
> $= k(k - 10) - 10(k - 10)$
> $= k^2 - 10k - 10k + 100$
> $= k^2 - 20k + 100$
>
> 5 $(5y - 3)(9y + 7)$
> $= 5y(9y + 7) - 3(9y + 7)$
> $= 45y^2 + 35y - 27y - 21$
> $= 45y^2 + 8y - 21$
>
> 6 $(3x - 4)^2$
> $= (3x - 4)(3x - 4)$
> $= 3x(3x - 4) - 4(3x - 4)$
> $= 9x^2 - 12x - 12x + 16$
> $= 9x^2 - 24x + 16$

With practice, you should be able to go straight to the penultimate line and then simplify to your final answer. Some learners like the visual aid of adding lines to show the expansion order, for example $2x$ multiplying the second bracket (blue lines) and then $+ 5$ multiplying the second bracket (red lines).

$$(2x + 5)(x + 1) = 2x^2 + 2x + 5x + 5$$

Exercise 11A

1 Expand and simplify:
   a) $(x + 2)(x + 3)$
   b) $(x + 6)(x + 2)$
   c) $(y + 7)(y + 1)$
   d) $(y + 9)^2$
   e) $(x + 4)(x - 2)$
   f) $(x + 5)(x - 4)$
   g) $(y + 6)(y - 7)$
   h) $(y + 2)(y - 8)$

2 Expand and simplify:
   a) $(x - 8)(x + 1)$
   b) $(x - 7)(x + 4)$
   c) $(y - 1)(y + 9)$
   d) $(y - 3)(y + 3)$
   e) $(x - 2)(x - 4)$
   f) $(x - 1)(x - 9)$
   g) $(y - 5)(y - 8)$
   h) $(y - 4)^2$
   i) $(x - 9)(x + 2)$
   j) $(x - 3)^2$
   k) $(y + 1)(y - 1)$
   l) $(y + 6)(y + 5)$

3 Expand and simplify:
   a) $(x + 4)(2x + 3)$
   b) $(x + 2)(3x + 9)$
   c) $(x + 7)(4x - 1)$
   d) $(x + 6)(5x - 2)$
   e) $(y - 8)(3y + 5)$
   f) $(y - 10)(2y + 5)$
   g) $(m - 3)(6m - 1)$
   h) $(10h - 7)^2$
   i) $(7x + 5)(3x + 4)$
   j) $(9m + 2)(2m + 3)$
   k) $(4p + 3)(5p - 2)$
   l) $(8k + 1)(k - 6)$
   m) $(3x - 4)(2x + 7)$
   n) $(5x - 2)(9x - 1)$
   o) $(7w - 2)^2$
   p) $(3a - 8)(4a - 1)$

Worked examples Expand and simplify:

1. $(3x + 4)(2x^2 + 7x - 8)$
    $= 3x(2x^2 + 7x - 8) + 4(2x^2 + 7x - 8)$
    $= 6x^3 + 21x^2 - 24x + 8x^2 + 28x - 32$
    $= 6x^3 + 29x^2 + 4x - 32$

2. $(x + 2)(x + 3) + (5x + 1)(x - 4)$
    $= x(x + 3) + 2(x + 3) + 5x(x - 4) + 1(x - 4)$
    $= x^2 + 3x + 2x + 6 + 5x^2 - 20x + x - 4$
    $= 6x^2 - 14x + 2$

Notice the pattern this time: $(3x + 4)(2x^2 + 7x - 8) = 6x^3 + 21x^2 - 24x + 8x^2 + 28x - 32$

3. $(3x + 5)(2x - 7) - (4x - 3)(x + 6)$
    $= 3x(2x - 7) + 5(2x - 7) - [4x(x + 6) - 3(x + 6)]$
    $= 6x^2 - 21x + 10x - 35 - [4x^2 + 24x - 3x - 18]$
    $= 6x^2 - 11x - 35 - [4x^2 + 21x - 18]$
    $= 6x^2 - 11x - 35 - 4x^2 - 21x + 18$
    $= 2x^2 - 32x - 17$

Be careful here! You MUST subtract all terms from the expansion of $(4x - 3)(x + 6)$. Introducing new brackets around this expansion will help. We have used square brackets to stand out from the round brackets everywhere else.

---

4. Expand and simplify:
    a) $(x + 5)(x^2 + 3x + 1)$
    b) $(x + 7)(x^2 + 2x + 3)$
    c) $(x + 9)(x^2 + 4x - 2)$
    d) $(x + 6)(x^2 - 5x + 4)$
    e) $(2x - 4)(x^2 - 2x + 8)$
    f) $(3x - 1)(x^2 + 9x + 2)$
    g) $(x^2 + 7x - 1)(5x - 3)$
    h) $(x^2 - 2x - 3)(7x - 8)$
    i) $(x^2 - 6x + 9)(4x - 5)$

5. Expand and simplify:
    a) $(2x + 1)(x + 6) + (3x - 1)(x + 2)$
    b) $(2x - 3)(x + 4) + (3x - 5)(x + 1)$
    c) $(3x - 7)(5x + 3) + (2x - 4)(x + 9)$
    d) $(3x - 1)(9x + 2) + (2x - 10)(4x - 1)$
    e) $(x - 4)(5x - 2) - (x + 1)(x + 5)$
    f) $(2x - 3)(x + 6) - (x + 4)(x - 7)$
    g) $(3x + 5)(2x - 6) - (2x - 1)(x + 8)$
    h) $(7x + 4)(x - 2) - (5x + 1)(3x - 5)$

6. For each of the following shapes, find an expression without brackets for their area. All lengths are in centimetres.

    a)
    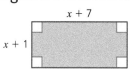
    $x + 7$
    $x + 1$

    b)
    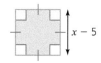
    $x - 5$

    c)
    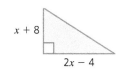
    $x + 8$
    $2x - 4$

    d)

    $x - 1$

7. The two rectangles below have the same area. All lengths are in centimetres.

    $x + 12$
    $2x - 3$
    $2x + 3$
    $x + 2$

    a) Obtain an expression without brackets for the area of each rectangle.

    b) Form an equation and solve it to find the dimensions of each rectangle.

▶ Factorising: common factor

Expanding $2x(x + 3)$ gives $2x^2 + 6x$. Starting with $2x^2 + 6x$ we can rewrite it as the product $2x(x + 3)$. Introducing brackets into an expression to write it like this is called **factorising**.

- Identify the **highest common factor (HCF)** of all terms.
- The highest common factor goes in front of the bracket.
- Determine what is inside the bracket by dividing the original terms in the expression by the highest common factor.
- Check your answer by expanding the bracket back out.

> **Worked examples**   Factorise fully:
>
> 1 $5x + 10$
>
> $= 5(x + 2)$
>
> The HCF of $5x$ and 10 is 5 so put 5 outside the bracket.
>
> Dividing each term in $5x + 10$ by 5 gives $x + 2$ so put $x + 2$ inside the bracket.
>
> 2 $12 + 18x - 24y$
>
> $= 6(2 + 3x - 4y)$
>
> The HCF of **every** term is 6 so put 6 outside the bracket.
>
> Dividing each term by 6 gives $2 + 3x - 4y$ so put $2 + 3x - 4y$ inside the bracket.
>
> For this question we could have written $2(6 + 9x - 12y)$ but this is not fully factorised because a factor remains in the bracket. Taking a further factor of 3 out gives $2 \times 3(2 + 3x - 4y) = 6(2 + 3x - 4y)$.
>
> 3 $xy + 4x$
>
> $= x(y + 4)$
>
> The HCF is x.
>
> Dividing each term by x gives $y + 4$. $x \times y = xy$ so $xy \div x = y$
>
> 4 $abc + ac$
>
> $= ac(b + 1)$
>
> The HCF is ac.
>
> Dividing each term by ac gives $b + 1$. $a \times b \times c = abc$ so $abc \div ac = b$
>
> $a \times c = ac$ so $ac \div ac = 1$
>
> 5 $y^2 - 3y$
>
> $= y(y - 3)$
>
> The HCF is y.
>
> Dividing each term by y gives $y - 3$. $y \times y = y^2$ so $y^2 \div y = y$

Exercise 11B

1 Factorise fully:

a) $3x + 12$ b) $2x + 10$ c) $7x + 14$ d) $5 + 15x$ e) $9 + 9x$

f) $2x - 8$ g) $4x - 12$ h) $6x - 36$ i) $40 - 8x$ j) $30 - 10x$

2 Factorise fully:

a) $4y + 10$ b) $8y + 20$ c) $20k - 15$ d) $6m + 9$ e) $18w - 4$

f) $12x + 42$ g) $14m + 35$ h) $32x - 24$ i) $40 - 16x$ j) $27 + 99x$

k) $2x + 8y - 4$ l) $3x - 9 + 18y$ m) $10y + 5x - 35$ n) $14x - 21 - 28y$ o) $8m - 12n + 16$

3 Factorise fully:

a) $xy + 6x$ b) $5y + xy$ c) $9c - cd$ d) $mn + 2n$ e) $kp - p$

f) $abc + 2bc$ g) $pqr - 9pr$ h) $3vw + vwx$ i) $5mnp - 6mn$ j) $7abc + 2ab$

➜

4 Fill in the blanks to make the following statements true:

a) $2pqr + 5r = \boxed{}(2pq + 5)$

b) $9ab + 11\boxed{} = ab(9 + 11c)$

c) $3x^2 + 10x = \boxed{}(3x + 10)$

d) $8\boxed{} - 5m = m(8m - 5)$

5 Factorise fully:

a) $x^2 + 6x$ b) $x^2 - 8x$ c) $2y + y^2$ d) $3p^2 + 10p$ e) $9h - 5h^2$

f) $x^2y - 15x^2$ g) $xy^2 + 13y^2$ h) $8m^2n + m^2$ i) $8x^3 + x^2y$ j) $wxy^2 + xy^3$

In the previous set of worked examples, the first two examples had a numerical common factor and the latter three an algebraic one. It is possible that the highest common factor of an expression comprises both a numerical and algebraic part.

Worked examples   Factorise fully:

1 $2x + 6xy$

$= 2x(1 + 3y)$

There is a common factor of 2 and a common factor of x so the HCF is $2x$.
Dividing each term by $2x$ gives $1 + 3y$.

2 $15x^2 + 20x$

$= 5x(3x + 4)$

The HCF is $5x$.
Dividing each term by $5x$ gives $3x + 4$.

3 $12abc + 6bc + 3c$

$= 3c(4ab + 2b + 1)$

The HCF is $3c$.
Dividing each term by $3c$ gives $4ab + 2b + 1$.

4 $12mn^2 + 18n^4$

$= 6n^2(2m + 3n^2)$

The HCF is $6n^2$.
Dividing each term by $6n^2$ gives $2m + 3n^2$.

Recall that $n \times n \times n \times n = n^4$ so $n^4 \div n^2 = n^2$

---

6 Factorise fully:

a) $3x + 15xy$ b) $4mn - 8n$ c) $6xy + 54x$ d) $5ac - 15c$ e) $18pq + 9p$

f) $14x - 49xy$ g) $24vy - 32v$ h) $28h - 40gh$ i) $40pq + 100q$ j) $88mn + 96n$

7 Factorise fully:

a) $2x^2 + 4x$ b) $5m^2 - 20m$ c) $9y - 3y^2$ d) $21k + 7k^2$ e) $50m^2 + 10m$

f) $12k^2 - 24k$ g) $18p + 27p^2$ h) $15w^2 + 20w$ i) $40h - 32h^2$ j) $27m^2 - 81m$

8 Factorise fully:

a) $5wxy + 15xy$ b) $7bc - 14abc$ c) $10pqr - 8pq$ d) $18fh + 6fh^2$

e) $2x^3 + 12x^2$ f) $5h^2 - 45h^3$ g) $8p^2 + 32p^4$ h) $70k^3 + 60k^4$

9 Factorise fully:

a) $2p^3 + 6p^2 + 8p$ b) $12a^3 + 9a^2 - 15a$ c) $50m^3 + 10m^2 + 15m$ d) $8p^4 - 12p^3 + 36p^2$

10 Fill in the blanks to make the following statements true:

a) $2x^2 + \boxed{}x = \boxed{}(x + 9)$ b) $20\boxed{} - 5m^3 = 5m^2\left(4 - \boxed{}\right)$ c) $\boxed{} - 24k^2 = 6k^2\left(5k - \boxed{}\right)$

▶ Factorising: a difference of two squares

If we expand double brackets with identical terms but different signs the resulting expression is a **difference of two squares** (one square subtracting another). Consider $(x + 3)(x - 3)$:

$$(x + 3)(x - 3)$$

$$= x^2 - 3x + 3x - 9 \quad \text{The middle terms cancel as they have opposite signs.}$$

$$= x^2 - 9 \quad \leftarrow \quad \text{This is called a difference of two squares.}$$

We can use this pattern to factorise any difference of two squares.

- Place the positive square root of the first term at the start of each bracket.
- Place the positive square root of the second term at the end of the bracket.
- One bracket will have a plus and the other a minus.
- Check your answer by expanding the brackets back out.

Worked examples   Factorise:

1   $x^2 - 4$

$= (x + 2)(x - 2)$

2   $1 - y^2$

$= (1 + y)(1 - y)$

3   $4x^2 - 9$

$= (2x + 3)(2x - 3)$

4   $y^2 - a^2$

$= (y + a)(y - a)$

5   $16k^2 - 25y^2$

$= (4k + 5y)(4k - 5y)$

6   $x^4 - 4x^2y^2$

$= (x^2 - 2xy)(x^2 + 2xy)$

Exercise 11C

1   Factorise:

a) $x^2 - 16$ b) $x^2 - 25$ c) $y^2 - 81$ d) $m^2 - 49$

e) $p^2 - 121$ f) $1 - x^2$ g) $64 - q^2$ h) $36 - n^2$

i) $100 - w^2$ j) $144 - y^2$ k) $9x^2 - 16$ l) $49x^2 - 25$

m) $64x^2 - 49$ n) $36y^2 - 1$ o) $121m^2 - 4$ p) $100 - 9y^2$

q) $25 - 4k^2$ r) $144 - 25r^2$ s) $400 - 9m^2$ t) $10000 - 81p^2$

2   Factorise:

a) $x^2 - m^2$ b) $x^2 - 16y^2$ c) $h^2 - 25m^2$ d) $9p^2 - k^2$

e) $100m^2 - t^2$ f) $49x^2 - 4y^2$ g) $25r^2 - 36t^2$ h) $100g^2 - 81h^2$

i) $121v^2 - 144w^2$ j) $900x^2 - 49y^2$ k) $225a^2 - 169b^2$ l) $2500p^2 - 289q^2$

3   Consider the expressions below. Sort them into two groups, those that are a difference of two squares and those that are not.

$h^2 - 10$ $81 - 25p^2$ $a^2b - 4c^2d$

$12x^2 - y^2$ $100p^2 - 49q^2$

$25a^2 + 121c^2$ $36g^2 - r^2h^2$ $196m^2 - n^2$

4   Factorise:

a) $x^2y^2 - 9$ b) $16 - m^2n^2$ c) $25x^2 - 49a^2b^2$

d) $64p^2q^2 - 81k^2$ e) $121a^2b^2 - 144c^2d^2$ f) $900x^2y^2 - 169v^2w^2$

5   Factorise:

a) $x^4 - 9m^2y^2$ b) $16x^2 - 25w^4y^4$ c) $1600x^4y^4 - 81v^4w^4$

▶ Factorising: trinomials

When we expand double brackets, we are multiplying a pair of factors. The resulting expression often has three terms $ax^2 + bx + c$, where $a, b, c \neq 0$. We call an expression like this a **trinomial**.

$(x+3)(x+6)$ ← pair of factors

$= x^2 + 6x + 3x + 18$

$= x^2 + 9x + 18$ ← trinomial

To factorise a trinomial, we must split it back into a pair of factors.

$x^2 + 9x + 18$ ← trinomial

$= (x+3)(x+6)$ ← pair of factors

To factorise we pay close attention to the **quadratic term** ax^2, **linear term** bx and **constant term** c. This investigation explores this interesting process and will build up your factorising skills.

Investigation

1   Expand the brackets:

    a) $(x + 2)(x + 3)$ **b)** $(x - 2)(x - 3)$ **c)** $(y + 1)(y + 6)$ **d)** $(y - 1)(y - 6)$

2   Use your answers to **Q1** to answer the following questions.

    a) The constant term in each trinomial is 6. What is the link between the numbers inside the brackets and the constant term in the trinomial?

    b) The linear term in each trinomial is different. For **1a)** the linear term is $+ 5x$. What is the link between the signs in the brackets and the sign of the linear term?

    c) What is the link between the numbers inside the brackets and the linear term in the trinomial?

3   Here are two attempts to factorise $x^2 + 6x + 8$ which give a constant term of $+ 8$:

$$(x + 1)(x + 8) \qquad \text{and} \qquad (x + 2)(x + 4).$$

    Which attempt gives the correct linear term $+ 6x$? Why?

4   Complete the brackets to factorise these trinomials. Check your answer by expanding.

    a) $x^2 + 9x + 20$ **b)** $x^2 + 7x + 12$ **c)** $x^2 - 8x + 15$ **d)** $x^2 - 7x + 10$

        $= (x \_ 4)(x \_ 5)$ $= (x \_ 4)(x \_ 3)$ $= (x \_ 3)(x \_ 5)$ $= (x \_ 5)(x \_ 2)$

5   Match the trinomials to their factorised form. Can you do it without expanding the brackets?

    $x^2 + 6x - 40$ $x^2 + 3x - 40$ $x^2 + 18x - 40$

      $(x + 20)(x - 2)$ $(x + 10)(x - 4)$ $(x - 5)(x + 8)$

6   In **Q5** the constant term for each trinomial is negative.

    a) What do you notice about the signs inside the brackets?

    b) What is the link between the numbers inside the brackets and the linear term in the trinomial?

7   By considering your answers so far, complete the brackets to factorise these trinomials. Check by expanding.

    a) $x^2 + 8x - 20$ **b)** $x^2 + 5x - 24$ **c)** $x^2 + 4x - 12$ **d)** $x^2 - 4x - 12$

        $= (x + 10)(x - \_)$ $= (x + 8)(x - \_)$ $= (x \_ 6)(x \_ 2)$ $= (x \_ 6)(x \_ 2)$

To factorise a trinomial of the form $x^2 + bx + c$:

- Set down the two brackets and begin by splitting the quadratic term.
- Consider factor pairs of the constant term, c (ignoring whether c is positive or negative).
- Find a factor pair that can be summed to give b, the coefficient of x, using these rules:
  - If the constant term is positive, the signs are the same (either both + or both −).
  - If the constant term is negative, the signs are different (one + and one −).
- Complete the brackets using the factor pair and signs.

Worked examples Factorise:

1 $x^2 + 3x + 2$

$= (x \,.....)(x \,.....)$

Consider the factors of 2.

Constant is +2 so the signs are the same.

Aim to make +3 from the factors.

	2
1	2

$x^2 + 3x + 2$

$= (x + 1)(x + 2)$

As +1 +2 = +3 we complete the brackets with + 1 and + 2.

2 $y^2 - 11y + 18$

$= (y \,.....)(y \,.....)$

Consider the factors of 18.

Constant is +18 so the signs are the same.

Aim to make −11 from the factors.

	18
1	18
2	9
3	6

$y^2 - 11y + 18$

$= (y - 2)(y - 9)$

As $- 2 - 9 = -11$ we complete the brackets with −2 and −9.

(Note that $-1 - 18 = -19$ and $-3 - 6 = -9$, so they are no good.)

The brackets can be written in either order: $(y - 2)(y - 9)$ is exactly the same as $(y - 9)(y - 2)$.

Exercise 11D

1 Factorise:

a) $x^2 + 6x + 5$ b) $x^2 + 4x + 3$ c) $x^2 + 8x + 7$ d) $x^2 + 14x + 13$

e) $x^2 + 6x + 8$ f) $x^2 + 7x + 10$ g) $x^2 + 8x + 15$ h) $x^2 + 9x + 20$

i) $x^2 + 7x + 12$ j) $x^2 + 15x + 14$ k) $x^2 + 10x + 16$ l) $x^2 + 11x + 18$

2 Factorise:

a) $x^2 - 3x + 2$ b) $x^2 - 8x + 7$ c) $x^2 - 2x + 1$ d) $x^2 - 12x + 11$

e) $x^2 - 5x + 6$ f) $x^2 - 6x + 9$ g) $x^2 - 14x + 24$ h) $x^2 - 11x + 30$

i) $x^2 - 9x + 8$ j) $x^2 - 7x + 10$ k) $x^2 - 8x + 12$ l) $x^2 - 17x + 16$

3 Factorise:

a) $x^2 + 5x + 6$ b) $x^2 + 9x + 18$ c) $x^2 - 13x + 30$ d) $x^2 - 18x + 45$

e) $y^2 - 20y + 36$ f) $n^2 + 15n + 50$ g) $m^2 - 22m + 72$ h) $k^2 + 29k + 100$

i) $p^2 + 22p + 121$ j) $q^2 - 23q + 60$ k) $h^2 - 20h + 51$ l) $a^2 - 20a + 84$

Worked examples Factorise:

1 $x^2 + x - 2$

$= (x \ .....)(x \ ......)$

Consider the factors of 2.

Constant is -2 so the signs are different.

Aim to make $+1$ from the factors.

	2	
1		2

$x^2 + x - 2$

$= (x - 1)(x + 2)$

As $-1 + 2 = +1$ we complete the brackets with -1 and $+2$.

(Note that $+1 - 2 = -1$ so the signs can't be this way.)

2 $m^2 - 6m - 27$

$= (m \ ......)(m \ ......)$

Consider the factors of 27.

Constant is -27 so the signs are different.

Aim to make -6 from the factors.

	27	
1		27
3		9

$m^2 - 6m - 27$

$= (m + 3)(m - 9)$

As $+3 - 9 = -6$ we complete the brackets with $+3$ and -9.

---

4 Factorise:

a) $x^2 + 2x - 3$ b) $x^2 + 6x - 7$ c) $x^2 + 4x - 5$ d) $x^2 + 12x - 13$

e) $x^2 + 8x - 9$ f) $x^2 + 3x - 10$ g) $x^2 + 4x - 12$ h) $x^2 + 2x - 8$

i) $x^2 + 5x - 14$ j) $x^2 + 2x - 24$ k) $x^2 + 7x - 30$ l) $x^2 + 12x - 45$

m) $x^2 + 14x - 32$ n) $x^2 + 23x - 50$ o) $x^2 + 21x - 72$ p) $x^2 + 24x - 81$

5 Factorise:

a) $x^2 - 2x - 3$ b) $x^2 - x - 2$ c) $x^2 - 10x - 11$ d) $x^2 - 16x - 17$

e) $x^2 - 5x - 6$ f) $x^2 - x - 12$ g) $x^2 - 6x - 16$ h) $x^2 - 11x - 26$

i) $x^2 - 2x - 15$ j) $x^2 - 11x - 42$ k) $x^2 - 5x - 36$ l) $x^2 - 5x - 50$

m) $x^2 - 6x - 40$ n) $x^2 - 2x - 63$ o) $x^2 - x - 56$ p) $x^2 - 8x - 84$

6 Factorise:

a) $x^2 - 4x - 5$ b) $x^2 + 10x - 11$ c) $x^2 - 6x - 7$ d) $x^2 - 7x - 18$

e) $x^2 + 6x - 16$ f) $x^2 - x - 20$ g) $y^2 - 4y - 32$ h) $y^2 - 7y - 44$

i) $k^2 + 10k - 56$ j) $a^2 - 18a - 63$ k) $c^2 + 6c - 72$ l) $d^2 + 21d - 100$

m) $m^2 + 17m - 84$ n) $q^2 - 16q - 80$ o) $h^2 + 10h - 75$ p) $p^2 - 29p - 96$

7 Factorise:

a) $x^2 - 9x + 18$ b) $x^2 + x - 12$ c) $x^2 + 13x - 14$ d) $x^2 - 20x + 19$

e) $x^2 - 25x - 26$ f) $x^2 - 13x + 22$ g) $x^2 + 7x - 30$ h) $x^2 + 11x + 28$

i) $x^2 - 16x - 36$ j) $x^2 + 6x - 55$ k) $x^2 - 25x + 46$ l) $x^2 + 5x - 50$

m) $y^2 - 9y - 52$ n) $n^2 + 19n + 60$ o) $q^2 - 15q + 54$ p) $p^2 + 29p - 62$

q) $h^2 + 3h - 70$ r) $r^2 + 16r + 64$ s) $t^2 - 14t - 72$ t) $v^2 + 12v - 85$

u) $w^2 - 18w + 81$ v) $g^2 + 20g + 99$ w) $x^2 - 21x + 110$ x) $y^2 - 17y - 200$

To factorise a trinomial $ax^2 + bx + c$ when a is an integer greater than one:

- Consider factor pairs of the constant term and factor pairs of ax^2, ensuring the x appears twice.
- Write an algebraic factor pair and a constant factor pair vertically and cross multiply to obtain their products (see working in blue in the worked examples).
- Use the same rule for signs as earlier on the resulting terms to try and get the coefficient of the x term. If the coefficient cannot be made, try another combination of factors.
- Check your answer by expanding the brackets back out.

Worked examples

Factorise:

1 $2x^2 + 7x + 5$ Constant term is $+5$ so the signs are the same. Aim to make $+7x$ from products.

The factor pairs are

$$\frac{2x^2}{x \quad 2x} \qquad \frac{5}{1 \quad 5}$$

Write the pairs vertically
$$\begin{array}{cc} x & 1 \\ 2x & 5 \end{array}$$
Cross multiply to get the products
$(x \times 5 = 5x$ and $2x \times 1 = 2x)$

$$\begin{array}{cc|c} x & 1 & 2x \\ 2x & 5 & 5x \end{array}$$

The aim is now to use $2x$ and $5x$ to make $+7x$. As $+2x + 5x = +7x$ this works.

$$\begin{array}{cc|l} x & 1 & + 2x \\ 2x & 5 & + 5x \\ \hline & & + 7x \end{array}$$
Take the $+$ signs to the left hand side to form the two brackets.

$$\begin{array}{cc|l} x + 1 & + 2x \\ 2x + 5 & + 5x \\ \hline & + 7x \end{array}$$

$2x^2 + 7x + 5$
$= (x + 1)(2x + 5)$

2 $5x^2 + 6x - 8$ Constant term is -8 so the signs are different. Aim to make $+6x$ from products.

The factor pairs are

$$\frac{5x^2}{x \quad 5x} \qquad \frac{8}{\begin{array}{cc} 1 & 8 \\ 2 & 4 \end{array}}$$
Choose one of the factor pairs for 8 to start with.

$$\begin{array}{cc|l} x & 1 & 5x \\ 5x & 8 & 8x \\ \hline & & ? \end{array}$$
We cannot make $+6x$ from $5x$ and $8x$.

Swap the order of the 1 and 8 and try again.

$$\begin{array}{cc|l} x & 8 & 40x \\ 5x & 1 & x \\ \hline & & ? \end{array}$$
We cannot make $+6x$ from $40x$ and x.

Now try the factor pair 2 and 4.
We can make $+6x$ this time.

$$\begin{array}{cc|l} x + 2 & + 10x \\ 5x - 4 & - 4x \\ \hline & + 6x \end{array}$$

$5x^2 + 6x - 8$
$= (x + 2)(5x - 4)$

3 $8x^2 - 2x - 15$ Constant term -15 so the signs are different. Aim to make $-2x$ from products.

The factor pairs are

$$\frac{8x^2}{\begin{array}{cc} x & 8x \\ 2x & 4x \end{array}} \qquad \frac{15}{\begin{array}{cc} 1 & 15 \\ 3 & 5 \end{array}}$$

No combination using the x and $8x$ gives $-2x$ from the products.

$$\begin{array}{cc|l} x & 1 & 8x \\ 8x & 15 & 15x \\ \hline & & ? \end{array} \qquad \begin{array}{cc|l} x & 15 & 120x \\ 8x & 1 & x \\ \hline & & ? \end{array} \qquad \begin{array}{cc|l} x & 3 & 24x \\ 8x & 5 & 5x \\ \hline & & ? \end{array} \qquad \begin{array}{cc|l} x & 5 & 40x \\ 8x & 3 & 3x \\ \hline & & ? \end{array}$$

Using $2x$ and $4x$ we find the combination that gives $-2x$.

$$\begin{array}{cc|l} 2x & 1 & 4x \\ 4x & 15 & 30x \\ \hline & & ? \end{array} \qquad \begin{array}{cc|l} 2x & 15 & 60x \\ 4x & 1 & 2x \\ \hline & & ? \end{array} \qquad \begin{array}{cc|l} 2x - 3 & - 12x \\ 4x + 5 & + 10x \\ \hline & - 2x \end{array}$$

$8x^2 - 2x - 15$
$= (2x - 3)(4x + 5)$

With enough practice you should hopefully be able to do these by inspection.

Exercise 11E

1 Factorise:

a) $2x^2 + 5x + 3$ b) $2x^2 - 5x + 3$ c) $2x^2 + 11x + 5$ d) $2x^2 + 13x - 7$

e) $3x^2 + 2x - 5$ f) $3x^2 + 10x - 8$ g) $3x^2 + x - 10$ h) $5x^2 + 46x + 9$

i) $5x^2 - 13x - 6$ j) $5x^2 - 17x - 12$ k) $7x^2 - 13x + 6$ l) $7x^2 - 54x - 16$

2 Factorise:

a) $4x^2 + 7x + 3$ b) $4x^2 + 8x - 5$ c) $6x^2 + x - 7$ d) $6x^2 - 11x + 3$

e) $8x^2 + 6x + 1$ f) $8x^2 - x - 7$ g) $8x^2 - 6x - 5$ h) $9x^2 - 6x + 1$

i) $10x^2 + 19x - 2$ j) $10x^2 + 13x - 3$ k) $12x^2 + 16x + 5$ l) $12x^2 - 11x + 2$

3 Factorise:

a) $4x^2 + 20x + 9$ b) $4x^2 + 16x + 15$ c) $6x^2 + 17x - 28$ d) $6x^2 - 13x - 8$

e) $8x^2 + 14x - 15$ f) $9x^2 - 18x + 8$ g) $12x^2 - x - 6$ h) $12x^2 - 44x - 45$

We can also use this technique to factorise $ax^2 + bx + c$ when a is a negative integer

Worked examples Factorise:

1 $7 + 6x - x^2$. The signs are different due to the $-x^2$. Aim to make $+ 6x$.

$$\begin{array}{cc} 7 \\ \hline 1 \quad 7 \end{array} \quad \begin{array}{cc} x^2 \\ \hline x \quad x \end{array} \quad \begin{array}{c|c} 1 + x & + 7x \\ 7 - x & - x \\ \hline & + 6x \end{array} \quad \begin{array}{l} 7 + 6x - x^2 \\ = (1 + x)(7 - x) \end{array}$$

Don't be tempted to write the x terms at the start of the brackets.

2 $15 + x - 2x^2$. The signs are different due to the $- 2x^2$. Aim to make $+ x$.

$$\begin{array}{cc} 15 \\ \hline 1 \quad 15 \\ 3 \quad 5 \end{array} \quad \begin{array}{cc} 2x^2 \\ \hline x \quad 2x \end{array} \quad \begin{array}{cc} 1 \quad x & 15x \\ 15 \quad 2x & 2x \\ \hline & ? \end{array} \quad \begin{array}{cc} 1 \quad 2x & 30x \\ 15 \quad x & x \\ \hline & ? \end{array} \quad \begin{array}{c|c} 3 - x & - 5x \\ 5 + 2x & + 6x \\ \hline & + x \end{array} \quad \begin{array}{l} 15 + x - 2x^2 \\ = (3 - x)(5 + 2x) \end{array}$$

4 Factorise:

a) $5 - 4x - x^2$ b) $3 + 2x - x^2$ c) $11 - 10x - x^2$ d) $13 + 12x - x^2$

e) $8 + 2x - x^2$ f) $10 - 3x - x^2$ g) $12 - x - x^2$ h) $16 + 6x - x^2$

i) $20 + x - x^2$ j) $18 + 7x - x^2$ k) $21 - 4x - x^2$ l) $24 - 2x - x^2$

5 Factorise:

a) $7 + 13x - 2x^2$ b) $5 - 3x - 2x^2$ c) $2 - 3x - 2x^2$ d) $1 + 4x - 5x^2$

e) $6 - x - 2x^2$ f) $8 + 18x - 5x^2$ g) $6 + 11x - 10x^2$ h) $8 - 6x - 9x^2$

i) $12 - x - 6x^2$ j) $15 + 14x - 8x^2$ k) $63 + 26x - 24x^2$ l) $-15x^2 - 8x + 12$

▶ Factorising: combining techniques

Factorising techniques can be combined in one question. In this case:

● Always take out the highest common factor first.

● Factorise the expression inside the bracket using the techniques for trinomials or a difference of two squares.

● Check your answer by expanding the brackets back out.

Worked examples Factorise fully:

1 $4x^2 - 100y^2$ HCF is 4
  $= 4(x^2 - 25y^2)$ A difference of two
  $= 4(x + 5y)(x - 5y)$ squares

2 $3x^2 + 12x - 15$ HCF is 3
  $= 3(x^2 + 4x - 5)$ Trinomial.
  $= 3(x - 1)(x + 5)$

3 $c^3 - 5c^2 + 4c$ HCF is c
  $= c(c^2 - 5c + 4)$ Trinomial
  $= c(c - 1)(c - 4)$

4 $8p^3 + 10p^2 - 12p$ HCF is $2p$
  $= 2p(4p^2 + 5p - 6)$ Trinomial
  $= 2p(p + 2)(4p - 3)$

Take care to always fully factorise an expression. For example, in Example 1 you could have $(2x + 10y)(2x - 10y)$ but this is not fully factorised. Each bracket still has a common factor of 2:
$(2x + 10y)(2x - 10y) = 2(x + 5y)2(x - 5y) = 4(x + 5y)(x - 5y)$.

Exercise 11F

1 Factorise fully these expressions which have a common factor and a difference of two squares.

   a) $2a^2 - 18$ b) $16 - 4q^2$ c) $10x^2 - 10$ d) $9y^2 - 36x^2$ e) $2h^2 - 242g^2$

   f) $x^3 - 16x$ g) $49y^3 - 25y$ h) $xy^2 - 4x$ i) $3q - 27q^3$ j) $32m^3 - 8m$

2 Factorise fully these expressions which have a common factor and a trinomial.

   a) $5x^2 + 15x + 10$ b) $2x^2 - 16x + 30$ c) $4k^2 - 16k - 20$ d) $3p^2 - 15p + 18$

   e) $x^3 + 13x^2 + 36x$ f) $q^3 - 7q^2 + 12q$ g) $h^3 + 4h^2 - 21h$ h) $2x^3 + 6x^2 - 36x$

   i) $12x^2 - 8x - 4$ j) $10y^2 - 34y + 12$ k) $6m^2 - 33m + 15$ l) $25x^2 + 45x + 20$

   m) $2x^3 + 7x^2 - 4x$ n) $6y^3 + 15y^2 + 6y$ o) $12k^3 + 8k^2 - 32k$ p) $18p^3 - 3p^2 - 6p$

3 Denise's attempt at fully factorising two expressions is given below. She has made one mistake in each attempt.

   A) $25x^2 - 100y^2$
     $= (5x + 10y)(5x - 10y)$ X

   B) $3a^2 - 15a - 42$
     $= 3(a^2 - 5a - 14)$
     $= (a - 7)(a + 2)$ X

   a) Describe the mistakes Denise has made.

   b) Write the correct answer for each question.

4 Factorise fully:

   a) $90 - 10x^2$ b) $2x^2 - 14x + 24$ c) $20y^2 + 36y + 16$ d) $72k^2 - 18$

   e) $5a^3 - 5a^2 - 100a$ f) $24t^3 + 42t^2 + 9t$ g) $98y^2 - 50x^2y^2$ h) $8p^4 - 50p^3 + 12p^2$

▶ Simultaneous equations: a graphical approach

Given a pair of equations that relate the same two variables it may be possible to find a solution which satisfies both of them. Here we will consider two equations which represent straight lines and find the solution by drawing them.

- Use the techniques from Chapter 8 to draw both straight lines on one set of axes.
- At the **point of intersection** (where the two lines meet) the x and y values of the coordinate point satisfy both equations and are the solution to the pair of equations.

Worked example

By sketching a graph, find the solution of $y = 2x - 9$ and $y = -x + 3$.

$y = 2x - 9$

x	0	1	2
y	−9	−7	−5

When $x = 0, y = 2(0) - 9 = -9$

When $x = 1, y = 2(1) - 9 = -7$

When $x = 2, y = 2(2) - 9 = -5$

Plot the points (0, −9), (1, −7) and (2, −5) and extend a line through them.

$y = -x + 3$

x	0	1	2
y	3	2	1

When $x = 0, y = -(0) + 3 = 3$

When $x = 1, y = -(1) + 3 = 2$

When $x = 2, y = -(2) + 3 = 1$

On the same axes, plot (0, 3), (1, 2) and (2, 1) and extend a line through them.

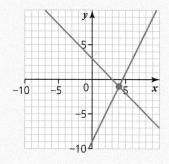

The point of intersection is (4, −1) so the solution to $y = 2x - 9$ and $y = -x + 3$ is $x = 4, y = -1$.

Exercise 11G

1 For the following pairs of equations:

    i) Form a table of values with $x = 0, 1$ and 2 and complete it by finding the y values.

    ii) Draw axes from −15 to 15 in both the x and y directions and sketch the straight lines.

    iii) Identify the point of intersection and write down the solution.

   a) $y = x - 3$
      $y = -x + 7$

   b) $y = 2x + 9$
      $y = -2x - 7$

   c) $y = 3x$
      $y = -x - 12$

   d) $y = x - 9$
      $y = -2x + 15$

2 Sketch to find the solution of:

   a) $y = 2x - 8$
      $y = -x + 4$

   b) $y = x - 5$
      $y = -2x + 13$

   c) $y = 2x + 6$
      $y = -x - 15$

   d) $y = 2x - 3$
      $y = -4x$

   e) $y = x - 14$
      $y = -3x + 6$

   f) $y = 2x + 2$
      $y = x - 1$

   g) $y = 4x + 6$
      $y = 2x + 5$

   h) $y = 1 - x$
      $y = -3x + 6$

3 What property must two straight lines share if you cannot find a point of intersection for them?

▶ Simultaneous equations: elimination

The first algebraic approach we will consider is called elimination. The aim of this method is to eliminate one of the variables to allow us to solve for the other.

- Write the two equations so that the variables are aligned vertically and label them (1) and (2).
- Add or subtract an entire equation from the other one to eliminate one of the variables.
- Solve for the remaining variable.
- Return to an original equation to solve for the variable you eliminated.
- Check your solutions using the other original equation.

Worked examples

Solve algebraically. The equations have been labelled (1) and (2) for you.

1 $x + 2y = 8$ ... (1)

$x + y = 5$ ... (2)

$y = 3$

Subtract (1) − (2)

If we subtract vertically (down the way) we get

$$\begin{array}{ccc} x & 2y & 8 \\ - \ x & - \ y & - \ 5 \\ \hline 0 & y & 3 \end{array}$$

This leaves $y = 3$. We have eliminated x.

$x + 2(3) = 8$

$x + 6 = 8$

$x = 2$

Substitute $y = 3$ into (1) and balance to get x

$x + y$

$= 2 + 3$

$= 5$ ✔

Check by substituting your values for x and y into the left-hand side of (2). If you get the right-hand side of (2) your values are correct. If you don't, you have made a mistake – go back and check your working!

Alternatively, we could do (2) − (1) but this would lead to negative numbers which can make the arithmetic harder.

2 $5x + 7y = 13$ ... (1)

$x − 7y = 11$ ... (2)

$6x \qquad = 24$

$x = 4$

Add (1) + (2)

If we add vertically we get

$$\begin{array}{ccc} 5x & 7y & 13 \\ + \ x & + \ (-7y) & + \ 11 \\ \hline 6x & 0 & 24 \end{array}$$

This leaves $6x = 24$. We have eliminated y.

$5(4) + 7y = 13$

$20 + 7y = 13$

$7y = -7$

$y = -1$

Substitute $x = 4$ into (1) to get y

$x − 7y$

$= 4 − 7(-1)$

$= 11$ ✔

Check using (2)

Notice that when the signs in front of a variable are the same, we subtract to eliminate the variable. When the signs are different, we add.

Exercise 11H

1   Solve each pair of equations simultaneously by subtracting the equations.

   a) $x + 4y = 9$
       $x + 3y = 7$

   b) $8x + y = 21$
       $6x + y = 17$

   c) $11x + 3y = 50$
       $2x + 3y = 14$

   d) $5x + 7y = 47$
       $5x + 4y = 44$

2   Solve each pair of equations simultaneously by adding the equations.

   a) $3x + y = 18$
       $4x - y = 3$

   b) $6x + 2y = 38$
       $3x - 2y = 7$

   c) $8x - 3y = 13$
       $4x + 3y = 11$

   d) $2x + 5y = 27$
       $-2x + 9y = 15$

3   Solve algebraically each pair of equations.

   a) $x + y = 5$
       $7x - y = 11$

   b) $x + y = 5$
       $x - y = 1$

   c) $6x + y = 36$
       $4x + y = 26$

   d) $8x + 3y = 83$
       $8x + y = 65$

   e) $7x + 4y = 68$
       $5x + 4y = 60$

   f) $6x - 5y = 8$
       $2x + 5y = 16$

   g) $9x + 7y = 53$
       $9x + 3y = 33$

   h) $4x - 5y = 13$
       $6x + 5y = 57$

   i) $3x + 5y = 28$
       $-3x + 4y = 17$

   j) $5x + 6y = 16$
       $7x - 6y = 8$

   k) $9x + 2y = 17$
       $15x - 2y = 7$

   l) $2x + 7y = 64$
       $-2x + y = 0$

   m) $5x + 7y = 13$
       $2x - 7y = 15$

   n) $9x + 2y = 2$
       $9x + 8y = 62$

   o) $5x - 2y = -25$
       $3x + 2y = 1$

   p) $2x - 5y = 8$
       $2x + 5y = -12$

4   Can you find a second method to eliminate a term in **3b)** and **3p)**?

5   Safian and Taylor both try the same question. Here are their attempts.

   Safian
   $4x - 3y = 9$
   $\underline{4x - 9y = 21}$
   $-12y = -12$
   $y = 1$

   $4x - 3(1) = 9$
   $4x - 3 = 9$
   $4x = 12$
   $x = 3$

   Taylor
   $4x - 3y = 9$
   $\underline{4x - 9y = 21}$
   $6y = -12$
   $y = -2$

   $4x - 3(-2) = 9$
   $4x + 6 = 9$
   $4x = 3$
   $x = \dfrac{3}{4}$

   a) Who obtained the correct solution?

   b) What mistake was made in the wrong attempt?

6   Ailish is asked to solve the following equations simultaneously:

   $4x + y = 9$

   $5x + 3y = 13$

   She realises she cannot eliminate a variable by adding or subtracting. Her teacher tells her she can multiply one of the equations to have the same amount of one variable in both equations.

   a) Which equation should she multiply and what should she multiply it by?

   b) Solve the equations simultaneously.

If we do not have the same number of either x or y (ignoring the signs) then we have to multiply either one or both of the equations to obtain the same number before eliminating a variable.

- Multiply an entire equation to match the coefficients of a variable (ignoring the signs).
- If an equation does not need to be changed, simply rewrite it. For any equation that is changed, write the new equation. Always keep the equations in pairs until you can eliminate a variable.
- Label the new pair of equations (3) and (4) and follow the steps from the previous exercise for equations (3) and (4).

Worked examples

Solve algebraically:

1

$$3x + 2y = 15 \quad ... (1) \qquad 4 \times (1) \text{ to get } 8y \to (3)$$
$$\underline{5x - 8y = -77} \quad ... (2) \qquad (2) \text{ unchanged} \to (4)$$
$$12x + 8y = 60 \quad ... (3)$$
$$\underline{5x - 8y = -77} \quad ... (4)$$
$$17x \quad\quad = -17 \qquad \text{Add } (3) + (4)$$
$$x = -1$$
$$3(-1) + 2y = 15 \qquad \text{Substitute } x = -1 \text{ into}$$
$$-3 + 2y = 15 \qquad (1) \text{ and balance}$$
$$2y = 18$$
$$y = 9$$
$$5x - 8y \qquad\qquad \text{Check using (2)}$$
$$= 5(-1) - 8(9)$$
$$= -77 \checkmark$$

2

$$3x - 2y = 11 \quad ... (1) \qquad (1) \text{ unchanged} \to (3)$$
$$\underline{x - 5y = -5} \quad ... (2) \qquad 3 \times (2) \text{ to get } 3x \to (4)$$
$$3x - 2y = 11 \quad ... (3)$$
$$\underline{3x - 15y = -15} \quad ... (4)$$
$$13y = 26 \qquad \text{Subtract } (3) - (4)$$
$$y = 2$$
$$3x - 2(2) = 11 \qquad \text{Substitute } y = 2 \text{ into (1)}$$
$$3x - 4 = 11 \qquad \text{and balance}$$
$$3x = 15$$
$$x = 5$$
$$x - 5y \qquad\qquad \text{Check using (2)}$$
$$= 5 - 5(2) = -5 \checkmark$$

Be very careful here! $-2y - (-15y) = -2y + 15y = 13y$

Exercise 11I

1 Solve each pair of equations simultaneously using the hint provided.

a) $x + 4y = 16$ (×2)
$2x + 3y = 17$

b) $x + 7y = 9$ (×5)
$5x + 4y = 14$

c) $2x - y = 1$ (×3)
$5x + 3y = 30$

d) $4x + 11y = 26$
$x + 5y = 11$ (×4)

e) $5x - 8y = 1$
$3x + 4y = 27$ (×2)

f) $7x - 6y = 16$
$3x + 2y = 16$ (×3)

2 Solve each pair of equations simultaneously.

a) $x + 4y = 23$
$2x + 3y = 21$

b) $3x + y = 9$
$11x + 7y = 43$

c) $5x - y = 1$
$2x + 5y = 22$

d) $2x + 3y = 16$
$6x + 5y = 40$

e) $7x - 6y = 10$
$4x + 3y = 25$

f) $2x + 3y = 7$
$18x + 11y = 15$

g) $5x + 2y = -3$
$10x + 3y = 3$

h) $3x - y = 6$
$2x + 5y = -64$

i) $7x - y = 20$
$13x - 3y = 36$

j) $12x - 5y = 39$
$2x - y = 5$

k) $6x - 7y = 27$
$x - 2y = 2$

l) $9x - 2y = 44$
$7x - 4y = 22$

In these examples, we will multiply both equations to then eliminate a variable.

Worked examples Solve algebraically:

1

$$3x + 2y = 5 \quad \text{...(1)}$$
$$2x + 5y = -4 \quad \text{...(2)}$$

$2 \times$ (1) $\rightarrow$ (3)
$3 \times$ (2) $\rightarrow$ (4)

$$6x + 4y = 10 \quad \text{...(3)}$$
$$6x + 15y = -12 \quad \text{...(4)}$$
$$11y = -22$$
$$y = -2$$

(4) − (3)

$$3x + 2(-2) = 5$$
$$3x - 4 = 5$$
$$3x = 9$$
$$x = 3$$

Substitute $y = -2$
into (1) and balance

$$2x + 5y$$
$$= 2(3) + 5(-2)$$
$$= -4 \checkmark$$

Check using (2)

2

$$5x + 8y = -57 \quad \text{...(1)}$$
$$7x - 9y = 1 \quad \text{...(2)}$$

$9 \times$ (1) $\rightarrow$ (3)
$8 \times$ (2) $\rightarrow$ (4)

$$45x + 72y = -513 \quad \text{...(3)}$$
$$56x - 72y = 8 \quad \text{...(4)}$$
$$101x = -505$$
$$x = -5$$

(3) + (4)

$$5(-5) + 8y = -57$$
$$-25 + 8y = -57$$
$$8y = -32$$
$$y = -4$$

Substitute $x = -5$
into (1) and balance

$$7x - 9y$$
$$= 7(-5) - 9(-4)$$
$$= 1 \checkmark$$

Check using (2)

Notice that in Example 1 we could have started with $5 \times$ (1) and $2 \times$ (2) to eliminate y and get the same answer. Similarly, in Example 2, we could have started with $7 \times$ (1) and $5 \times$ (2) to eliminate x.

Exercise 11J

1 Solve each pair of equations simultaneously using the hints provided.

a) $5x + 2y = 9$ (×3)
$\quad 2x + 3y = 8$ (×2)

b) $3x + 4y = 18$ (×3)
$\quad 5x - 3y = 1$ (×4)

c) $4x - 9y = 3$ (×2)
$\quad 3x + 2y = 11$ (×9)

d) $2x + 7y = 30$ (×3)
$\quad 3x + 4y = 19$ (×2)

e) $3x + 5y = -9$ (×2)
$\quad 7x - 2y = 20$ (×5)

f) $3x + 8y = 10$ (×7)
$\quad 7x + 6y = -2$ (×3)

g) $6x + 5y = 7$ (×3)
$\quad 7x - 3y = 17$ (×5)

h) $11x - 7y = 75$ (×5)
$\quad 2x + 5y = 45$ (×7)

i) $9x - 8y = -11$ (×3)
$\quad 7x - 6y = -9$ (×4)

2 Solve algebraically:

a) $2x + 9y = 15$
$\quad 3x + 2y = 11$

b) $3x + 2y = 19$
$\quad 5x + 3y = 31$

c) $4x + 7y = 18$
$\quad 6x - 5y = -4$

d) $7x - 2y = -4$
$\quad 4x + 5y = 53$

e) $2x + 11y = 5$
$\quad 3x + 4y = -5$

f) $7x + 3y = 36$
$\quad 5x + 2y = 26$

g) $4x + 3y = 5$
$\quad 10x - 7y = 27$

h) $11x - 2y = 37$
$\quad 6x + 5y = 8$

i) $9x + 5y = 3$
$\quad 7x - 2y = -33$

j) $8x + 3y = 28$
$\quad 5x - 4y = 41$

k) $3x - 13y = 7$
$\quad 5x - 2y = -8$

l) $6x - 7y = 10$
$\quad 8x - 3y = -12$

m) $5x - 3y = 13$
$\quad 2x + 11y = -7$

n) $3x - 8y = -25$
$\quad 5x + 7y = -1$

o) $9x + 3y = -51$
$\quad 11x + 2y = -54$

p) $4x - 5y = -34$
$\quad 6x - 7y = -48$

▶ Simultaneous equations: substitution

This method is useful if you are given or can find at least one equation explicitly (i.e. $y = ...$ or $x = ...$).

- Take one equation where you know a variable explicitly and substitute into the other equation.
- Solve for the remaining variable.
- Return to an original equation to solve for the second variable. Check your answer.

Worked examples   Solve algebraically:

1   $y = 2x - 4$ In this example, y is given explicitly in both cases.
    $y = x + 1$ This means that the two expressions for y must be equal to each other.

    $2x - 4 = x + 1$ Equate the expressions and balance to solve.
       $2x = x + 5$
        $x = 5$

    $y = x + 1$ Substitute $x = 5$ into one of the equations (the second one is quicker here).
      $= 5 + 1$
      $= 6$

    $y = 2x - 4$ Check using the equation you did not use for the substitution (the first one).
      $= 2(5) - 4$
      $= 6\ ✓$

2   $x = 5y + 7$ In this example, x is given explicitly in both cases. This means that the two
    $x = 3y + 5.$ expressions for x must be equal to each other.

    $5y + 7 = 3y + 5$ Equate the expressions and balance to solve.
      $5y = 3y - 2$
      $2y = -2$
       $y = -1$

    $x = 5y + 7$ Substitute $y = -1$ into one of the equations.
    $= 5(-1) + 7$
    $= 2$

    $x = 3y + 5$ Check using the equation you did not use for the substitution.
    $= 3(-1) + 5$
    $= 2\ ✓$

3       $y = x + 3$ In this example, only y is given explicitly. This means that
   $5x + 2y = 62.$ whenever y appears in the second equation it is equal to $x + 3$.

   $5x + 2(x + 3) = 62$ Substitute $x + 3$ where y appears in the second equation and
    $5x + 2x + 6 = 62$ balance to solve. You have to multiply the whole of $x + 3$ by 2 so
       $7x + 6 = 62$ take care to put brackets around $x + 3$ and then expand correctly.
         $7x = 56$
          $x = 8$

    $y = x + 3$ Substitute $x = 8$ into the first equation (as it is known explicitly).
     $= 8 + 3$
     $= 11$

   $5x + 2y = 5(8) + 2(11) = 62\ ✓$ Check using the second equation.

4

$x = 7 - 2y$
$4y - 3x = 29.$

$4y - 3(7 - 2y) = 29$
$4y - 21 + 6y = 29$
$10y - 21 = 29$
$10y = 50$
$y = 5$

In this example, only x is given explicitly. This means that whenever x appears in the second equation it is equal to $7 - 2y$.

Substitute $7 - 2y$ where x appears in the second equation and balance to solve. You have to multiply the whole of $7 - 2y$ by 3 so take care to put brackets around $7 - 2y$ and then expand correctly.

$x = 7 - 2y$
$ = 7 - 2(5)$
$ = -3$

Substitute $y = 5$ into the first equation (as it is known explicitly).

$4y - 3x = 4(5) - 3(-3) = 29$ ✓   Check using the second equation.

Exercise 11K

1   Solve by substituting for y.

a) $y = 3x + 9$
    $y = x + 15$

b) $y = x + 5$
    $y = 6x - 15$

c) $y = 10 + x$
    $y = 10 - x$

d) $y = 6 + 7x$
    $y = 13x - 6$

2   Solve by substituting for x.

a) $x = 3y + 12$
    $x = y + 20$

b) $x = 5y - 9$
    $x = y + 7$

c) $x = 14 + y$
    $x = 7y + 2$

d) $x = 5 + 2y$
    $x = 20 - 3y$

3   Solve by substituting for y. Remember to include brackets where necessary.

a) $y = x + 7$
    $3x + y = 11$

b) $y = x - 1$
    $5x + y = 23$

c) $y = 3x$
    $2x + y = 15$

d) $y = 2x$
    $x + 4y = 18$

e) $y = x + 6$
    $x + 2y = 27$

f) $y = 4x + 1$
    $2x + 3y = 17$

g) $y = 2x - 5$
    $4x + 3y = 45$

h) $y = 6x - 2$
    $10x - y = 22$

4   Solve by substituting for x. Remember to include brackets where necessary.

a) $x = y + 8$
    $x + y = 12$

b) $x = 2y - 3$
    $x + 6y = 37$

c) $x = 6y$
    $x + 3y = 9$

d) $x = 4y$
    $x - 2y = 8$

e) $x = 6 + y$
    $2x + y = 21$

f) $x = y + 4$
    $2x + 3y = 38$

g) $x = 2y - 6$
    $4x - 3y = 11$

h) $x = 2 - 3y$
    $2y - 5x = 7$

5   Solve algebraically:

a) $x = 8 + y$
    $x = 16 - y$

b) $y = 7x$
    $10x + y = 17$

c) $x = y - 9$
    $y + 1 = 3x$

d) $y = 3x - 4$
    $2y + x = 6$

e) $y = x + 40$
    $y = 7x - 26$

f) $x = 5y + 4$
    $x = 9y - 20$

g) $2y = x + 7$
    $y = 4x - 35$

h) $y = 7 - x$
    $3x - 2y = 1$

i) $x - 2y = 14$
    $y = 3x + 3$

j) $y = 6 - x$
    $y = 18 - 2x$

k) $y = 2x + 3$
    $5y = 4x + 3$

l) $x + y = -5$
    $y = x - 1$

6   Find algebraically the point of intersection of $y = 3x + 1$ and $y = 2x - 5$.

▶ Simultaneous equations in context

It may be necessary to form the equations before then solving them simultaneously.

Worked example

Three bottles of juice and four cartons of milk cost £5·30. Five bottles of juice and two cartons of milk cost £4·40. Find the total price of two bottles of juice and three cartons of milk.

Let j = juice and m = milk. We can then form two equations from the information given in the question: $3j + 4m = 5·3$ and $5j + 2m = 4·4$.

There is no need for units in the equations, just remember to put them back into your final answer.

$3j + 4m = 5·3 \dots (1)$ $(1) \rightarrow (3)$

$\underline{5j + 2m = 4·4 \dots (2)}$ $2 \times (2) \rightarrow (4)$

$3j + 4m = 5·3 \dots (3)$

$\underline{10j + 4m = 8·8 \dots (4)}$

$7j \quad\quad = 3·5$ $(4) - (3)$

$j = 0·5$

Substitution into (1) gives

$3(0·5) + 4m = 5·3$

$1·5 + 4m = 5·3$

$4m = 3·8$

$m = 0·95$

As j stands for juice and m for milk we have found that a bottle of juice costs 50p and a carton of milk costs 95p.

We can now calculate the cost of two bottles of juice and three cartons of milk as

$2j + 3m = 2(0·5) + 3(0·95)$

$= £3·85$ The total price of two bottles of juice and three cartons of milk is £3·85.

Exercise 11L

1   The cost of five adults and one child to visit a museum is £28.

   a)   Write down an equation to represent this information. Use a for adults and c for children.

   The cost of three adults and four children is £27.

   b)   Write down an equation to represent this information. Use a for adults and c for children.

   c)   Solve the two equations simultaneously to calculate the cost of one adult ticket and the cost of one child ticket for the museum.

2   Three sandwiches and two wraps cost £13·40.

   a)   Write down an equation to represent this information. Use s for sandwich and w for wrap.

   Five sandwiches and three wraps cost £21·35.

   b)   Write down an equation to represent this information. Use s for sandwich and w for wrap.

   c)   Solve the two equations simultaneously to calculate the total cost of three sandwiches and one wrap.

3   Four inner circle tickets and one standing ticket for a concert cost £95.

   a)   Write down an equation to represent this information. Use i for inner and s for standing.

   Two inner circle tickets and five standing tickets cost £115.

   b)   Write down an equation to represent this information. Use i for inner and s for standing.

   c)   Find the total price of seven inner circle and four standing tickets.   ➔

4   The cost to hire two adult bikes and seven child bikes is £110.

   a) Write down an equation to represent this information.

   The cost to hire six adult bikes and five child bikes is £170.

   b) Write down an equation to represent this information.

   c) Bez estimates it will cost £300 to hire nine adult bikes and ten child bikes. Has he overestimated or underestimated the hire cost and if so by how much?

5   The cost of printing three canvases and five photos is £17·50.

   a) Write down an equation to represent this information.

   The cost of printing four canvases and three photos is £21·50.

   b) Write down an equation to represent this information.

   c) Calculate the total cost of printing eight canvases and four photos.

You may use a calculator for **Q6–Q14**.

6   A school enterprise group sell badges and mugs to raise money for charity. Nine badges and two mugs raise £10·80.

   a)   Write down an equation to represent this information.

   Four badges and three mugs raise £11·45.

   b) Write down an equation to represent this information.

   c) How much would the group raise if they sold nine badges and ten mugs?

7   Stirling gets paid a different rate for working during the week compared to working at the weekend. In one week, he works 20 weekday hours and 13 weekend hours and gets paid £454·25.

   a)   Write down an equation to represent this information. Be careful with your choice of letters!

   The next week, he gets paid £408·25 for working 10 weekday hours and 17 weekend hours.

   b) Write down an equation to represent this information.

   c) What is the difference in Stirling's weekend and weekday rates?

8   Three glasses and four bottles have a combined capacity of 2·9 litres. Eight glasses and five bottles have a combined capacity of 4·9 litres. How much liquid will two glasses and two bottles hold altogether?

9   Four small pots and seven large pots have a mass of 10·968 kg. Nine small pots and six large pots have a mass of 13·953 kg. Find the mass of three small pots and four large pots.

10 It is more expensive to print in colour than it is in black and white. Morag prints 16 sheets in colour and 35 in black and white. It costs her £1·66. Claire prints 31 colour and 56 black and white sheets at a cost of £2·98. David wants to print 49 colour and 55 black and white. He thinks it will cost him £4. Is he correct?

11 Two numbers, x and y, have a sum of 67 and a difference of 17. Find the two numbers.

12 Two numbers add to 2. Twice the first number plus three times the second number is 1. Find the two numbers.

13 The difference between two numbers is 2. The sum of five times the first number and four times the second number is –62. Find the product of the two numbers.

14 A savings tin contains some 10p and some 20p coins. There are 48 coins in the tin and the total value of the coins is £6·50. How many of each type of coin are in the tin?

Check-up

1 Expand and simplify:

   a) $(x + 4)(x + 1)$ b) $(x + 7)(x + 2)$ c) $(x + 8)(x - 10)$ d) $(x + 9)(x - 11)$

   e) $(x - 4)(x + 6)$ f) $(x - 5)(x + 8)$ g) $(v - 9)(v - 3)$ h) $(v - 10)^2$

2 Expand and simplify:

   a) $(2x + 3)(x + 5)$ b) $(3y + 4)(2y + 9)$ c) $(4m + 5)(6m - 1)$ d) $(7k + 6)(8k - 3)$

   e) $(5w - 2)(3w + 7)$ f) $(6p - 1)(2p + 5)$ g) $(10q - 3)(6q - 5)$ h) $(9x - 4)^2$

3 Expand and simplify:

   a) $(x + 4)(x^2 + 3x + 1)$ b) $(2x + 5)(x^2 - 2x + 6)$

   c) $(4x - 9)(x^2 + 7x - 5)$ d) $(3x - 2)(x^2 - 4x - 1)$

4 Expand and simplify:

   a) $(x + 2)(x + 3) + (x + 5)(x - 6)$ b) $(x + 6)^2 + (2x - 1)(3x + 5)$

   c) $(7x + 1)(x - 2) - (3x + 4)(x + 5)$ d) $(x - 6)(2x - 5) - (x - 3)(2x - 1)$

5 Obtain an expression without brackets for the area of the following shapes.

   a)   b)   c)

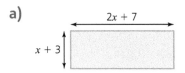 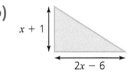 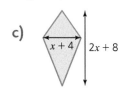

6 Factorise:

   a) $4t + 16$ b) $6a - 18$ c) $8m + 10$ d) $15 + 6k$ e) $45x - 55$

   f) $xy + 9x$ g) $8p - pq$ h) $10mn + n^2$ i) $4abc + 5ab$ j) $pqr^2 - 7r^2$

7 Factorise fully:

   a) $4x + 12xy$ b) $6mn + 30m$ c) $10pq - 25q$ d) $18h + 21gh$ e) $49cd - 84d$

   f) $9x^2 + 3x$ g) $24y + 40y^2$ h) $8y^2 - 6y^3$ i) $5pqr + 35pq$ j) $18m^2n - 27m^3n^2$

8 Factorise:

   a) $y^2 - 9$ b) $4 - x^2$ c) $81p^2 - 25$ d) $16 - 49a^2$

9 Factorise:

   a) $x^2 + 12x + 11$ b) $x^2 + 8x + 12$ c) $x^2 - 18x + 17$ d) $x^2 - 9x + 20$

   e) $x^2 + 16x - 17$ f) $x^2 + 7x - 18$ g) $x^2 - 22x - 23$ h) $x^2 - 6x - 16$

10 Factorise:

   a) $2x^2 - 7x + 3$ b) $5x^2 + 9x - 2$ c) $7x^2 + 78x + 11$ d) $7x^2 - 31x - 20$

   e) $11x^2 + 97x - 18$ f) $6x^2 - 7x - 5$ g) $10x^2 - 3x - 1$ h) $16x^2 - 62x + 21$

   i) $17 - 16x - x^2$ j) $6 - x - x^2$ k) $5 - 2x - 3x^2$ l) $4 + 4x - 3x^2$

11 Factorise fully:

   a) $10m^2 - 40n^2$ b) $242p^2 - 32q^2$ c) $8a^2 + 16a - 64$ d) $5x^2 + 10x + 5$

   e) $8y^2 - 20y + 12$ f) $3k^3 - 3k^2 - 18k$ g) $6m^3 + 18m^2 - 24m$ h) $4y^3 + 22y^2 + 10y$

12 Draw accurate sketches to find the solution of:

a) $y = 4x$
   $y = 2x + 6$

b) $y = x - 2$
   $y = 3x - 12$

c) $y = 2x - 4$
   $y = -x - 10$

d) $y = 5x + 3$
   $y = -2x - 11$

13 Solve algebraically:

a) $8x + y = 44$
   $x - y = 1$

b) $6x + 7y = 54$
   $3x + 7y = 48$

c) $8x + 5y = 1$
   $8x + 3y = 7$

d) $8x - 3y = 20$
   $9x + 3y = -3$

14 Solve algebraically:

a) $x + 6y = 15$
   $2x + 7y = 20$

b) $3x + y = 11$
   $2x + 5y = 42$

c) $5x + 8y = 49$
   $3x + 2y = 21$

d) $3x + 7y = 1$
   $9x - 4y = 53$

15 Solve algebraically:

a) $3x + 2y = 7$
   $5x + 7y = 19$

b) $11x - 5y = 7$
   $2x + 3y = 13$

c) $5x - 2y = 40$
   $4x + 3y = 55$

d) $5x + 3y = 22$
   $2x + 5y = 24$

e) $6x - 5y = 16$
   $7x + 4y = -1$

f) $4x - 7y = -32$
   $3x + 2y = 5$

g) $7x + 8y = 10$
   $9x + 10y = 12$

h) $5x - 6y = 3$
   $8x - 9y = 3$

16 Solve algebraically:

a) $y = 2x + 5$
   $y = x + 9$

b) $x = 34 - y$
   $x = 6y - 8$

c) $y = 3x + 1$
   $5x + 4y = 21$

d) $y = 1 - 5x$
   $3x - y = 23$

17 Five milkshakes and two fruit smoothies cost £15.

a) Write down an equation to represent this information.

Six milkshakes and one fruit smoothie cost £14·50.

b) Write down an equation to represent this information.

c) Calculate the total price of three milkshakes and four fruit smoothies.

18 A train fare for three adults and two children is £21·50. For four adults and five children the same journey costs £34·50. Calculate the train fare for two adults and three children.

19 At a school fun day, house points are awarded for every ball successfully thrown into a pot. Each person is given 10 balls and there are two pots, a large one and a small one. You get different points for each pot. Richard successfully threw six balls in the large pot and one in the small pot. He scored 23 points. Stewart gained 25 points after throwing five balls in the large pot and two in the small pot.

How many points would you get if you threw eight in the large pot and two in the small one?

20 Angus continually loses his gel pens and pencils. One month, he replaced five pencils and three gel pens at a cost of £5·35. The next month, the prices of the gel pens and pencils were still the same and twelve pencils and seven gel pens cost him £12·60.

His friend Sean bought a multipack containing twenty pencils identical to the ones Angus bought. Sean paid £5. How much did Sean save per pencil compared to Angus?

21 Two numbers sum to 99. The difference between the numbers is 37. What are the two numbers?

12 Circle

▶ Triangles in circles

A line inside a circle which has two points of contact with the circumference but does not pass through the centre is called a chord.

A triangle can be formed by joining the ends of a chord to the centre of the circle. The two lines which have been added are radii therefore they have the same length and the triangle is isosceles.

Recall that angles in a triangle add up to 180° and that two angles within an isosceles triangle are equal.

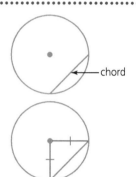
chord

Worked examples   Calculate the size of the unknown angles.

1

Equal angles at the circumference end of the equal sides.

$$\begin{array}{r} 70 \\ +70 \\ \hline 140 \end{array} \qquad \begin{array}{r} 180 \\ -140 \\ \hline 40 \end{array} \quad x = 40°$$

2

$$\begin{array}{r} 180 \\ -120 \\ \hline 60 \end{array}$$

Find half of the remaining angle as the angles are equal.

$$\begin{array}{r} 30 \\ 2\overline{)60} \end{array} \quad y = 30°$$

Exercise 12A

1   Calculate the size of the unknown angles.

a)

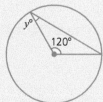

b)

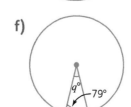

c)

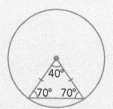

d)

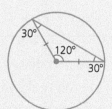

e)

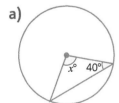

f)

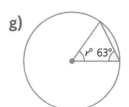

g)

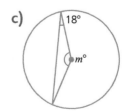

h)

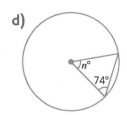

2 Calculate the size of the unknown angles.

a)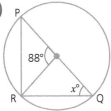

b)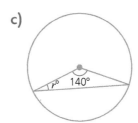

c)

d)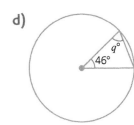

e)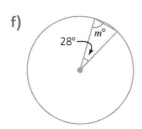

f)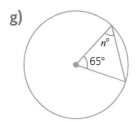

g)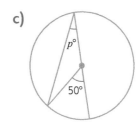

h)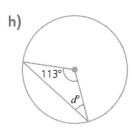

3 Copy the diagrams and fill in as many angles as you can to help you find the size of the unknown angles.

a)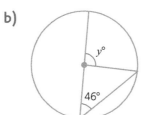

b)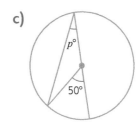

c)

d)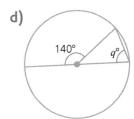

e)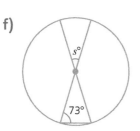

f)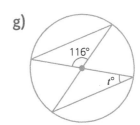

g)

h)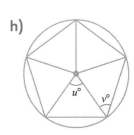

4 There are two isosceles triangles in each of these questions and PQ is a straight line. Copy the diagrams and fill in as many angles as you can to help you find the size of the unknown angles.

a)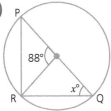

b)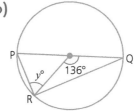

c)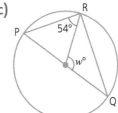

d)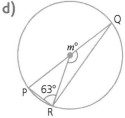

5 Look at the diagrams in Q4. Two angles, one from each of the isosceles triangles, meet at point R on the circumference. Find these two angles and add them together. What do you notice?

▶ Angle in a semicircle

Given a circle with diameter AB, we can pick a third point on the circumference, C and form a triangle. When we draw a second diameter CD, we can add two chords to form the quadrilateral ACBD. As the diagonals AB and CD are equal in length and **bisect** each other (cut each other in half), ACBD must be a rectangle.

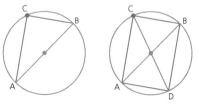

Therefore, if a triangle is formed by connecting the end points of a diameter to another point on the circumference of a circle, the triangle is right-angled and the diameter is the hypotenuse.

This is often referred to as an angle in a semicircle as the triangle is contained within one half of the circle.

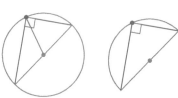

Hopefully you verified this result at the end of the previous exercise.

- To calculate unknown angles or sides you may have to use angle properties, trigonometry, or Pythagoras' theorem.

Worked examples

1   Calculate the size of angle x.

a)

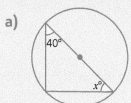

Angle in a semicircle so the triangle is right-angled.

$$\begin{array}{r} 90 \\ +\,40 \\ \hline 130 \end{array} \qquad \begin{array}{r} 180 \\ -\,130 \\ \hline 50 \end{array} \quad x = 50°$$

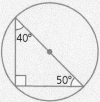

b)

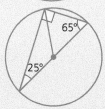

Angle in a semicircle within the large triangle so it is right-angled.

$$\begin{array}{r} 90 \\ +\,25 \\ \hline 115 \end{array} \qquad \begin{array}{r} 180 \\ -\,115 \\ \hline 65 \end{array} \quad x = 65°$$

You could have used two isosceles triangles here to get the same answer – it's just more work!

2   Calculate the length of the unknown sides.

a)

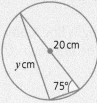

$$\sin x = \frac{O}{H}$$

$$\sin 75 = \frac{y}{20}$$

$$y = 20 \sin 75$$
$$= 19\cdot31\ldots$$
$$= 19\cdot3\,\text{cm to 3 s.f.}$$

b)

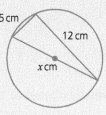

$$c^2 = a^2 + b^2$$
$$x^2 = 5^2 + 12^2$$
$$x^2 = 169$$
$$x = \sqrt{169}$$
$$x = 13\,\text{cm}$$

Exercise 12B

Throughout this exercise, where necessary, round your answers to three significant figures.

1 Calculate the size of the unknown angles.

a) 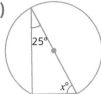   b) 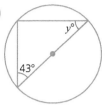   c) 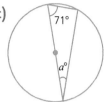   d)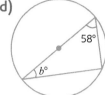

2 Calculate the size of the unknown angles.

a) 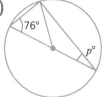   b) 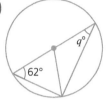   c) 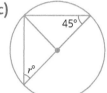   d)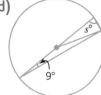

3 Use trigonometry to calculate the length of the unknown sides.

a) 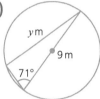   b) 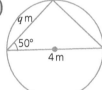   c)   d)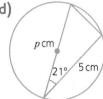

4 Use Pythagoras' theorem to calculate the length of the unknown sides.

a) 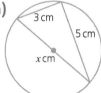   b)   c) 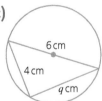   d)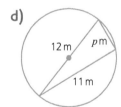

5 A club logo consists of a blue circle of diameter 8 cm, with a red isosceles triangle inside it. One side of the triangle is the diameter of the circle. Calculate:

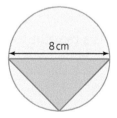

a) the height of the red triangle

b) the area of the red triangle

c) the area of the blue part of the logo.

The club decide to design a new logo. It will remain a blue circle of diameter 8 cm, however, there will be a yellow triangle inside it, as shown in the diagram. One side of the triangle is the diameter of the circle. Find:

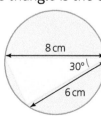

d) the height of the yellow triangle

e) the area of the yellow triangle

f) the percentage increase in the area of the blue part from the old to the new logo.

▶ Tangent to a circle

A **tangent** to a circle is a straight line which has only one point of contact with the circumference of the circle.

 Not a tangent as there are two points of contact.

 Not a tangent as there is no point of contact.

 Tangent as there is only one point of contact. A right angle is formed with the radius.

- Where a radius of a circle meets a tangent, a right angle is formed at the point of contact.
- Sketch the diagram and fill in as many angles as you need to in order to find the one asked for.

Worked examples

1 The line PQ is a tangent to the circle. Calculate the size of angle x.

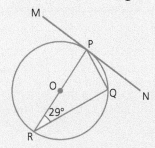

A right angle is formed where the radius meets the tangent.

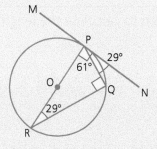

$$\begin{array}{cc} 90 & 180 \\ +52 & -142 \\ \hline 142 & 38 \end{array}$$

$x = 38°$

2 Given MN is a tangent, find the size of angle QPN.

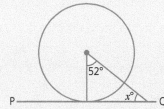

Angle in a semicircle, so triangle PQR is right-angled.

A right angle is formed where the radius meets the tangent.

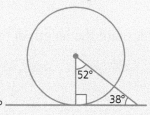

$$\begin{array}{cc} 90 & 180 \\ + 29 & - 119 \\ \hline 119 & 61 \end{array}$$

$\angle OPQ = 61°$

$$\begin{array}{c} 90 \\ - 61 \\ \hline 29 \end{array} \quad \angle QPN = 29°$$

3 PQ is a tangent to the circle. Calculate the size of angle x.

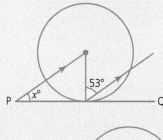

Arrows indicate parallel lines so corresponding angles (F angles) exist. There is a right angle at the tangent.

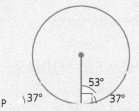

$$\begin{array}{c} 90 \\ -53 \\ \hline 37 \end{array}$$

$x = 37°$

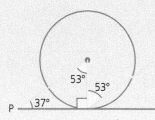

Using alternate angles (Z angles) with the 53° and angles in a triangle add to 180° would give the same answer.

Exercise 12C

1   For each circle, AB is a tangent to the circle. Find the size of the unknown angles.

a)
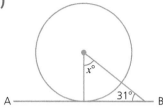

b)
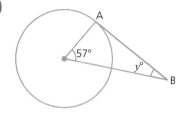

c)
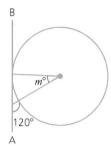

2   Line PQ is a tangent to the circle. Calculate the size of the unknown angles.

a)
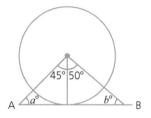

b)
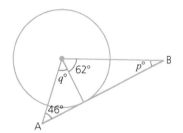

c)
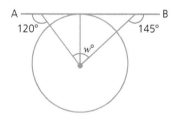

3   Given AB is a tangent to the circle, find the size of the unknown angles.

a)
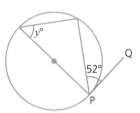

b)
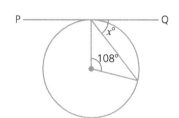

c)
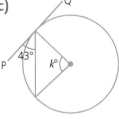

4   The line PQ is a tangent to the circle. Find the size of the unknown angles.

a)
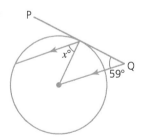

b)
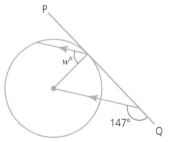

c)
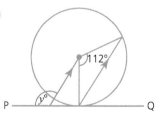

5   Evana sketches the following diagram and believes the line AB is a tangent to the circle. Use the converse of Pythagoras' theorem to determine whether Evana is correct.

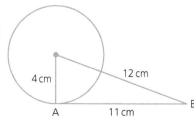

Worked example

AC is a tangent to the circle at point B and FD is a diameter. Find the size of ∠EFB.

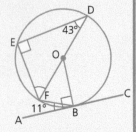

Angle in a semicircle so triangle DEF is right-angled.

AC is a tangent so there is a right angle at B.

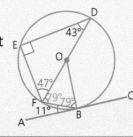

Within triangle DEF:

$$\begin{array}{rr} 90 & 180 \\ +\ 43 & -133 \\ \hline 133 & 47 \end{array}$$

∠EFD = 47°

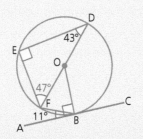

Triangle OFB is isosceles. Using the tangent we get

$$\begin{array}{r} 90 \\ -11 \\ \hline 79 \end{array}$$

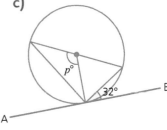

∠EFB

= 47° + 79°

= 126°

---

6 Find the size of the unknown angles. The line AB is a tangent to the circle.

a)

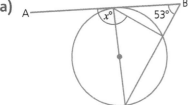

b)

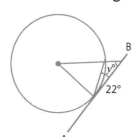

c)

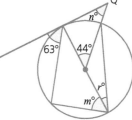

7 The lines PQ, AB and BC are tangents to the circle. Find the size of the unknown angles.

a)

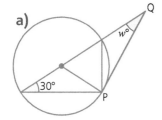

b)

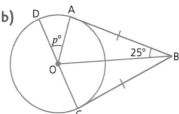

c)

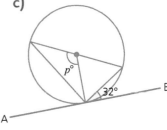

8 MN is a tangent to the circle. Use trigonometry to find the size of angle x to the nearest degree. (Hint: consider the length of sides first!)

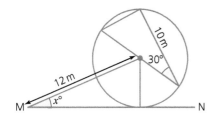

▶ Symmetry in the circle

Recall that a line inside a circle which has two points of contact with the circumference but does not pass through the centre is called a chord.

- When a radius meets a chord at a right-angle, the chord is **bisected** (cut in half).
- When a radius bisects a chord, it does so at a right-angle.
- If a line bisects a chord at right angles (a **perpendicular bisector**) it will pass through the centre of the circle.

To find unknown lengths associated with chords, form a right-angled triangle **from the centre of the circle** and use Pythagoras' theorem.

 Form a right-angled triangle from the centre of the circle. The radius will be the hypotenuse.

Note that without the section below the chord, the diagram would look like this:

Worked examples

1. Chord PQ is bisected by a line OR which is 7 cm long. The radius of the circle is 10 cm. Calculate the length of chord PQ. Give your answer to three significant figures.

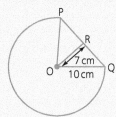 Right angles at R as PQ is bisected and OR is from the centre of the circle.

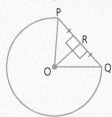

$$c^2 = a^2 + b^2$$
$$10^2 = 7^2 + x^2$$
$$x^2 = 10^2 - 7^2$$
$$= 51$$
$$x = \sqrt{51}$$
$$x = 7 \cdot 141\ldots$$

The line RQ is half the length of chord PQ so

$$PQ = 2 \times RQ$$
$$= 2 \times 7 \cdot 141\ldots$$
$$= 14 \cdot 282\ldots$$
$$= 14 \cdot 3 \text{ cm to 3 s.f.}$$

2. Chord AB is 6 cm long and is part of a circle of radius 5 cm. Calculate the shortest distance from the centre of the circle to the chord.

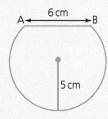 Adding a perpendicular bisector from the centre (shown in dark blue) forms a right angle and halves the length of the chord.

Adding a radius to point A (shown in red) forms a right-angled triangle.

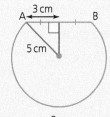

$$c^2 = a^2 + b^2$$
$$5^2 = 3^2 + x^2$$
$$x^2 = 5^2 - 3^2$$
$$= 16$$
$$x = \sqrt{16}$$
$$x = 4 \text{ cm}$$

Exercise 12D

Throughout this exercise, give your answers to three significant figures.

1 Calculate the length of each chord.

a)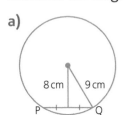

b)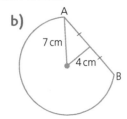

c)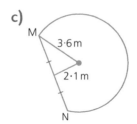

d)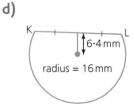

2 Calculate the length of the unknown side.

a)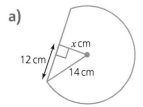

b)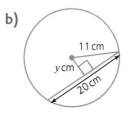

c)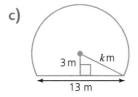

d)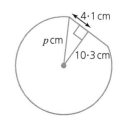

3 Copy each diagram and form a right-angled triangle to find the shortest distance from the centre of the circle to the chord (a line from the centre that bisects the chord).

a)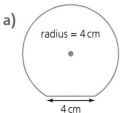

b)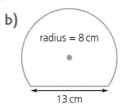

c)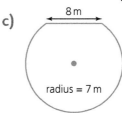

d)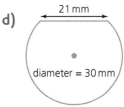

4 For the diagrams shown, calculate the
  i) length of the side x cm
  ii) height, h cm.

a)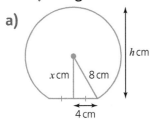

b)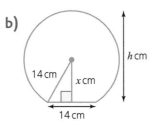

c)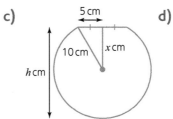

d)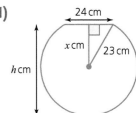

5 For the diagrams shown, calculate the
  i) radius of the circle
  ii) length marked x cm.

a)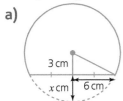

b)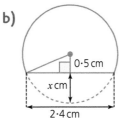

c)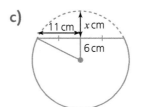

d)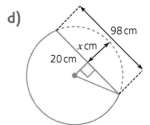

Worked example

Water is still in a horizontal cylindrical pipe of radius 8 cm.

The diagram shows a cross-section of the pipe and PQ represents the surface of the water.

Calculate the maximum depth of water in the pipe when the surface of the water is 14·2 cm wide. Give your answer to three significant figures.

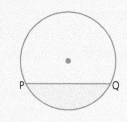

Add a line from the centre to bisect the chord. Each half of the chord has length 7·1 cm.

Add a radius to complete a right-angled triangle. This line is 8 cm long as it is a radius.

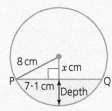

$$c^2 = a^2 + b^2$$
$$8^2 = 7{\cdot}1^2 + x^2$$
$$x^2 = 8^2 - 7{\cdot}1^2$$
$$= 13{\cdot}59$$
$$x = \sqrt{13{\cdot}59}$$
$$x = 3{\cdot}686\ldots \text{ cm}$$

The depth of the water can be found by subtracting x from the radius, so

depth = 8 − 3·686

= 4·314

= 4·31 cm to 3 s.f.

---

6   Liquid is transported in a cylindrical tanker of radius 2 metres. The diagram shows a cross-section of the tanker and PQ represents the surface of the liquid.

Form a right-angled triangle from the centre of the circle to calculate the maximum depth of liquid in the tanker when the surface of the liquid is 3 metres wide.

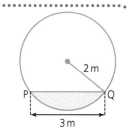

7   A trophy is part of a circle with centre O and a horizontal base. A cross-section of the trophy is shown in the diagram.

The circle has radius 9 cm and the base of the trophy is 16 cm long. Calculate:

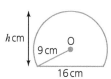

   a)   the shortest distance from O to the base of the trophy

   b)   the height, h cm, of the trophy.

8   A microphone consists of a cylinder with part of a sphere attached to it. A cross-section of the microphone is shown in the diagram. The sphere has a radius of 4 cm and the height of the circular part is 7 cm. For this cross-section:

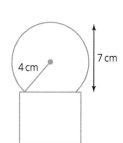

   a)   sketch a diagram and add a right-angled triangle where the sphere attaches to the cylinder

   b)   find the length of the chord where the sphere attaches to the cylinder.

9   A garden area consists of a lawn and a patio. The lawn is part of a circle of radius 8 metres and it meets the rectangular patio as shown in the diagram. A wall is built along the edge where the two sections meet (shown in red).

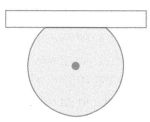

   The shortest distance from the centre of the circle to the patio is 6·2 metres.

   Calculate the length of the wall.

▶ Arcs and sectors

When a section is removed from the centre of a circle, it is called a **sector**. The curved edge of a sector is known as an **arc**. It is part of the circumference and we call its length the **arc length**.

We can split this circle into two sectors.

The smaller one is called the **minor sector** and the larger one the **major sector**.

The arcs are shown in red and the blue straight sides of the sectors are radii of the original circle

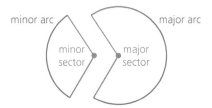

The measurements in a sector are in direct proportion to the measurements in the full circle.

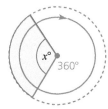

The angle inside a sector, $x°$, is from one complete revolution of 360°.

Fraction: $\dfrac{\text{angle}}{360}$

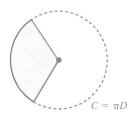

The arc length of a sector is from the circumference of a full circle $C = \pi D$.

Fraction: $\dfrac{\text{arc}}{\pi D}$

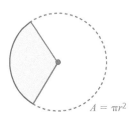

The area of a sector area is from the area of a full circle $A = \pi r^2$.

Fraction: $\dfrac{\text{sector}}{\pi r^2}$

Each ratio is from the same circle, so they are all equal and can be written as

$$\frac{\text{angle}}{360} = \frac{\text{arc}}{\pi D} = \frac{\text{sector}}{\pi r^2}.$$

You will only use two of these fractions at any one time.

To calculate a measurement for a sector:

- Tick what you know and what you want to know on the numerator of the fractions to help you decide which two fractions you are going to use.

- Substitute in values and rearrange to obtain the final answer.

▶ Length of an arc

Calculate the arc length of each sector. Give your answers to three significant figures.

1

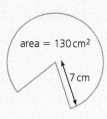

$$\frac{\text{angle}^{\checkmark}}{360} = \frac{\text{arc}^{\checkmark}}{\pi D} = \frac{\text{sector}}{\pi r^2}$$

We know the angle and want to know the arc length so tick and use these fractions.

$$\frac{100}{360} = \frac{\text{arc}}{\pi \times 10}$$

Substitute in values: if $r = 5$ then $D = 10$.
Balance by multiplying both sides by $\pi \times 10$.

$$\text{arc} = \frac{100 \times \pi \times 10}{360}$$
$$= 8{\cdot}726\ldots$$
$$= 8{\cdot}73\,\text{cm to 3 s.f.}$$

2

area = 130 cm²

7 cm

$$\frac{\text{angle}}{360} = \frac{\text{arc}^{\checkmark}}{\pi D} = \frac{\text{sector}^{\checkmark}}{\pi r^2}$$

We do not use the angle fraction this time.

$$\frac{\text{arc}}{\pi \times 14} = \frac{130}{\pi \times 7^2}$$
$$\text{arc} = \frac{130 \times \pi \times 14}{\pi \times 7^2}$$
$$= 37{\cdot}14\ldots$$
$$= 37{\cdot}1\text{cm to 3 s.f.}$$

Exercise 12E Throughout this exercise, round your answers to three significant figures.

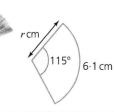

1 Calculate the arc length of each sector.

a)

x cm
65°
6 cm

b)

y cm
140°
13 cm

c)

k mm
235°
27 mm

d)

30°
9 cm
p cm

Be careful here!
What angle is in the sector?

2 Calculate the arc length of each sector.

a)

x cm
area = 55 cm²
8 cm

b)

y cm
area = 960 cm²
21 cm

c)

2·4 mm
r mm
area = 2 mm²

d)

q m
area = 1·5 m²
91 cm

Think about units.

3 A fan is opened out to form a sector with an angle of 150°.
If the radius of the fan is 18 cm, calculate the length of its arc.

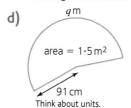

4 A sector is formed from a circle with radius r cm.
The arc length of the sector is 6·1 cm and the angle in the sector is 115°.
Calculate the radius of the circle.

r cm
115°
6·1 cm

▶ Area of a sector

Worked examples   Calculate the area of each sector.

1

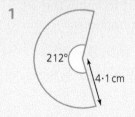

$$\frac{\text{angle}^✓}{360} = \frac{\text{arc}}{\pi D} = \frac{\text{sector}^✓}{\pi r^2}$$

$$\frac{212}{360} = \frac{\text{sector}}{\pi \times 4 \cdot 1^2}$$

$$\text{sector} = \frac{212 \times \pi \times 4 \cdot 1^2}{360}$$

$$= 31 \cdot 09\ldots$$

$$= 31 \cdot 1 \,\text{cm}^2 \text{ to 3 s.f.}$$

2

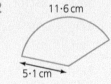

$$\frac{\text{angle}}{360} = \frac{\text{arc}^✓}{\pi D} = \frac{\text{sector}^✓}{\pi r^2}$$

$$\frac{11 \cdot 6}{\pi \times 10 \cdot 2} = \frac{\text{sector}}{\pi \times 5 \cdot 1^2}$$

$$\text{sector} = \frac{11 \cdot 6 \times \pi \times 5 \cdot 1^2}{\pi \times 10 \cdot 2}$$

$$= 29 \cdot 58$$

$$= 29 \cdot 6 \,\text{cm}^2 \text{ to 3 s.f.}$$

Exercise 12F

Throughout this exercise, round your answers to three significant figures.

1 Calculate the area of each sector.

  a)
area = x cm²

  b)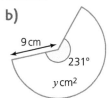

  c)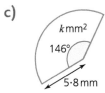

  d)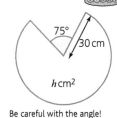
Be careful with the angle!

2 Given the following arc lengths, calculate the area of the sectors.

  a)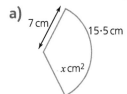

  b)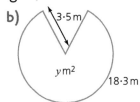

  c)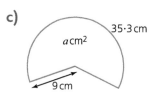

  d)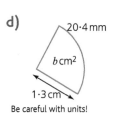
Be careful with units!

3 A kitchen work surface is to be covered in a protective material.

The work surface is a rectangle with a sector on the end, as shown in the diagram.

Calculate, to the nearest square centimetre, the area of protective material required to cover the top of the work surface.

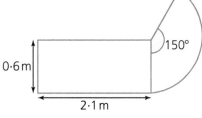

4 A sector is formed from a circle with radius r cm as shown in the diagram.

The area of the sector is 105 cm² and the size of angle n is 20°.

Calculate the length of the radius of the circle.

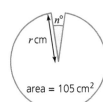

▶ Angle in a sector

Worked examples   Calculate the size of angle x in each sector.

1

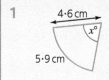

4·6 cm

$x°$

5·9 cm

$$\frac{\text{angle}}{360} = \frac{\text{arc}}{\pi D} = \frac{\cancel{\text{sector}}}{\cancel{\pi r^2}}$$

$$\frac{\text{angle}}{360} = \frac{5·9}{\pi \times 9·2}$$

$$\text{angle} = \frac{5·9 \times 360}{\pi \times 9·2}$$

$$= 73·48...$$

$$= 73·5° \text{ to } 3 \text{ s.f.}$$

2

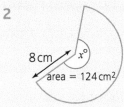

8 cm

$x°$

area = 124 cm²

$$\frac{\text{angle}}{360} = \frac{\cancel{\text{arc}}}{\cancel{\pi D}} = \frac{\text{area}}{\pi r^2}$$

$$\frac{\text{angle}}{360} = \frac{124}{\pi \times 8^2}$$

$$\text{angle} = \frac{124 \times 360}{\pi \times 8^2}$$

$$= 222·0...$$

$$= 222° \text{ to } 3 \text{ s.f.}$$

Exercise 12G

Throughout this exercise, round any decimal answers to three significant figures.

1   Given the arc length, calculate the size of the angle marked in each sector.

a)
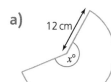
12 cm
$x°$
49 cm

b)
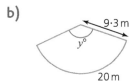
9·3 m
$y°$
20 m

c)

6·7 cm
$w°$
5·5 cm

d)
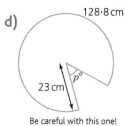
128·8 cm
23 cm
$p°$

Be careful with this one!

2   Calculate the size of the unknown angle inside each sector.

a)
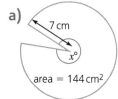
7 cm
$x°$
area = 144 cm²

b)
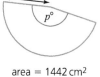
$y°$
5·9 m
area = 26 m²

c)

31 cm
$p°$
area = 1442 cm²

d)
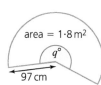
area = 1·8 m²
$q°$
97 cm

Be careful with units!

3   A toy chest is 50 cm wide. When it is fully open, the bottom of the lid travels through an arc of 70 cm.

A cross-section of the opening is shown in the diagram.

What angle does the lid reach when the chest is fully open?

70 cm
50 cm

4   For each of the following sectors, calculate the unknown measure.

a)
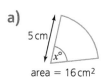
5 cm
$x°$
area = 16 cm²

b)
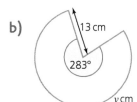
13 cm
283°
y cm

c)
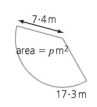
7·4 m
area = p m²
17·3 m

d)
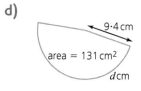
9·4 cm
area = 131 cm²
d cm

▶ Arcs and sectors problem solving

A pin badge is cut from a sector of a circle as shown by the shaded part of the diagram.

a) Calculate the area of the front of the badge.

The designer wants to add a trim around the perimeter of the badge.

b) What length of trim is required?

Give your answers to 3 significant figures.

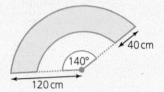

a) We have to calculate the area of the large sector then subtract the area of the small one.

Large sector

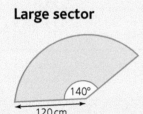

$$\frac{140}{360} = \frac{sector}{\pi \times 120^2}$$

$$sector = \frac{140 \times \pi \times 120^2}{360}$$

$$= 17\,592\cdot\ldots\, cm^2$$

Small sector

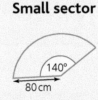

$$\frac{140}{360} = \frac{sector}{\pi \times 80^2}$$

$$sector = \frac{140 \times \pi \times 80^2}{360}$$

$$= 7819\cdot\ldots\, cm^2$$

The area of the front of the badge is $17\,592\cdot\ldots - 7819\cdot\ldots = 9773\cdot\ldots = 9770\, cm^2$ to 3 s.f.

b) To find the perimeter we calculate the arc lengths of the large and small sectors and then add them to the two 40 cm sections at either side.

Large arc

$$\frac{140}{360} = \frac{arc}{\pi \times 240}$$

$$arc = \frac{140 \times \pi \times 240}{360}$$

$$= 293\cdot2\ldots\, cm^2$$

Small arc

$$\frac{140}{360} = \frac{arc}{\pi \times 160}$$

$$arc = \frac{140 \times \pi \times 160}{360}$$

$$= 195\cdot4\ldots\, cm^2$$

The total perimeter is $293\cdot2 + 195\cdot4 + 40 + 40 = 568\cdot6\ldots = 569\, cm$ to 3 s.f.

Throughout this exercise, round your answers to three significant figures.

1 The diagrams below are formed by removing a small sector from a larger one within concentric circles. Find the area of each shaded area.

a)

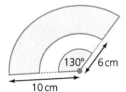

b)

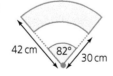

c)

2 The diagrams below are formed by removing a small sector from a larger one within concentric circles. Find the perimeter of each shaded area.

a)

b)

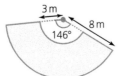

c)

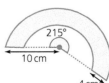

3   Sectors from two concentric circles, one with radius 15 cm and the other with radius 8 cm, are shown in the diagram opposite. Calculate:

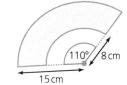

a)   the area of the shaded shape

b)   the perimeter of the shaded shape.

4   Lauren loves her swing and has an adapted seat to stop her from falling off it. The back of Lauren's seat is 40 cm high and the furthest away point of her seat from the top of the swing is 1·5 m. The shaded area in the diagram shows the area covered by the back edge of her seat when Lauren swings through an angle of 120°. Calculate the shaded area.

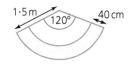

5   A club badge is a square of side 3·2 cm. It has three sectors on it which are all the same size.

Each sector has an arc length of 2·5 cm and a radius of 1·4 cm. Calculate:

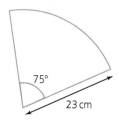

a)   the area of one sector

b)   the area of the badge which does not contain a sector.

6   A window decoration has five identically sized coloured sectors on it. One sector is shown in the diagram. The perimeter of each sector is to be painted black.

a)   Calculate the total length to be painted black on the sectors.

Each sector is 8 mm thick.

b)   Calculate the volume of one sector.

Try **Q7** and **Q8** without a calculator. Hint: simplify fractions!

7   A computer graphic is in the shape of a sector as shown.

a)   Calculate the exact area of the graphic.

A blue border is introduced around the graphic.

b)   Calculate the exact length of the border.

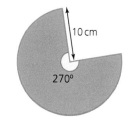

8   The sector shown is folded to make a cone. Find the exact value of:

a)   the surface area of the cone

b)   the circumference of the base of the cone

c)   the base radius of the cone

d)   height of the cone.

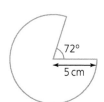 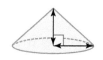

12 Circle

Check-up

1 Calculate the size of the unknown angles.

a)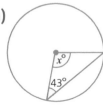

b)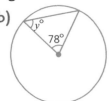

c)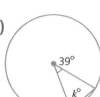

2 Calculate the size of the unknown angles.

a)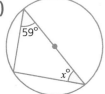

b)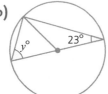

c)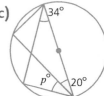

3 Calculate the length of the unknown sides to three significant figures.

a)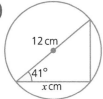

b)

c)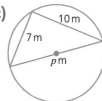

4 AB is a tangent to each circle. Find the size of the unknown angles.

a)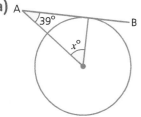

b)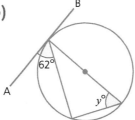

c)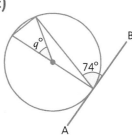

5 Find each unknown length. Round your answers to three significant figures.

a)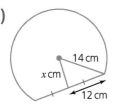

b)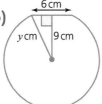

c)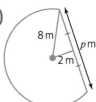

6 Two concentric circles with centre O are shown in the diagram.

The line PQ is 32 cm long, is a tangent to the smaller circle and forms a chord within the larger circle

The radius of the smaller circle is 12 cm. Calculate the radius of the larger circle.

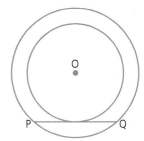

7 Calculate the arc length of each sector. Round any decimal answers to three significant figures.

a)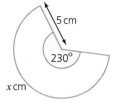

5 cm
230°
x cm

b)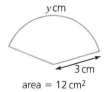

y cm
3 cm
area = 12 cm²

c)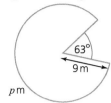

63°
9 m
p m

8 Calculate the area of each sector to three significant figures.

a)

10 cm
135°
area = *x* cm²

b)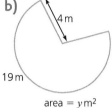

4 m
19 m
area = *y* m²

c)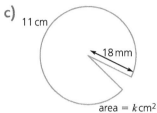

11 cm
18 mm
area = *k* cm²

9 Calculate the angle in each sector to the nearest degree.

a)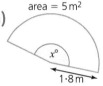

area = 5 m²
x°
1·8 m

b)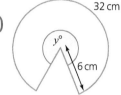

32 cm
y°
6 cm

c)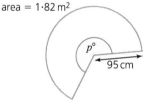

area = 1·82 m²
p°
95 cm

10 Calculate the unknown measure for each sector. Round any decimal answers to three significant figures.

a)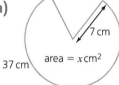

7 cm
37 cm
area = *x* cm²

b)

y m
115°
8 m

c)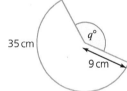

35 cm
q°
9 cm

11 A jelly mould is a prism which is 8 cm tall.

The cross-section of the mould is a sector with radius 6 cm and arc length 14 cm, as shown in the diagram.

Calculate the volume of the mould correct to three significant figures.

8 cm
14 cm
6 cm

12 A school enterprise group have made up a game to raise money for charity. Each person pays £1 to play and if they chip a golf ball into the shaded area they win a prize.

The shaded area is formed by taking sectors from two concentric circles.

The larger circle has radius 5 m and the smaller circle has radius 3·8 m.

a) Calculate the area available to land in to win a prize.

The group decide to build a border around the entire prize area to stop the golf balls running off.

b) What length is the border? Give your answers to three significant figures.

5 m
3·8 m
93°

▶ Mean and range

An **average** is a single central value which represents a whole data set. It gives an idea of what the values in the data set are like. Measures of **spread** assess how close together the numbers in the set are and indicate how well the average represents the data set.

To calculate the **mean**:

- Add up the numbers in the data set to find the total.
- Divide by the **sample size** n, the number of items in the data set.

$$\text{mean} = \frac{\text{total}}{n}$$

The **range** is the difference between the highest and lowest numbers. It is a basic measure of spread. A higher range indicates a more spread out set of data.   range = highest − lowest

---

Worked example

Two teams of S3 pupils are running a cross country race. Calculate the mean and range for each team. Make two valid comments about the results.

Team 1 Race times (minutes): 17, 14, 15, 16, 13

$17 + 14 + 15 + 16 + 13 = 75$

5 items

$\text{mean} = \dfrac{\text{total}}{n}$

$= \dfrac{75}{5} = 15$ minutes

range = highest − lowest

$= 17 - 13 = 4$ minutes

Team 2 Race times (minutes): 9, 18, 14, 13, 10, 9·5

$9 + 18 + 14 + 13 + 10 + 9·5 = 73·5$

6 items

$\text{mean} = \dfrac{\text{total}}{n}$

$= \dfrac{73·5}{6} = 12·25$ minutes

range = highest − lowest

$= 18 - 9 = 9$ minutes

Typically, Team 2 ran the race faster as their mean time was lower ($12·25 < 15$).

Team 2's race times were more varied as their range is higher ($9 > 4$).

Exercise 13A   Where necessary, round to 2 decimal places.

1   Calculate the mean of each data set. Try **a)**–**d)** without a calculator.

   a)   3, 4, 5, 6, 7   b)   102, 110, 115, 93   c)   1·5, 2, 2·5, 3, 4, 2   d)   −4, 3, 1, −5, −2, 1, 2, −4
   e)   500, 400, 420, 480, 410, 530, 520, 600, 300, 150   f)   20·4, 24·6, 20·9, 18·1, 27·2, 29·3, 30
   g)   58, 65, 62, 57, 52, 51, 50, 61, 64, 62, 57, 51, 50   h)   −3, 5, 4, −9, 6, −2, −1, 4, 8, −2, 3, −5, −2, 3

2   Calculate the range for each data set in **Q1**.

3   The table shows the maximum wind speed recorded each day one week in Lerwick and Falkirk.

		Mon	Tue	Wed	Thur	Fri	Sat	Sun
Windspeed km/h	Lerwick	31	21	18	19	11	10	25
	Falkirk	18	16	10	12	14	8	18

a) Calculate the mean and range for Lerwick and Falkirk. Copy and complete the table of results.

b) Which town was windier that week on average?

c) Which town's weather was more varied that week? Give evidence from the table of results to justify your answer.

	Mean	Range
Lerwick		
Falkirk		

4 A price comparison site checks the price of an 800 g loaf of bread in 20 shops. The data is shown:

Supermarkets (£s)	1·10	0·90	1·15	1·10	0·95	1·20	1·05	1·00	0·85	0·55
Convenience shops (£s)	1·25	1·60	1·35	1·20	1·80	1·10	1·85	1·90	1·40	0·95

a) Calculate the mean and range for each kind of shop. Copy and complete the table of results.

b) Where is a loaf likely to be cheaper?

c) Which kind of shop has more varied prices? Justify your answer.

	Mean	Range
Supermarket		
Convenience		

5 A train operator keeps a record of the ages of people doing different jobs.

Trainee driver	18, 19, 20, 19, 18, 21, 24, 23
Ticket examiner	24, 26, 65, 28, 22, 59, 21, 19, 23, 25, 28, 22
Revenue manager	38, 40, 42, 48

	Mean	Range
Trainee driver		
Ticket examiner		
Revenue manager		

a) Calculate the mean age and age range for each job. Copy and complete the table.

b) Which job has the lowest mean age?

c) Which job has the largest spread of ages?

d) The HR manager thinks one of the means doesn't represent the workforce very well. Which do you think it is? Explain why it does not represent the workforce well.

6 The mean weight of five parcels is 3 kg. What is the total weight of all five parcels?

7 The mean height of 4 people is 180 cm.

a) What is the total height of the four people?

b) The heights of 3 of the people are 150 cm, 190 cm and 195 cm. How tall is the 4th person?

8 The mean age of six people is 29. Five of the people are 17, 35, 28, 34 and 20. How old is the 6th?

9 There are 129 Members of the Scottish Parliament (MSPs). Suppose their mean age is 50 before some changes. Three MSPs aged 47, 78 and 69 leave. Two more are elected who are 24 and 21 years old. After the 3rd new MSP is elected the mean age is 49. How old is the 3rd new MSP?

10 Here is the formula for the mean using mathematical notation $\bar{x} = \dfrac{\Sigma x}{n}$

The symbol for the mean is $\bar{x}$: we read this as '*x* bar'

Σ is a Greek letter 'Sigma'. In mathematics this tells us to find the sum of (add up) whatever follows it.

a) A manager calculates the sum of their employees ages, $\Sigma x = 2000$. There are 50 employees. Calculate the mean age, $\bar{x}$.

b) A factory weighs 8000 cakes and calculates the mean weight, $\bar{x} = 130$ g. Calculate Σx, the total weight of the cakes produced.

c) An air quality monitoring station records the amount of nitrogen dioxide every 30 seconds for 1 hour. The total, Σx, recorded was 1670 μg/m³. Calculate the mean amount, $\bar{x}$.

d) A customer review site records a total of 120 stars, $\Sigma x = 120$, with $\bar{x} = 3$ stars. Calculate *n*, the sample size or number of reviews.

▶ Standard deviation

Standard deviation is a very robust measure of how spread out a data set is. A low value indicates the data is **consistent**, i.e. the numbers are close to the mean. A higher value indicates the data is **varied**, i.e. the numbers are more widely spread from the mean.

One formula to calculate the standard deviation of a sample is $s = \sqrt{\dfrac{\sum(x - \bar{x})^2}{n-1}}$, where n is the sample size.

Worked example A speed camera records the speed of passing cars. The data set shows the speeds of a sample of 5 of the cars that passed during rush hour. **Speeds** (mph) 14, 16, 21, 24, 25

a) Calculate the mean and standard deviation.

$\text{mean} = \dfrac{\text{total}}{n}$ 'x bar' is the mean.

$= \dfrac{100}{5}$ $\bar{x} = 20$

$= 20 \, \text{mph}$

x	$x - \bar{x}$	$(x - \bar{x})^2$
14	−6	36
16	−4	16
21	1	1
24	4	16
25	5	25
	Total	94

● Add up the final column. This number is crucial for the next step. This is the **sum of the squared deviations**.

$\sum(x - \bar{x})^2 = 94$

This is the numerator in our formula.

$s = \sqrt{\dfrac{\sum(x - \bar{x})^2}{n-1}}$

$= \sqrt{\dfrac{94}{5-1}}$

$= \sqrt{\dfrac{94}{4}}$

$= \sqrt{23 \cdot 5}$

$= 4 \cdot 847 \ldots$

$= 4 \cdot 85$ to 2 d.p.

To calculate the standard deviation:

● Calculate the mean.

● Set up a table with headings as shown.

x	$x - \bar{x}$	$(x - \bar{x})^2$

● Fill out the first column with the data set.

● To complete the middle column, $x - \bar{x}$, take each data point and subtract the mean.

$14 - 20 = -6$

$16 - 20 = -4$ and so on

These are the '**deviations**' from the mean. If we added these up the answer should be 0. Carry out a quick check.

● To complete the last column, $(x - \bar{x})^2$, square all the deviations.

$(-6)^2 = 36$

$(-4)^2 = 16$

$1^2 = 1$ and so on

Every number in this column will now be positive. Adding these up will give a positive measure of the total deviation from the mean.

● Now use the formula $s = \sqrt{\dfrac{\sum(x - \bar{x})^2}{n-1}}$.

For a sample from a larger population we divide by $n - 1$, one less than the sample size. This accounts for sampling differences.

Divide the fraction first. Keep the answer on your calculator, do not round at this stage.

Take the square root and round your final answer carefully.

The mean speed was 20 mph and the standard deviation of the speeds is 4·85 to 2 decimal places.

b) The same camera recorded a sample of speeds from cars during the night. The table includes the mean and standard deviation of these. Make two valid comments comparing the speed of cars at rush hour and at night-time.

	Mean	Standard deviation
Rush hour	20	4·85
Night-time	29	6·21

Typically, rush hour traffic was slower as the mean is lower (20 < 29 mph).

Rush hour traffic had more consistent speeds as the standard deviation is lower (4·85 < 6·21) *or*

Night-time traffic had more varied speeds as the standard deviation is higher (6·21 > 4·85).

Exercise 13B

In each question, the data set is a sample of a larger population. Round standard deviations to 2 decimal places.

Standard deviation formula:

$$s = \sqrt{\frac{\sum (x - \bar{x})^2}{n - 1}}$$

where n is the sample size

1 The data set shows a sample of six S1 pupils' heights.
Heights (cm): 144, 155, 138, 160, 168, 147

a) Copy and complete to calculate the mean and standard deviation of the heights.

$$\bar{x} = \frac{\text{total}}{n}$$

$$= \frac{}{6}$$

$$= \text{cm}$$

x	$x - \bar{x}$	$(x - \bar{x})^2$
144		
155		
138		
160		
168		
147		
Total		

$$s = \sqrt{\frac{\sum (x - \bar{x})^2}{n - 1}}$$

$$= \sqrt{\frac{}{6 - 1}}$$

$$= \sqrt{\frac{}{5}}$$

$$= \sqrt{}$$

$$= \quad \text{unrounded}$$

$$= \quad \text{to 2 d.p.}$$

b) A sample of S3 heights was also taken. The table shows the mean and standard deviation of these. Copy and complete the table with S1s results.

	mean	s
S1		
S3	172	7·05

c) Which year group was taller? Give evidence for your answer.

d) Which year group had more varied heights? Justify your answer.

2 A researcher is checking the price of beans. The price of seven, own brand, 400 g tins is shown.
Prices (p): 40, 85, 95, 110, 150, 100, 50

a) Calculate the mean and standard deviation of the own-brand prices.

The prices for branded beans are also checked. The table shows the mean and standard deviation for the branded prices.

b) Copy and complete the table with your results.

c) Make two valid comments comparing the prices of own-brand and branded beans.

	mean	s
Own-brand		
Branded	125	15·20

3 A study records the age at which a sample of 6 teachers learned to swim. Age: 5, 5, 4, 2, 6, 8

   a) Calculate the mean and standard deviation. Do not use a calculator.

   The researchers then asked a sample of lifeguards how old they were when they learned to swim.

   b) Copy and complete the table of results.

   c) Make two valid comments about how old the teachers and life guards were when they learned to swim.

	mean	s
Teachers		
Lifeguards	4	1·14

4 A hotel gathers data on the number of nights a sample of guests stay for. The results are shown. Number of nights: 2, 3, 5, 1, 6.

   a) Calculate the mean and standard deviation of the number of nights.

   The hotel installs a new spa facility. After the spa opens the mean number of nights guests stay for is 6·2 and the standard deviation is 3·1.

   b) Copy and complete the table of results.

   c) Make two valid comments about the opening of the spa and guest stays.

	mean	s
No Spa		
Spa		

5 Sports scientists measure the amount of oxygen a person can use during intense exercise. This measure of fitness is called VO_2 Max and is measured in millilitres of oxygen used in one minute per kilogram of body weight (ml/kg/min). The table shows the VO_2 Max levels of athletes before and after a training programme.

VO_2 Max	Before programme	33·2	25·8	26·1	31·7	27·6	30·2
ml/kg/min	After programme	38·5	30·4	39·1	40	31·5	37·1

   a) Calculate the mean and standard deviation for each group of athletes.

   b) Make two valid comments comparing the fitness of athletes before and after the programme.

6 The standard deviation of 5, 8, 11, 14, 19 and 21 is 6·23.

   Write down the standard deviation of 105, 108, 111, 114, 119 and 121.

7 Two classes take a maths test at the start of the year and another 2 months later. The means and standard deviations of their results are shown.

	Test 1		Test 2	
	mean	s	mean	s
Class 1·1	55	4	52	8
Class 1·2	50	10	75	2

   Which of the following statements are true?

   A Both classes improved their scores.

   B Class 1·1's scores were more varied in the second test.

   C Class 1·2's scores were more consistent at the start of the year.

8 Mary has calculated the mean and standard deviation of her data set but has smudged ink on three numbers in her data set. She remembers they are whole numbers. Can you find the smudged numbers?

   Data set: ▮, 4, ▮, 1, ▮ $\bar{x} = 4$ $s = \sqrt{5}$

▶ Standard deviation: an alternative formula

$$s = \sqrt{\dfrac{\sum x^2 - \dfrac{(\sum x)^2}{n}}{n-1}}$$

9 Copy and complete to find the standard deviation of 3, 4, 5 and 12 using this formula.

	x	x^2
	3	9
	4	
	5	
	12	
Total	24	194

$\uparrow$ $\sum x$ $\qquad \uparrow$ $\sum x^2$

$$s = \sqrt{\dfrac{\sum x^2 - \dfrac{(\sum x)^2}{n}}{n-1}}$$

$$= \sqrt{\dfrac{194 - \dfrac{(24)^2}{4}}{3}}$$

$$= \sqrt{\dfrac{50}{3}}$$

$$=$$

$$=$$

$\sum x^2$, square the data points and then add them up.

$(\sum x)^2$ add up the data points and square the total

Use your calculator carefully.

$$194 - \dfrac{576}{4}$$
$$= 194 - 144$$
$$= 50$$

$$\sqrt{\dfrac{50}{3}}$$
$$= \sqrt{16 \cdot 66 \ldots}$$

10 Use the alternative formula to calculate the mean and standard deviation for each data set:

   a) Temperatures (°C): 3, 4, 6, 8, 9   b) Lengths (m): 3, 6, 9, 12, 13, 17

11 The table shows the mean and standard deviation of the prices of an 800 g loaf of bread from **Q4** in Exercise 13A. Make two valid comments about the price of bread in supermarkets and convenience shops.

	mean	s
Supermarket	0·99	0·19
Convenience	1·44	0·33

12 The standard deviation of 4, 5, 6, 7 and 8 is 1·58.

   Write down the standard deviation of 104, 105, 106, 107 and 108.

In statistics the mean and standard deviation are a rich source of information. For certain data sets, like the ones in the example and **Q13** and **Q14** below, 95% of the data in a set is less than 2 standard deviations away from the mean.

Worked example   A geologist studies a sample of pebbles on a shore. They find the mean weight of a pebble is 35·2 g and the standard deviation is 4·5 g. Write a sentence about the weight of 95% of the pebbles.

$4 \cdot 5 \times 2 = 9$ $2 \times s = 9$ The geologist knows that 95% of the pebbles on the
$35 \cdot 2 - 9 = 26 \cdot 2$ shore weigh between 26·2 g and 44·2 g.
$35 \cdot 2 + 9 = 44 \cdot 2$ mean ± 9

13 A study of people at the cinema finds the mean age is 36 years and the standard deviation is 4 years. Write a sentence about the age of 95% of the people at the cinema.

14 A study finds the mean beep test score for an age group is 5·5 with a standard deviation of 2·1. Someone in the age category scores 9·8. What can you say about their fitness?

▶ The median

The **median** is a different kind of average. To find the median:

- List the numbers in ascending order.
- Find the middle number. If the list is even, take the mean of the two middle numbers.

Worked examples Find the median of each data set.

1 14, 15, 17, 19, 20, (23), 25, 25, 37, 48, 50

This data set is in order.

There are eleven numbers, 11 ÷ 2 = 5 r 1.

There are 5 numbers on each side and 1 in the middle. The middle number is the median.

median = 23

2 37, 5, 37, 11, 36, 35, 32, 31

First, write the numbers in order:
5, 11, 31, 32, | 35, 36, 37, 37

8 ÷ 2 = 4. There are 4 numbers each side. The median is halfway between 32 and 35.

$$\frac{32 + 35}{2} = 33.5$$

median = 33·5

Exercise 13C

1 These data sets are already listed in order. Find the median of each list:

a) 4, 5, 6, 7, 8, 9, 11   b) 28, 30, 30, 38, 39, 40, 41, 45, 49   c) 10, 14, 18, 22, 26, 30

d) 10, 20, 30, 50, 60, 80, 90, 100   e) −12, −8, −7, −3, −2, 0, 1, 1, 2, 5

f) 0·25, 0·3, 0·45, 0·5, 0·55, 0·6, 0·66   g) 2350, 2475, 2918, 3105, 3190, 3207, 3291, 3300

2 Two friends have the same phone app, which records the total number of times they pick up their phone each day. Their data for a particular week is shown below.

	Mon	Tue	Wed	Thurs	Fri	Sat	Sun
Annie	70	32	36	30	45	8	10
Joseph	47	45	38	41	34	50	80

a) Write out Annie's data set in ascending order. Find Annie's median number of phone pick-ups.

b) Write out Joseph's data set in ascending order. Find Joseph's median number of phone pick-ups.

c) Which friend typically picked up their phone more often?

3 A drive-through coffee shop records the time taken for customers to complete their purchase.

Time to complete purchase (minutes): 2·5, 2, 3·3, 2·2, 4, 3·1, 14·5, 6·2, 2·1, 2·3, 2·8, 3

a) Find the median time taken to complete a purchase.

b) Calculate the mean waiting time. Which average do you think better represents the data set? Write a sentence to explain your choice.

4 A company lists employee salaries.

Annual salary (£1000s): 22, 115, 20, 20·5, 21, 24·5, 21, 16, 24, 23, 26, 98, 17·5

The company says that typically employees earn more than £30 000. Which average do you think they are using? Which average would you choose to represent this data? Explain why.

▶ Interquartile range

The median splits a data set written in ascending order, in half. If we split the data evenly again this makes four parts or quartiles.

- To find the lower quartile, find the median of the lower half.
- To find the upper quartile, find the median of the upper half.
- To calculate interquartile range: $IQR = Q_3 - Q_1$

Q_1	lower quartile
Q_2	median
Q_3	upper quartile

The IQR is a measure of spread. A low value indicates the data is consistent and close to the median. A high value indicates the data is varied and spread more widely from the median.

Worked example

Running club data is shown.

5k run time (minutes)	
Feb Meet	13, 14, 14·5, 15, 15, 16, 16·2, 17, 18, 18·5, 20, 21, 22, 34, 42
July Meet	9, 9, 10, 10·2, 10·5, 12, 13, 14, 14·5, 15, 15·5, 16, 18, 22·3

a) Find the median, upper and lower quartiles for each race. Calculate the IQR for each race.

b) Make two valid comments about the performance in February and July's 5K races.

a) **Feb meet**

13, 14, 14·5, (15) 15, 16, 16·2, (17) 18, 18·5, 20, (21) 22, 34, 42

Q_2 median

lower quartile Q_1, median of first half   upper quartile Q_3, median of upper half

$Q_1 = 15$ $Q_2 = 17$ $Q_3 = 21$

$$IQR = Q_3 - Q_1 = 21 - 15 = 6$$

July meet

9, 9, 10, (10·2) 10·5, 12, 13, 14, 14·5, 15, (15·5) 16, 18, 22·3

$Q_1 = 10·2$ $Q_2 = \dfrac{13 + 14}{2} = 13·5$ $Q_3 = 15·5$

The original list was even. The median falls between two numbers. The lower half includes 13.

$IQR = 15·5 - 10·2 = 5·3$ The upper half includes 14.

b) The runners were typically faster in July as the median time is lower ($13·5 < 17$).

The runners had more consistent times in July as the IQR is lower ($5·3 < 6$).

Exercise 13D

1. Find Q_1, Q_2 and Q_3 for each data set:

a) Heart rates (bpm): 55, 58, 59, 60, 61, 62, 64, 65, 67, 70, 71

b) Sleep duration(hrs): 5·5, 6, 6, 6·5, 7, 7·5, 7·5, 8, 8·5, 9, 9·5, 10, 10·5, 11, 11·5

c) Books read: 2, 4, 5, 6, 7, 8, 10, 11, 12

d) Age: 38, 41, 42, 44, 45, 46, 48, 50, 52, 54, 55, 56, 57

e) Puppy weights (g): 350, 400, 410, 415, 420, 455, 500, 510, 520, 524, 530, 531, 535

f) Beep test scores: 6, 9·5, 6·5, 6·9, 8, 8·2, 7·8, 7·9, 6·9, 7·2, 9, 6·5 (Remember to order these!)

g) Wages (£1000s): 24, 28, 13·5, 12, 35, 33, 33·9, 38, 24, 25, 21·8, 51, 48·5, 22

➜

2 Find the IQR for each of the data sets in **Q1**.

3 A streaming service records how long people watch pilot episodes of two new comedies before switching off. Both programmes last 30 minutes. The results are shown:

The Call Centre	5, 14, 23, 18, 21, 30, 22, 26, 19, 21, 20·5, 24, 22, 19, 11, 20, 21
Monday Morning	6, 10, 9, 8, 29, 27, 30, 26, 4, 30, 2, 30, 28, 29, 1, 27, 26, 28, 30, 4, 5, 2, 29, 30

a) Find the median, upper and lower quartiles for each comedy.

b) Calculate the IQR for each.

c) Make two valid comments about the audience reaction to the two comedies.

4 A biotech company gathers data on the number of days a stitching material takes to dissolve. The company develops a new material and a second data set is gathered.

	Time taken to dissolve (days)
Original material	10, 18, 35, 40, 41, 60, 70, 95, 12, 15, 100, 105, 110, 120, 150
New material	5, 8, 10, 14, 25, 30, 32, 40, 50, 85, 85, 90, 90, 110

a) Find the median, upper and lower quartiles for each material.

b) Calculate the IQR for each material.

c) Make two valid comments about the time taken for the materials to dissolve.

Worked example

Display the results of the biotech material study in **Q4** in a parallel box plot.

A **box plot** is a diagram which shows quartiles 1, 2 and 3 in a box, with whiskers extending to the lowest and highest values. A **parallel box plot** displays more than one data set on the same axes so that they can be easily compared.

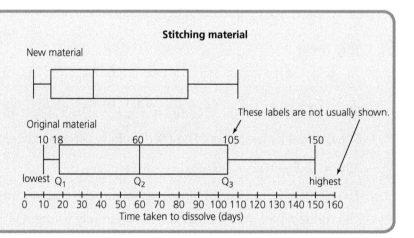

Stitching material

5 The midday temperatures in two towns were recorded over a nine-day period. The results are shown:

	Temperature °C
Bethton	1, 2, 4, 4, 5, 6, 7, 9, 10
Billford	3, 4, 6, 6, 7, 9, 10, 13, 15

a) Find the highest and lowest values and the median, lower and upper quartiles for the temperatures in each town.

b) Draw a parallel box plot to display the data.

c) Calculate the length of the box for each town. This has the same value as the interquartile range.

d) Make two valid comments about the temperatures in Bethton and Billford.

▶ The mode

The mode is a third kind of average. The mode is the most common item in a data set. This average is useful for categorical data sets. The mode does not have a measure of spread associated with it. When there is more than one most popular item there will be more than one mode. If all items are equally popular then there is no mode.

> **Worked example**   State the mode for each data set.
>
> a) Cards picked at random: H, H, D, D, D, S, C, C, H, D, S, C, S, D   The **modal** category is D with 5.
>
> b) Shoe sizes: 5, 6, 7, 6, 6, 5, 7, 8, 5, 9, 4, 10   The data set is bimodal (has two modes).
>   The modal values are size 5 and size 6 with 3 of each.
>
> c) Fruits: apple, banana, banana, orange, apple, orange   No mode (all equal)

Exercise 13E

1   Find the modal value in each data set:

   a) Flavours: Sweet, Sweet, Savoury, Sour, Savoury, Sour, Sour, Savoury, Sweet, Sour

   b) Colours: R, R, B, B, B, Y, B, R, G, G, G, R, R, R, G, B, B, Y, G, Y, R

   c) Sizes: M, M, S, XS, L, L, XL, XL, XS, XS, M, M, S, S, XL, XL

   d) VO$_2$ Max (nearest whole number): 26, 25, 24, 26, 26, 27, 25, 28, 27, 24, 25, 24, 25

   e) Preferred time for meeting:1 pm, 1.30 pm, 2 pm, 2 pm, 1 pm, 1.30 pm, 1.30 pm, 2 pm, 1 pm

2   Sarah-Anne makes a summary note about averages and measures of spread

   Complete the table choosing from:
   mean, standard deviation, IQR, median and mode.

Data	Type of average	Measure of spread
Numerical data, no unusually high or low values		
Numerical data, unusually high or low values		
Categorical data		

3   Which measures of location and spread would you choose for each situation?

   a) A survey of earnings with a relatively small number of very high earners

   b) A study of which colours are used in the flags of the world

   c) A study of insect numbers in a location with no unusually high or low numbers

4   Find the mean, median and mode for each data set. Decide which best represents the data.

   a) Prices for a model of phone: (£): 120, 102, 99, 100, 110, 105, 115, 95, 95, 109

   b) House prices (£1000s): 105, 90, 98, 112, 88, 125, 110, 126, 635, 810

   c) Shoes sold (size): 5, 5, 6, 6, 4, 7, 7, 5, 5, 5, 8, 11

5   Find a set of seven positive whole numbers with mode = 5, mean = 8 and median = 9.

6   How many sets of five positive whole numbers can you find with mode = 4, mean = 6, median = 5?

7   Why can a set of five positive whole numbers not have mode = 4, mean = 5, median = 3?

▶ Stem and leaf diagrams

Stem and leaf diagrams organise data into ascending order and display the **distribution** of the data. This means that they show how spread out or clustered the data is. To make a stem and leaf diagram, include five elements:

- Stem, leaves, title, sample size and key.

Worked example

A graphic designer records how long it takes to create some elements of a computer game.

Time (minutes): 45, 56, 62, 71, 80, 44, 63, 60, 61, 59, 54, 67, 48, 88, 63

a) Draw a stem and leaf diagram to display the results.

b) Find the median time taken.

a) Make a draft, recording each value on the correct branch.

In this example:

the **stem** records the first digit

4 | 4 5 8

the **leaves** record the second, grouping the data.

This branch records 80 and 88

In your final diagram, rewrite the branches in ascending order.

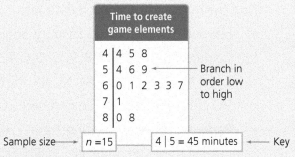

Draft	
4	5 4 8
5	6 9 4
6	2 3 0 1 7 3
7	1
8	0 8

Time to create game elements

4	4 5 8
5	4 6 9	← Branch in order low to high
6	0 1 2 3 3 7
7	1
8	0 8

Sample size ⟶ $n = 15$ 4 | 5 = 45 minutes ⟵ Key

b) The data is now listed in ascending order. The sample size is 15.

15 ÷ 2 = 7 r 1: the median is the 8th value. Counting along to the 8th value, the median time is 61 minutes.

Notice that with a different **key** this diagram could represent very different data. e.g. 4|5 = £4·5 million.

Exercise 13F

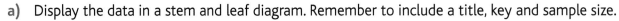

1 Amina is doing a sponsored run. She keeps a record of the donations.

Donations (£s): 23, 30, 15, 28, 10, 21, 13, 19, 30, 35, 41, 15, 34, 39, 48, 5

a) Draw a stem and leaf diagram to display the data.

b) Find the median amount donated.

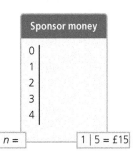

Sponsor money

0	
1	
2	
3	
4	

$n =$ 1 | 5 = £15

2 Taylor is having a problem with noise from a building site beside her house. She uses an app on her phone to record an estimate of the noise level every day at 8 am for 12 days.

Noise level (Db, decibels): 70, 85, 79, 88, 90, 95, 98, 86, 80, 100, 95, 92

a) Display the data in a stem and leaf diagram. Remember to include a title, key and sample size.

b) Find the modal (most common) noise level recorded.

c) Find the median noise level recorded.

d) The council says noise levels above 90 Db are unacceptable. What fraction of the days recorded had unacceptable noise levels?

3   Gemma sells antiques at auction. She records the sale price of each item at one sale.

Sale price (£s): 310, 330, 200, 210, 450, 560, 380, 470, 310, 440, 570, 690

a) Using the key 3|1 = £310, draw a stem and leaf diagram to display the data.

b) Find the median sale price.

c) What percentage of the sales were over £500?

4   A courier delivery service records the number of deliveries each of its sixteen employees made in a week.

Number of deliveries made (1 week)

```
 9 | 2 5 9
10 | 0
12 | 5
13 |
14 |
15 | 0 1 1 2 5 8 9
16 | 0 0 8 9
```

n = 16   9 | 2 = 92 deliveries

a) How many couriers made fewer than 120 deliveries?

b) What is the probability that a courier picked at random made fewer than 120 deliveries?

c) The couriers are paid £2 per delivery. How many couriers made at least £300?

d) What is the probability that a courier picked at random made at least £300?

5   A botanist is growing saplings in different soil conditions and records the results in the **back to back stem and leaf** diagram shown.

The heights on the left are for saplings grown in peat soil.

The heights on the right are for saplings grown in clay soil.

This branch reads 4·0, 4·5, 4·6 cm

Each set of results has its own sample size.

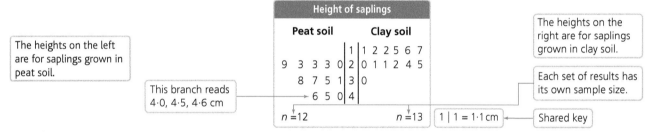

Height of saplings

Peat soil		**Clay soil**
	1	1 2 2 5 6 7
9 3 3 3 0	2	0 1 1 2 4 5
8 7 5 1	3	0
6 5 0	4	

n = 12   *n* = 13   1 | 1 = 1·1 cm ← Shared key

a) Find the range of heights for each soil type.   b) Find the median height for each soil type.

c) Which of the soil types would you recommend to grow the tallest saplings?

6   Maria Lyle is a parasport athlete from Dunbar. At age 14 Maria set a world record in the 200 m sprint. The data below is based on the results of the 200 m sprint from two international competitions.

	Finishing time (seconds)						
Rio Paralympics 2016	28·2	28·8	**29·4**	32·7	33·1	33·9	
World Championship 2017	28·5	28·6	**29·9**	32·0	32·9	32·7	33·7

a) Create a back to back stem and leaf diagram to display the data. The times in bold are Maria's.

b) Which event had a faster median time and by how much?

7   A charity records the annual cost of different development projects for two causes.

Cost of project (£ millions)	
Clean water	1·8, 2·5, 3, 3·6, 5, 4·2, 1·9, 2, 2·4, 2·8, 2·9, 3·4, 3·9, 3·1, 3, 3·6, 2·1, 6, 6·7
Education	0·5, 2·4, 3·8, 5, 0·4, 1·1, 5·6, 1·7, 6·2, 1·3, 7·8, 5, 6, 1·5, 0·4, 1·7, 5·3, 4

a) Display the data in a back to back stem and leaf diagram. Remember to include a shared key, sample sizes and title.

b) Find the interquartile range for each kind of development project.

c) Which type of project has more consistent development costs?

▶ Scatter graph and line of best fit

Scatter graphs display possible correlations between numerical data sets. We can add a line of best fit, which follows the trend of the points. The equation of the line of best fit gives us a formula for the connection between the data. This is called a linear model. To find the equation of the line of best fit:

- Use two points on the line of best fit to find its gradient.
- Use $y = mx + c$ to find the equation of the line.
- Replace y and x with the correct variables for the context.

Worked example The scatter graph shows how many supported study sessions learners attended (N) and their exam score (E).

a) Add a line of best fit to the scatter graph.

b) Find the equation of the line in terms of E and N.

c) Use your model to estimate the exam mark for someone who attended all six sessions.

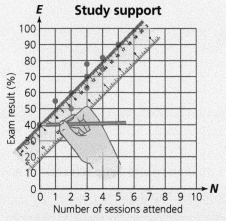

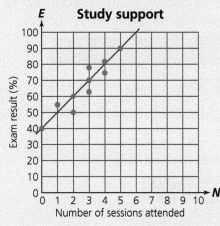

a) The line of best fit follows the trend of the points. Aim to have as many points close to the line as you can. Try to have roughly the same number of points above and below the line. Extend the line to the y-axis. It is easier to calculate the gradient if your line of best fit goes through two clear coordinate points.

b) $(0,40) \quad (2,60)$

$(x_1, y_1) \quad (x_2, y_2)$

$$m = \frac{y_2 - y_1}{x_2 - x_1}$$
$$= \frac{60 - 40}{2 - 0}$$
$$= \frac{20}{2}$$
$$= 10$$

The y-intercept is $(0, 40)$, $c = 40$

$y = mx + c$

$y = 10x + 40$

Now replace y and x with the correct variables for the context. Exam score E is on the y-axis, so replace y with E. Number of sessions N is on the x-axis, so replace x with N.

$E = 10N + 40$

c) For six supported study sessions $N = 6$, $E = 10N + 40$
$$= 10 \times 6 + 40$$
$$= 100$$

The model predicts an exam score of 100% for attending all six sessions.

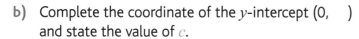

Exercise 13G

1   A courier company records the data shown. A line of best fit has been added to the graph for you.

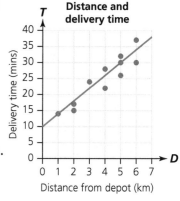

   a)   The points (1, 14) and (5, 30) lie on the line of best fit. Use these points and the gradient formula to calculate the gradient m, of the line of best fit.

   b)   Complete the coordinate of the y-intercept (0,   ) and state the value of c.

   c)   Using $y = mx + c$, write down the equation of the line of best fit.

   d)   State the equation of the line of best fit in terms of T and D, $T = mD + c$.

   e)   Use your model to estimate the time to deliver a parcel 8 km from the depot, $D = 8$.

2   A vegetable box delivery service pays for adverts on local radio. They run different numbers of adverts and record the number of new customers each day. The table shows the results.

Number of adverts per day (A)	1	2	3	4	5	6	7	8
Number of new customers (C)	2	5	6	7	12	10	14	16

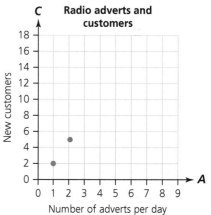

   a)   Draw a scatter graph to display the data.

   b)   Add a line of best fit to your scatter graph.

   c)   Find the gradient of the line of best fit and state the y-intercept.

   d)   State the equation of the line of best fit in terms of C and A.

   e)   Use your model to predict how many new customers the service would have if they ran 15 adverts per day.

3   A family records the daily minimum outside temperature T (°C) and their energy usage E (KWh).

Temperature T(°C)	10	9	1	7	2	8	7	3	6	4	5
Energy usage E(KWh)	15	16	45	35	35	20	25	38	21	33	30

   a)   Draw a scatter graph, display temperature T on the x-axis and energy usage E on the y-axis.

   b)   Add a line of best fit to your scatter graph.

   c)   Find the equation of the line of best fit in terms of E and T.

   d)   Use your formula to predict the energy usage if the minimum outside temperature is −2 °C.

4   A group of friends recorded the data shown about the number of weeks they studied for an exam.

   a)   Draw a scatter graph and add a line of best fit.

   b)   Find a formula which will predict exam results based on weeks of study.

   c)   One friend scored 70% in the exam. Estimate how many weeks they studied for.

Weeks of study (W)	1	2	3	4	5	4	3	6	5	8
Exam result (R)	32	45	50	61	53	65	48	85	80	95

▶ Data analysis

In this exercise, you will interpret a range of charts and diagrams.

1   The graph shows the medals won by four countries at the 2016 Olympic Games.

a)   Which of the countries shown won the most gold medals?

b)   How many medals did Brazil win in total?

c)   How many more gold medals than bronze medals did Great Britain win?

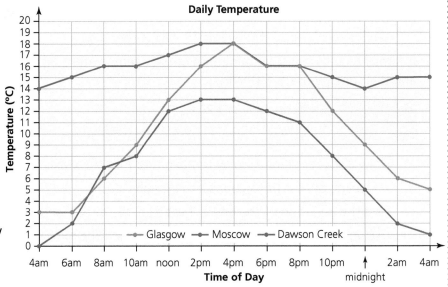

2016 Olympic Games

Key
Gold
Silver
Bronze

Number of medals — Country: Great Britain, China, Russia, Brazil

d)   How many silver medals did the countries shown win in total?

e)   Which of the countries shown won the most medals overall?

f)   What fraction of Russia's medals were bronze?

2   The graph shows the temperature over 24 hours in May in three cities which all sit on the same global line of latitude.

a)   What was the maximum temperature in Dawson Creek, Canada, during the period shown?

b)   It was the same temperature in Glasgow and Moscow for a while. How long did this last?

Daily Temperature

Temperature (°C) — Time of Day: 4am, 6am, 8am, 10am, noon, 2pm, 4pm, 6pm, 8pm, 10pm, midnight, 2am, 4am

Glasgow — Moscow — Dawson Creek

c)   Between which times was Dawson Creek warmer than Glasgow?

d)   What was the range of temperatures in Glasgow during the period shown?

e)   What was the range of temperatures in Moscow during the period shown?

f)   Describe the differences between the overall trend in daily temperature in Moscow compared to both Glasgow and Dawson Creek.

3 The pie chart shows the different types of music that Charlie has listened to over a month. The data is gathered by her streaming service.

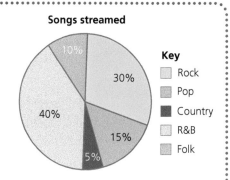

Songs streamed

Key
- Rock
- Pop
- Country
- R&B
- Folk

a) Charlie streamed 54 country songs during this time. How many folk songs did she stream?

b) How many R & B songs did Charlie stream?

c) How many songs did she stream altogether?

d) What should the angle at the centre of the 'rock' slice measure?

e) What should the angle at the centre of the 'pop' slice measure?

The stacked bar graph shows how many of the R&B and country songs that Charlie streamed had the lyrics 'life' and 'love'.

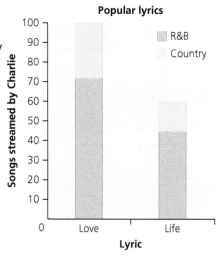

Popular lyrics

f) Charlie streamed 72 R&B songs with the lyric 'love'. How many country songs did she stream with this lyric?

g) How many R&B songs did Charlie stream with the lyric 'life'?

h) Charlie thinks this graph means that the words 'love' and 'life' are always used more often in R&B than country songs. Do you agree?

4 The chart below is called a dot plot. This dot plot shows nutritional information about 12 savoury items for sale in a takeaway bakery shop. Each dot represents one item. The table shows information about the government food labelling system.

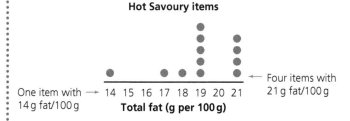

Hot Savoury items

One item with → 14g fat/100g
Four items with 21g fat/100g

Total fat (g per 100g)

	Low	Medium	High
Fat (g per 100g)	≤ 3g	3 < fat ≤ 17.5g	> 17.5g

a) How many of the items contain a high level of fat?

b) What fraction of the items contain a medium level of fat?

c) What is the modal amount of fat in the savoury items?

d) Find the median amount of fat in an item.

e) What is the mean amount of fat?

The same shop also sells cold sandwiches.

f) Complete the table and use it to make two valid comments about the fat content of savoury snacks and cold sandwiches in this shop.

	Mean (g fat per 100g)	Standard deviation
Savoury snacks		2·04
Cold sandwiches	5·65	2·99

▶ Displaying data

In this exercise you will choose how to display the data in each question. Q1 forms a summary and the **red keywords** will help you choose an appropriate graph or diagram throughout this exercise. You may wish to use technology. Remember to add titles, axes labels and keys as you would when creating graphs by hand.

1   Complete the last column of the table, naming each type of graph or chart, choosing from:
    scatter graph, compound bar graph, pie chart, compound line graph and stem and leaf diagram.

Graph or chart	Used to display...	Type of graph or chart
	how a total is **shared** into different categories.	a)
	the frequency of different **categories** with clusters or stacks of bars showing **groups** in each category.	b)
	trends in data over a continuous numerical scale with a key to identify groups.	c)
	possible **correlation** between two numerical data sets.	d)
	numerical data listed in **ascending order**.	e)

2   Three friends carry out an experiment to see who has the best takeaway coffee cup.
    They all buy the same drink at the same time and take the temperature of their drink every 3 minutes. The results are shown in the table.

	A	B	C	D
1		Temperature °C	Temperature °C	Temperature °C
2	Time since collection (minutes)	Disposable	Bamboo	Steel Insulated
3	0	72	75	70
4	3	65	72	68
5	6	62	62	68
6	9	55	54	67
7	12	49	47	66·5
8	15	45	41	66
9	18	41	36	66
10	21	39	32	64
11	24	37	29	63·5
12	27	36	28	63
13	30	35	27	62·5

→

a) Create a graph to display the **trends** in how the temperatures change over time for each cup. The data for all three cups should be shown on the same axes with a key to identify groups.

b) Which drink cooled the most in the first 3 minutes?

c) Use your graph to predict the temperature of each drink after 20 minutes.

d) Describe any differences you notice between the disposable and the insulated cups.

e) The owner of the insulated cup claimed they had only lost 5% of the heat from their drink in 30 minutes. Were they correct?

3 Aaron keeps a note of how he **shares** out his time one day.

a) What fraction of his full day does Aaron spend on schoolwork?

b) Create a suitable chart to show how he divides up his full day.

Activity	Time (hours)
Sleep	8
Schoolwork	8
Exercise	2
Meals	2
Hobbies	4

4 The table opposite shows the designation of land use in Scotland and England. The figures have been rounded to the nearest 1000 square miles.

a) Which country has the largest area of rural land?

	Area of Land (Square miles)	
	Scotland	England
Rural	29000	45000
Urban	1000	5000

b) Choose a graph to display this information. Show the area of land in Scotland and England and how this **category** is broken down into the **groups** rural and urban land.

c) Are you surprised by the data? Describe anything you notice about it.

5 A cloud computing expert gives different apps a cybersecurity score. A higher score is better.

Score: 1·5, 0·5, 2·0, 2·5, 1·0, 1·5, 3·5, 6·5, 5·5, 1·2, 1·8, 2·0, 1·1, 2·2, 2·5, 3·0, 2·1, 1·9, 4·0, 2·0, 1·3, 1·9

a) What were the highest and lowest scores?

b) Create a diagram to display this data in **ascending order**.

c) Find the median score.

6 The spreadsheet shows information about a call centre.

a) Draw a graph to display **trends** in the number of incoming calls over time. Use only the first two columns of data to do this.

b) Describe the trends in the number of incoming calls over time.

c) Is there a relationship between the number of incoming calls, N, and the mean customer wait time, W? Draw a graph to investigate a possible **correlation**. Use the second and third columns of data to do this.

	A	B	C
1	Time	Number of Incoming Calls	Mean Wait Time (minutes)
2	09:00	45	5
3	09:10	48	6
4	09:20	55	7
5	09:30	50	5
6	09:40	45	4.5
7	09:50	40	4
8	10:00	30	3.5
9	10:10	28	4
10	10:20	25	2.5
11	10:30	22	2
12	10:40	20	2
13	10:50	25	3.5
14	11:00	24	3

d) Find a formula that predicts the mean customer wait time W, based on the number of incoming calls N.

e) Predict the mean customer wait time if the number of incoming calls is 60.

▶ Probability and expectation

Recall that when there are several equally likely outcomes the **probability** of an event is calculated using the formula.

$$\text{probability (favourable outcome)} = \frac{\text{number of favourable outcomes}}{\text{total number of outcomes}}$$

We build probability as a fraction but it can also be expressed as a decimal and a percentage.

Probabilities lie between 0 (no chance) and 1 (certain).

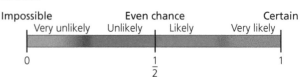

Worked example

A raffle has 200 tickets, 2 big prizes and 8 smaller prizes.

a) Calculate the probability of winning a big prize.

$$P(\text{big}) = \frac{2}{200}$$
$$= \frac{1}{100}$$
$$\frac{1}{100} = 1\% \text{ chance}$$

b) Calculate the probability of winning any prize.

$$P(\text{any}) = \frac{10}{200}$$
$$= \frac{1}{20}$$
$$\frac{1}{20} = 5\% \text{ chance}$$

c) Ella buys 40 tickets. Estimate how many prizes she might win.

$$\frac{1}{20} \text{ of } 40 = 2$$

d) What is the probability of not winning a prize at all?

$$1 - \frac{1}{20}$$
$$= \frac{20}{20} - \frac{1}{20} = \frac{19}{20}$$

Exercise 13J

1 Fraser picks a day of the week at random. What is the probability that he will choose:

   a) a weekday
   b) a weekend day
   c) a day that starts with T
   d) Wednesday
   e) any day except Wednesday?

2 A raffle has 50 tickets, 1 big prize and 4 smaller prizes. What is the probability of:

   a) winning the big prize
   b) winning a smaller prize
   c) winning any prize
   d) not winning at all?

3 Which event is more likely?

   a) Rolling a 6 on a die or tossing a coin to land on heads
   b) Rolling an even number on a die or picking a month that starts with J
   c) Winning a raffle with 800 tickets and 50 prizes or winning a raffle with 300 tickets and 10 prizes

4 A wheel of fortune has 8 sections. The wheel is set up so that when it spins an arrow will always land inside one of the sections and each one is equally likely. ➔

The wheel is spun. Calculate the probability that it will:

a) land on the big prize

b) land on a small prize

c) land on a spin again

d) land on you lose

e) land on any prize or a spin again

f) not land on the big prize.

5 In the wheel of fortune game above, the probability of winning a small prize is $\frac{1}{4}$. Imagine you are running this game. How many small prizes would you expect to give out if the wheel is spun:

a) 16 times **b)** 40 times **c)** 256 times **d)** 500 times?

As the number of spins increases the estimates become more accurate. We can't predict where the very next spin will land but we do know what to expect over time.

e) How many prizes would you expect to give out after 900 spins?

6 A class are doing probability experiments.

a) Helen rolls a die 180 times. Estimate how many times you would expect her to roll a six.

b) Evie rolls a die 70 times. Estimate how many times she should expect to roll an even number.

c) Sarah rolls a die 51 times. Estimate how many times you would expect her to roll at least a 5.

d) Rayaan rolls a die 39 times. Estimate how many times he should expect to roll a square number.

e) Which of these predictions do you think is most likely to be accurate?

7 A traffic light sequence is shown.

a) Calculate the total length of the light sequence.

A car draws up to the lights. What is the probability that the lights are:

b) red only **c)** green

d) red and amber **e)** showing any amber light?

	Red	Red/amber	Green	Amber
Time (secs)	120	10	150	20

f) a driver passes these lights at different times of day 720 times a year. How many times should she expect the lights to be green when she draws up?

8 Antonio is at a charity fete. There are two games of chance. There is a raffle with 200 tickets and 70 small prizes. There is also a wheel of fortune with 18 equally likely sections and 6 small prizes.

a) What is the probability of Antonio winning a small prize in the raffle?

b) What is the probability of Antonio winning a small prize in the wheel of fortune?

c) Which game of chance is Antonio more likely to win? Compare the values clearly.

9 Stacey-Anne has an 8-sided die and a bag of marbles which contains 2 red, 1 green and 4 blue marbles. Which of these outcomes are more likely?

a) Rolling a prime number on the die or choosing a blue marble?

b) Rolling at least six on the die or choosing a red or green marble?

10 Siobhan runs a 'hook the duck' stall at a charity fete. She has 50 floating ducks marked 1 to 50. Choose the more likely outcome from each pair.

a) Hooking a duck marked with a square number or one marked with a multiple of 6?

b) Hooking a duck marked with a prime number or one with a factor of 216?

▶ Probability from relative frequency and expectation

We can also use experience or the results of a **trial** or experiment to determine a probability.

probability (event) $= \dfrac{\text{number of events}}{\text{total number of trials}}$. This is the **relative frequency** of the event.

Although we cannot predict the next event with certainty, we can multiply by the probability to estimate the number of events we expect to happen over time.

Worked example

On a day with 150 pupils a teacher notices 21 learners ask for extra homework.

a) What is the probability that one of her pupils chosen at random wanted extra work?

$$P(\text{extra}) = \frac{21}{150} = \frac{7}{50}$$

b) There are 1200 pupils in the whole school. How many would you expect to want extra homework?

$$\frac{7}{50} \times 1200 = 168$$

$$0 \cdot 14 \times 1200 = 168$$

Exercise 13K

1 In a trial, 4 out of 50 people clicked on an online advert.

   a) Calculate the probability that someone will click on the advert.

   b) How many clicks could the advertiser expect with 2500 adverts?

2 A factory takes a sample of 90 zips and finds 3 are faulty.

   a) What is the probability that a zip is faulty?

   b) How many faulty zips would you expect in an order of 5400 zips?

3 An insurance company keeps the data shown on the probability that different models of car will be stolen.

Model	Probability
4 × 4	0·002
minicar	0·0035
estate	0·0007

   a) Which model is most likely to be stolen?

   b) The insurer has 10 000 clients with estate cars. How many would they expect to be stolen?

   c) The insurer has 500 clients with 4 × 4s. How many would they expect to be stolen?

   d) In one year the insurer had 200 000 clients with minicars and 800 got stolen. Was this unusual?

4 The probability of an expectant mother having triplets is estimated to be $\dfrac{1}{2500}$. Every year there are approximately 50 000 births in Scotland. How many sets of triplets would you expect?

5 The probability of someone living in Scotland having red hair is around 5% or $\dfrac{1}{20}$. There are approximately 5 500 000 people in Scotland. Estimate the number of people with red hair.

▶ Further probability

When an event is more complex a table can be useful to display all the possible outcomes.

> **Worked example**   What are the possible outcomes if we toss two coins?
>
> Copy and complete the table of all the possible outcomes. H = heads, T = tails
>
> | | H | T |
> |---|---|---|
> | H | HH | HT |
> | T | TH | TT |
>
> What is the probability of:
>
> a)   two tails $P(TT) = \dfrac{1}{4}$ We can see from the table that tails and tails is one possibility out of four.
>
> b)   both coins landing the same way? $P(HH \text{ or } TT) = \dfrac{2}{4} = \dfrac{1}{2}$

Exercise 13L

1   Two 6-sided dice are rolled and the scores are added. For example, rolling 2 and 4 gives a total of 6.

a)   Copy and complete the table to show all the possibilities for the sum of the two dice.

	1	2	3	4	5	6
1	2					
2				6		
3						
4						
5						
6						

What is the probability of:

b)   scoring less than 6

c)   scoring at least 6?

d)   Which scores are least likely?   e)   Which score is the most likely?

f)   In this game, you can score an 'easy eight' or a 'hard eight'. What do you think this means?

g)   The game is played 720 times. How many times would you expect to roll a total of 6?

As long as the results of events don't have an impact on each other we can calculate the probability of a combination of them by multiplying the individual probabilities together.

The probability of **independent successive events** P(A and B) = P(A) × P(B).

> **Worked example**   Find the probability of:
>
> a)   a coin landing on H twice in a row. $P(H \text{ and } H) = P(H) \times P(H) = \dfrac{1}{2} \times \dfrac{1}{2} = \dfrac{1}{4}$
>
> b)   rolling a 6 on a die and then an even number on a die $P(6 \text{ and even}) = P(6) \times P(\text{even}) = \dfrac{1}{6} \times \dfrac{1}{2} = \dfrac{1}{12}$

2   What is the probability of:

a)   a coin landing on tails and then rolling a 3 on a die

b)   a random number generator from 1–10 generating the same digit twice in a row

c)   picking a day starting with S from the days of the week and picking a month starting with M

d)   tossing a coin and it landing on tails three times in a row (multiply 3 fractions together)

e)   a Scottish expectant mother having triplets (0·000 4) and having red hair (0·05)

f)   the next three Scots who try this question all having red hair?

Check-up

1. Find the mean of each data set:
   a) 24, 25, 21, 20, 19, 28, 17   b) 70, 71, 80, 85, 60, 55, 70, 90, 95, 84

2. Find the range for each of the data sets in Q1.

3. Here are the same data sets written in ascending order. Find the median of each:
   a) 17, 19, 20, 21, 24, 25, 28   b) 55, 60, 70, 70, 71, 80, 84, 85, 90, 95

4. A researcher is testing voice recognition software. They play 10 male voices and note how many are 'understood' correctly. The results of a sample of 6 runs are shown: 9, 7, 5, 6, 7, 8.
   a) Find the mean and standard deviation of the results.
   b) The software is also tested using female voices. The results are shown. Make two valid comments about how well the software is recognising male and female voices.

	mean	s
Male		
Female	5	1·96

5. Find the median and lower and upper quartile for each data set below:
   a) 101, 104, 108, 115, 120, 122, 128, 130, 139, 140, 144, 147, 149, 150, 155
   b) −9, −6, −3, −3, −2, −1, 0, 3, 5, 6, 7, 8, 11, 13, 15, 17, 18
   c) (careful!) 82 000, 80 000, 72 000, 90 000, 68 000, 70 500, 92 500, 81 250

6. Calculate the interquartile range for each data set in Q5.

7. Find the mode from each data set:
   a) 0, 1, 3, 1, 0, 3, 3, 3, 1, 3
   b) H, C, D, S, H, C, C, S, H, D, H, C

8. The ages of people at a party are: 20, 48, 15, 19, 21, 18, 21, 20, 21, 20, 49, 67, 69, 20, 20, 21, 18
   a) Draw a stem and leaf diagram to display this data. Remember a title, sample size and key.
   b) How old were the youngest and oldest people at the party? What was the range?
   c) What was the modal age of the people at the party?
   d) Find the median age.
   e) Calculate the mean age.
   f) What fraction of the people at the party were in their 20s?
   g) Which average do you think best represents the people at the party?

9. A sample of people were asked how old they were when they chose their career.
   The data is shown below.

   Age : 23, 26, 25, 25, 22, 28, 26

   Calculate the mean and standard deviation of the ages. You could try this one without a calculator.

10. A TV talent show holds an audition in a big city. A total of 1500 people queue up to audition.
    From that queue, 75 people get to audition for the judges and 3 are chosen.
    What is the probability of:
    a) someone picked at random from the queue auditioning for the judges
    b) someone who auditions for the judges getting picked to go on TV
    c) someone picked at random from the queue being chosen to go on TV?

11 A psychologist does an experiment asking volunteers not to eat a marshmallow from a bowl. He found that 21 of the 56 volunteers ate the marshmallow.

   a) What is the probability that a volunteer picked at random would eat a marshmallow?

   b) The psychologist went on to test 296 volunteers. How many would you expect to eat the marshmallow?

The rest of the check-up questions on this page form a case study about a small study of eight smartphone users. The smartphone users were unhappy with their smartphone use and agreed to try a programme of exercise and offline creative tasks and puzzles for 30 days instead.

12 The study participants were asked to rate their wellbeing on a scale from 0 (low) to 6 (high). The table shows the results for day 1.

   a) Calculate the mean wellbeing score on day 1.   **Day 1 wellbeing scores** | 3, 3, 3, 1, 1, 4, 0, 1

   b) Calculate the standard deviation.

   c) The users rated their wellbeing again on day 30. The mean score was 5 with a standard deviation of 1. Do you think the programme improved wellbeing? Make two comments.

13 The scattergraph shows some further data from the study.

   a) The study had 8 participants. Can you explain why the scatter graph shows 16 data points?

   b) What was the maximum screen time recorded on day 1?

   c) What was the maximum screen time recorded on day 30?

   d) How many participants had at least 3 hours of screen time on day 1?

   e) Use a ruler to identify a line of best fit for the scattergraph. Do not write on the book. Calculate the gradient of the line.

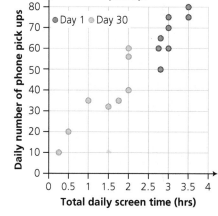

   f) Find the equation of the line of best fit in terms of N and T.

   g) Use your equation to predict the number of times a person with 5 hours of daily screen time would pick up their phone.

   h) Do you think the programme of exercise and offline activities was successful in reducing screen time? Give evidence from the scattergraph to explain your answer.

14 The same study recorded the data shown below about the participants' smartphone use. Describe 3 differences in how the participants changed their phone use.

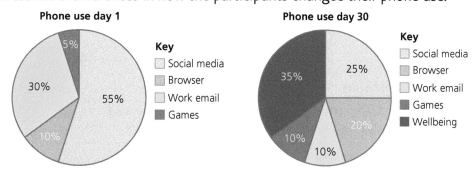

Answers

1 Number work

Exercise 1A

1. a) 40 000 b) 7000 c) 300 d) 700
   e) 80 f) 6 g) 0·8 h) 0·04
   i) 0·05 j) 0·009
2. 4000 3 700 000
4. a) 64 000 b) 4900 c) 640 d) 810
   e) 21 f) 8·6 g) 0·60 h) 0·092
   i) 0·31 j) 0·0040
5. £240 000
6. a) AU$193·597 b) AU$190
7. a) 14 900 b) 8270 c) 932 d) 51·9
   e) 2·16 f) 0·0346 g) 0·721 h) 0·0463
   i) 0·759 j) 0·300
8. 0·003 47 metres per month
9. 5 000 000 to 1 s.f., 5 500 000 to 2 s.f.,
   5 460 000 to 3 s.f.
10. a) 2 b) 3 c) 1 d) 3 e) 2
    f) 3 g) 4 h) 2 i) 1 j) 2
11. a) 9
    b) i) 3·14 ii) 3·142 iii) 3·1416
12. a) 0·44 b) 0·84 c) 0·576
    d) 0·261 e) 0·6096 f) 0·478 48
13. a) £54 875 b) £50 000 c) £55 000
    d) Pupils' own answer. It does not make sense
       to round as you want the prize money to
       be shared exactly.
    e) There would not be enough money to give
       each winner a quarter share.

Exercise 1B

1. a) 18 cm, 14 cm b) 16 mm, 6 mm
   c) 46 g, 14 g d) 76·5 cm, 75·5 cm
   e) 251 mm, 229 mm f) 452·5 g, 447·5 g
   g) 64·5 cm, 61·5cm h) 2·7 cm, 2·3 cm
2. 71 mm, 75 mm, 72 mm 3 3·3 cm and 2·9 cm

4. 25 mm, 24 mm, 26 mm, 24 mm, 23 mm,
   25 mm, 22 mm
5. a) 416 mm, 424 mm, 415 mm, 412 mm
   b) $\frac{3}{5}$
6. a) maximum perimeter: 1748 mm, minimum
      perimeter: 1732 mm
   b) 3480 mm^2
7. She is wrong as the length 154 mm is in both
   tolerances and you would not know if it was
   acceptable or not.
8. a) (12 ± 2) cm b) (84 ± 4) mm
   c) (150 ± 50) mm d) (79 ± 6) g
   e) (1·5 ± 0·5) mm f) (13·5 ± 1·5) mm
   g) (42·4 ± 0·4) g h) (6·7 ± 0·4) km
9. a) (27 ± 3)° b) decrease
10. (6·05 ± 0·25) mm
11. a) (422 ± 11) litres per minute
    b) 417 litres per minute and 433 litres
       per minute
12. (−0·04 ± 0·3) °C

Exercise 1C

1. a) 49 b) 81 c) 36 d) 144 e) 324
   f) 8 g) 216 h) 3375 i) 81 j) 625
   k) 100 000 l) 64
2. a) 25 b) 64 c) 100 d) 121
   e) 400 f) −8 g) −27 h) −64
   i) 10 000 j) −243 k) 1 000 000 l) −1
3. a) The power is even b) The power is odd
4. a) 617 b) 180 c) 2124
5. a) 1849 b) 8281 c) 12 326 391
   d) 205 962 976 e) 6 436 343 f) 815 730 721
6. 1 048 576 which is 2^{20}
7. a) Power of 3 b) 18 × 19 × 20

Exercise 1D

1. a) 5 and −5 b) 2 and −2 c) 7 and −7
   d) 10 e) 9 f) 12
2. a) Yes. 10^3 = 1000 so 10 is a cube root of 1000.
   b) No. $(−10)^3$ = −1000 so −10 cannot be the
      cube root of 1000.

3 −5

4 a) $\sqrt[3]{729} = 9$ b) $\sqrt[3]{343} = 7$ c) $\sqrt[4]{6561} = 9$
   d) $\sqrt{9} = 3$ e) $\sqrt[3]{512} = 8$ f) $\sqrt[4]{20736} = 12$

5 a) 2 b) 5 c) 4
   d) 3 e) 2 f) 10

6 a) False b) True c) False
   d) True e) True f) True

7 a) 7·07 and −7·07 b) 10·8 and −10·8
   c) 6·80 and −6·80 d) 8·35
   e) 3·32 f) 1·43 g) 4·58
   h) 4·14 i) 1·26

8 a) Fourth root b) 35 × 36 × 37 × 38

Exercise 1E

1 a) 42 b) 41 c) 4 d) 1 e) 15 f) 14
   g) 24 h) 32 i) 61

2 a) 19 b) 191 c) 11 d) 90 e) 2 f) 41

3 a) 33 b) 1 c) 38 d) 97 e) 15 f) 2

4 a) 6 b) 10 c) 28 d) 14 e) 4 f) 40
   g) 27 h) 35 i) 1

5 a) $(2 + 5) \times 8 - 2 + 3 = 57$,
      $2 + 5 \times (8 - 2) + 3 = 35$,
      $2 + 5 \times 8 - (2 + 3) = 37$
   b) $6^2 - (3 - 2) \times 5 + 1 = 32$,
      $6^2 - 3 - (2 \times 5 + 1) = 22$,
      $6^2 - 3 - 2 \times (5 + 1) = 21$
   c) $\left(18 - \sqrt{100}\right) \div 2 - 1 + 9 = 12$
      $18 - \left(\sqrt{100} \div 2 - 1\right) + 9 = 23$
      $18 - \sqrt{100} \div (2 - 1 + 9) = 17$
   d) $\frac{3}{4}$ of $\left(4^3 - 2^3\right) + 15 - \sqrt{9} = 54$
      $\frac{3}{4}$ of $4^3 - \left(2^3 + 15 - \sqrt{9}\right) = 28$
      $\frac{3}{4}$ of $\left(4^3 - 2^3 + 15 - \sqrt{9}\right) = 51$

6 a) 6 b) $\frac{1}{3}$ c) −6
   d) 1 e) $\frac{1}{2}$ f) $\frac{5}{8}$

Exercise 1F

1 a) 4×10^2 b) 7×10^4 c) 9×10^3
   d) 6×10^5 e) 7.5×10^3 f) 8.9×10^2
   g) 6.2×10^4 h) 2.3×10^5 i) 5.9×10^5
   j) 1.4×10^4 k) 2.9×10^3 l) 8.5×10^6

2 The number of digits after the first digit when the number is written in full is equal to the size of the index.

3 $4.3 \times 10^4 \text{kg}$

4 a) 4.51×10^4 b) 7.23×10^5
   c) 1.93×10^6 d) 2.31×10^8
   e) 8.506×10^5 f) 1.072×10^6
   g) 3.7601×10^8 h) 5.1023×10^7

5 1.98×10^8

6 a) 1×10^6 b) 3×10^7
   c) 4.2×10^8 d) 1×10^9

7 a) $6 \times 10^5 \text{cm}$ b) $8.3 \times 10^5 \text{cm}$
   c) $9.1 \times 10^6 \text{cm}$ d) $5.07 \times 10^7 \text{cm}$

8 £1.28×10^5

Exercise 1G

1 a) 6×10^{-2} b) 7×10^{-3} c) 9×10^{-5}
   d) 2×10^{-4} e) 3.2×10^{-3} f) 4.1×10^{-2}
   g) 6.7×10^{-4} h) 8.5×10^{-6} i) 9.9×10^{-4}
   j) 5.3×10^{-3} k) 8×10^{-6} l) 2.1×10^{-8}

2 The number of zeros at the start of the number written in full is equal to the size of the index.

3 a) 1.53×10^{-2} b) 6.27×10^{-3}
   c) 5.34×10^{-5} d) 8.48×10^{-4}
   e) 1.62×10^{-4} f) 7.19×10^{-6}
   g) 3.05×10^{-3} h) 6.407×10^{-9}

4 a) The index should be −3, not 3
   b) The zero between the 3 and the 9 should be in the decimal part of the answer: 3·09.

5 a) $1.4 \times 10^{-1} \text{g}$ b) 6×10^{-2} litres
   c) $2.304 \times 10^{-4} \text{m}$ d) $8.75 \times 10^{-3} \text{g}$
   e) $4 \times 10^{-7} \text{km}$ f) 1×10^{-6} tonnes

Exercise 1H

1. a) 600 b) 7000 c) 350 000
   d) 81 000 e) 4390 f) 682 000
   g) 7 040 000 h) 95 380 000
2. 384 400 km
3. a) 0·08 b) 0·002 c) 0·41
   d) 0·000 79 e) 0·005 62 f) 0·000 008 3 5
   g) 0·000 207 h) 0·000 009 34
4. a) £106 500 000 b) £110 000 000
5. 0·0244 cm per day
6. a) 0·000 000 007 3 m b) 7.3×10^{-7} cm
   c) 7.3×10^{-6} mm
7. Minimum = 0·000 000 38 m,
   maximum = 0·000 000 45 m

Exercise 1I

1. a) 1.02×10^{8} b) 3.71×10^{9}
   c) 8.22×10^{-12} d) 2.24×10^{-10}
   e) 2.95×10^{20} f) 8.87×10^{11}
2. a) $\$2.50 \times 10^{12}$ b) $\$1.63 \times 10^{12}$ c) 1·99
3. a) 3.87×10^{-5} m b) 5.54×10^{-7} m
4. a) 4.94×10^{9}
   b) January, November, December
   c) 1.05×10^{9}
5. 2.87×10^{12} m

Exercise 1J

1. a)

Number of games	Cost (£)
8	152
1	19
5	95

b)

Time (h)	Cost (£)
5	185
1	37
12	444

c)

Episodes	Time (mins)
4	28
2	14
10	70

2. £33 3. £102 4. 45 minutes
5. a) No, they are not in direct proportion. If they were, you would expect 3000 people to buy 60 t-shirts.
   b) No, they are not in direct proportion. If they were, you would expect 5 loaves to cost £4·50.
   c) Yes they are in direct proportion. One walk costs £15.
   d) No, they are not in direct proportion. If they were, you would expect 11 people to have blue eyes on a bus of 55 passengers.
   e) No, they are not in direct proportion. If they were, you would expect 344 books on 8 bookshelves.

Exercise 1K

1. 10 hours 2. 20 boxes 3. 10 days
4. 144 ml 5. £67·50 6. Direct: 40 hours
7. Direct: 35 square metres
8. Inverse: 25 bottles 9. Direct: 540 papers
10. Inverse: 20 cm 11. Inverse: 7 boxes

Check-up

1. a) 6000 b) 7 c) 0·003 d) 0·001
2. £6 000 000
3. a) 37 000 b) 44 c) 8·4 d) 0·0019
4. 59 000
5. a) 62 800 b) 19·0 c) 1·01 d) 0·009 12
6. a) 37 mm, 31 mm b) 154 cm, 146 cm
   c) 187°, 183° d) 303 mm, 265 mm
7. a) Small: 120 g, 133 g, 136 g
      Medium: 160 g, 190 g, 193 g, 200 g, 199 g, 176 g
      Large: 204 g, 234 g
   b) Because 140 g is in both the small and medium category.
8. (723 ± 11) mm
9. a) 49 b) 16 c) −125 d) 17
10. a) 4 b) 10 c) 3 d) 13
11. a) 14·1 b) 4·24 c) 5·52 d) 1·52
12. a) 133 b) 11 c) 118 d) 2

13 a) 5.7×10^4 b) 6.49×10^6

     c) 7.08×10^7 d) 9.143×10^8

14 6.048×10^5

15 a) 4×10^{-2} b) 3.7×10^{-3}

     c) 9.12×10^{-6} d) 5.03×10^{-5}

16 $7 \times 10^{-6}\,\text{m}$

17 a) 9000   b) 860 000

     c) 0.0003   d) 0.000 002 05

18 0.000 56 m

19 a) The decimal is not between 1 and 10.

     b) 5.21×10^{-7}

20 a) 7.04×10^{11} b) 1.48×10^{-24}

     c) 7.67×10^8 d) 1.24×10^{-4}

21 0.225 cm   22 1.53×10^8

23 a) 3   b) Mercury, Mars, Venus, Earth

     c) $1.701 \times 10^8\,\text{km}$ d) 18 times

     e) 6.7×10^7 miles

24 £2.88   25 19

26 24 minutes   27 £180

2 Fractions, decimals and percentages

Exercise 2A

1 a) 4.8   b) 2.7   c) 0.28   d) 0.25

   e) 0.12   f) 0.115   g) 0.416   h) 0.287

   i) 0.0874   j) 0.1581   k) 0.0024   l) 0.032

2 a) 7.8   b) 11.6   c) 0.74   d) 5.22

   e) 28.35   f) 9.476   g) 15.708   h) 6.5533

3 a) 8×10^{-6} b) 4.5×10^{-7} c) 1.8×10^{-8}

4 a) 2.3   b) 9.988   c) 6.56   d) 2.012

   e) 6.15   f) 1.29

5 a) 40   b) 160   c) 24   d) 16

   e) 0.2   f) 4.7   g) 0.3   h) 0.15

6 a) 150   b) 400   c) 270   d) 52.2

   e) 2000   f) 90   g) 50.2   h) 5.175

7 85 lengths

8 55 full hems

9 a) $1.2 \div 0.4 = 2.7 \div 0.9, 0.96 \div 0.03 = 6.4 \div 0.2,$
     $4.9 \div 0.7 = 0.42 \div 0.06$

   b) $0.6 \times 0.3 = 0.072 \div 0.4, 0.15 \times 0.8 = 0.06 \div 0.5,$
     $0.02 \times 0.05 = 0.0007 \div 0.7$

10 a) 2.5   b) 0.53   c) 148.76

Exercise 2B

1 76%   2 85%   3 95%

4 a)

Food	Calories (kCal)	Carbohydrate (g)	Total Fat (g)	Protein (g)
Banana	6.05%	11.92%	0.71%	2.8%
Egg	3.55%	0.15%	7.14%	12%
Pasta (100 g)	6.5%	9.62%	1.43%	10%
Bagel	18%	30.77%	2.86%	28%
Cashew nuts (100 g)	27.65 %	11.54%	62.86%	36%

   b) Pupils' own answer

Exercise 2C

1 a) £240   b) 756 km   c) 1336 miles

   d) £386.75   e) 67.27 m   f) £14.35

2 £1.62   3 £285 000

4 a) £360   b) 75 mg   c) 693 km

   d) 5.46 kg   e) 54 km   f) £8.32

5 £407.54   6 £721.38

7 a) i) £143 630   ii) £155 120.40

     iii) £133 403.54

   b) The increase is $1 \times 1.06 \times 1.08 = 1.1448$.
     Decreasing by 14% gives $1.1448 \times 0.86 =$
     0.98..., which is less than the initial value of 1.

Exercise 2D

1 a) £224.97   b) £757.70   c) £1738.91

   d) £2315.25   e) £3721.78   f) £3747.99

   g) £5826.83

2 a) £24.97   b) £57.70   c) £238.91

   d) £315.25   e) £773.78   f) £297.99

   g) £826.83

3 £18 701.30   4 £7088.92

5 £190 816.80   6 6711 bacteria   7 346 insects

8 a) 8.3 billion   b) 8.6 billion

9 a) i) £31 679.43   ii) £35 737.94

   b) 18 years

10 5 years

Exercise 2E

1. a) 20% b) 50% c) 40% d) 25%
   e) 11·1% f) 56·7% g) 37·8% h) 5·38%
2. 20% 　3 110% 　4 11·3%
5. a) 10% b) 55% c) 40% d) 25%
   e) 27·2% f) 17·2% g) 65% h) 21·5%
6. 23·1% 　7 33·3% or $33\frac{1}{3}$% 8 25%
9. 15% 　10 9% 　11 30%
12. 20·6% 　13 20%
14. a) 4·43% b) 93·0%
   c) i) 192% ii) 42·9% iii) 49·4%
15. No, the newsreader's claim is incorrect as it is an increase of 100%.
16. a) In the second year, it is 6% of 95%, not the original amount.
    b) 10·7%

Exercise 2F

1. a) 120% b) 145% c) 119% d) 90%
   e) 70% f) 90·5%
2. £40 　3 750 ml 　4 £24 000
5. 2500 people 6 £90 　7 £230
8. £150 000 　9 £16 　10 25 workers
11. £395 　12 7800 people 　13 £35
14. £400 　15 700 tickets
16. £125 　17 £24 000
18. a) Required: A, D, F; Not required: B, C, E.
    b) A: £95 B: £49 C: £15·33 D: £11·62 E: 625 ml F: 480 ml

Exercise 2G

1. a) $6\frac{2}{3}$ b) $5\frac{4}{5}$ c) $11\frac{7}{9}$ d) $15\frac{4}{7}$
   e) $4\frac{3}{5}$ f) $7\frac{5}{7}$ g) $5\frac{4}{9}$ h) $1\frac{1}{3}$
2. a) $7\frac{5}{8}$ b) $11\frac{7}{9}$ c) $3\frac{5}{6}$ d) $8\frac{11}{15}$
   e) $2\frac{3}{8}$ f) $4\frac{3}{10}$ g) $6\frac{1}{12}$ h) $1\frac{29}{42}$

   i) $14\frac{2}{5}$ j) $8\frac{2}{3}$ k) $14\frac{11}{20}$ l) $11\frac{1}{18}$
3. $3\frac{3}{4}$ hours 　4 $5\frac{7}{18}$ laps
5. a) $7\frac{1}{3}$ b) $11\frac{2}{5}$ c) $9\frac{4}{9}$ d) $6\frac{1}{8}$
   e) $9\frac{1}{6}$ f) $15\frac{3}{20}$ g) $7\frac{11}{21}$ h) $5\frac{17}{24}$
6. a) $6\frac{1}{2}$ b) $3\frac{4}{5}$ c) $4\frac{1}{3}$ d) $5\frac{1}{4}$
   e) $1\frac{2}{3}$ f) $3\frac{3}{5}$ g) $3\frac{7}{8}$ h) $5\frac{15}{28}$
7. a) $13\frac{1}{2}$ b) $12\frac{23}{36}$ c) $1\frac{5}{6}$ d) $8\frac{37}{40}$
8. a) $21\frac{23}{63}$ m 　b) $5\frac{34}{63}$ m

Exercise 2H

1. a) $9\frac{3}{10}$ b) $5\frac{1}{4}$ c) $7\frac{3}{5}$ d) $5\frac{3}{4}$
   e) 8 f) $21\frac{2}{3}$ g) $6\frac{5}{12}$ h) 6
2. a) $16\frac{1}{2}$ b) $25\frac{1}{3}$ c) 98 d) $49\frac{1}{2}$
3. $8\frac{4}{5}$ cm² 　4 100 cm
5. a) 17 b) $5\frac{7}{11}$ c) $2\frac{23}{32}$

Exercise 2I

1. a) $\frac{9}{26}$ b) $\frac{28}{45}$ c) $\frac{14}{33}$ d) $1\frac{5}{9}$
   e) $2\frac{1}{10}$ f) $\frac{15}{19}$ g) $2\frac{1}{2}$ h) $2\frac{1}{3}$
2. a) $1\frac{7}{13}$ b) $1\frac{9}{55}$ c) $\frac{18}{35}$ d) $1\frac{2}{19}$
3. $1\frac{3}{5}$ m 　4 $1\frac{1}{9}$ litres 　5 25 minutes
6. a) $\frac{1}{2}$ b) $2\frac{8}{11}$ c) 63

Check-up

1. a) 0·8   b) 0·56   c) 0·288   d) 0·5022
   e) 50   f) 9·4   g) 160   h) 1·8
2. a) 10·28   b) 13·982   3 30%
4. a) 67%   b) 2%   c) 16%
5. a) 98·44 g   b) 643·2 kg   c) 226·25 mm
6. a) 6·72 kg   b) 315·9 cm   c) 35·96 miles
7. 1740 tickets   8 81p
9. a) £579·64   b) £1297·92   c) £3183·62
   d) £875·16   e) £932·69   f) £4985·88
   g) £668·39
10. a) £79·64   b) £97·92   c) £183·62
    d) £155·16   e) £32·69   f) £785·88
    g) £18·39
11. £1930·46   12 56 ml
13. 50%   14 26·9%   15 15%
16. 13·1%   17 80 litres   18 £220
19. 24 umbrellas   20 £8
21. £300   22 £25
23. a) $6\frac{11}{12}$ b) $10\frac{5}{9}$ c) $8\frac{1}{4}$ d) $6\frac{4}{15}$
    e) $3\frac{11}{24}$ f) $6\frac{1}{2}$ g) $1\frac{13}{14}$ h) $1\frac{39}{40}$
24. a) $5\frac{1}{10}$ b) $14\frac{2}{3}$ c) $29\frac{3}{4}$ d) $23\frac{2}{3}$
    e) $1\frac{11}{16}$ f) $1\frac{1}{5}$ g) $3\frac{1}{3}$ h) $5\frac{3}{5}$
25. a) $26\frac{13}{30}$ kg   b) $\frac{23}{30}$ kg
26. a) 11   b) $1\frac{3}{5}$ c) $54\frac{2}{5}$ d) $16\frac{1}{2}$
27. a) 1350 square inches   b) 81 inches

3 Time

Exercise 3A

1. a) 6 hours 5 minutes   b) 8 hours 22 minutes
   c) 5 hours 45 minutes   d) 10 hours 41 minutes
   e) 8 hours 44 minutes   f) 9 hours 54 minutes

2. a) 103 hours   b) 195 hours   c) 336 hours
3. a) 2 days, 2 hours   b) 6 days, 13 hours
   c) 1 week, 1 day, 5 hours
   d) 2 weeks, 8 hours
4. 8 days, 3 hours, 18 minutes
5. a) 7 days, 14 hours, 35 minutes
   b) 3 days, 1 hour, 55 minutes
   c) 2 days, 18 hours, 40 minutes
   d) 4 days, 7 hours, 45 minutes
6. a) 2 pm   b) 7:15 pm
   c) 1:30 am Saturday   d) 04:05 Tuesday
7. a) 04:00   b) 3:15 pm
   c) 10:45 pm Monday   d) 18:40 Sunday
8. a) 08:18   b) 10:31
9. a) 1 hour 49 minutes   b) 2 hours 15 minutes
   c) 5 hours 35 minutes   d) 4 hours 5 minutes
   e) 3 hours 30 minutes   f) 1 day, 7 hours
   g) 18 hours 30 minutes   h) 23 hours 20 minutes

Exercise 3B

1. a) 4 hours 30 minutes   b) 3 hours 15 minutes
   c) 6 hours 45 minutes   d) 12 hours 15 minutes
   e) 30 minutes   f) 6 hours 6 minutes
   g) 10 hours 48 minutes   h) 7 hours 12 minutes
   i) 7 hours 24 minutes   j) 9 hours 36 minutes
   k) 3 hours 3 minutes   l) 12 hours 51 minutes
2. a) 2·5 hours   b) 3·25 hours   c) 6·75 hours
   d) 1·2 hours   e) 9·4 hours   f) 12·6 hours
3. a) 0·5 hours   b) 0·7 hours   c) 0·3 hours
   d) 0·35 hours   e) 1·1 hours   f) 2·2 hours
   g) 2·8 hours   h) 3·9 hours   i) 5·55 hours
   j) 6·8 hours   k) 2·6 hours   l) 7·1 hours

Exercise 3C

1. a) 75 miles   b) 52 km   c) 95 miles
   d) 40 km   e) 139·2 miles   f) 77 miles
2. 100 km   3 12·5 miles   4 28 km
5. 11·75 m   6 26·25 miles   7 48 miles

Exercise 3D

1   a) 50 mph   b) 40 km/h   c) 256 mph
    d) 20 km/h   e) 30 mph   f) 15 km/h
2   50 km/h
3   a) 9·5 mph   b) 2·5 mph faster
4   150 mph   5   20 km/h
6   a) 8044 seconds, 42195 metres
    b) 5·25 m/s   c) 18·9 km/h

Exercise 3E

1   a) 1 hour 30 minutes   b) 1 hour 15 minutes
    c) 3 hours 45 minutes d) 1 hour 12 minutes
    e) 3 hours 6 minutes   f) 4 hours 48 minutes
2   10 minutes   3   1 hour 12 minutes
4   3 hours 36 minutes   5   16 minutes
6   4 days, 21 hours

Check-up

1   a) 15 hours 25 minutes
    b) 7 hours 13 minutes
    c) 9 hours 26 minutes
    d) 11 hours 34 minutes
    e) 10 hours 36 minutes
    f) 8 hours 8 minutes
2   a) 48 minutes   b) 1 hour 2 minutes
    c) 3 hours 22 minutes
    d) No – he will arrive at 17:30
3   a) 3 days, 15 hours, 35 minutes b) £10 100
4   5 hours 45 minutes
5   a) 07:40   b) 4 hours 55 minutes
6   a) 7:15 am   b) 3 hours 5 minutes
7   a) i) 04:20   ii) 14 hours 35 minutes
    b) i) 06:55   ii) 9 hours 25 minutes
    c) i) 00:20 4 April   ii) 12 hours 45 minutes
    d) i) 09:30   ii) 14 hours 55 minutes
8   a) 02:50   b) 01:40 on 14 December
9   a) 3 days, 8 hours, 45 minutes
    b) Munich: 8 hours, Bucharest: 4 hours,
       Vienna: 24 hours
10 a) 3 hours 12 minutes b) 6 hours 24 minutes
    c) 12 hours 51 minutes d) 39 minutes

11 a) 2·75 hours   b) 5·25 hours
    c) 3·1 hours   d) 9·45 hours
12 a) 45 km   b) 24 miles   c) 90 km
13 a) 80 km/h   b) 34 mph   c) 30 km/h
14 a) 1 hour 30 minutes b) 2 hours 24 minutes
    c) 48 minutes   d) 4 hours 42 minutes
15 1 hour 48 minutes
16 a) 1 hour 24 minutes b) 8·75 km/h
17 18·8 km/h

4 Measurement

Exercise 4A

1   a) esf = 2, $a = 60$ cm   b) esf = 4, $b = 8$ m
    c) esf = 3, $c = 30$ mm   d) rsf = $\frac{1}{2}$, $d = 9$ cm
    e) rsf = $\frac{1}{3}$, $e = 15$ m   f) esf = $\frac{3}{2}$, $f = 9$ cm
    g) rsf = $\frac{1}{6}$, $g = 5$ mm   h) esf = $\frac{4}{3}$, $h = 16$ cm
    i) esf = $\frac{5}{4}$, $i = 15$ cm
2   a) $a = 40$ cm   b) $b = 45$ cm
    c) $c = 10$ cm, $d = 7·5$ cm
    d) No, the print could not be scaled in
       proportion

3
	small	medium	large
Waist (in)	28	35	43·75
Length (in)	16	20	25

4   a) The rsf has been rounded too soon,
       rsf = $\frac{1}{3} = 0·\dot{3}, 0·333...$
    b) The esf has been calculated instead of the rsf
    c) $x = 2$ cm

5   a)
	Full size	Ride on	Toy
Length	4·8 m	80 cm	15 cm
Height	1·92 m	32 cm	6 cm

    b) esf = $\frac{16}{3}$

Exercise 4B

1. a) $a = 8\,cm$ b) $b = 24\,mm$ c) $c = 12\,cm$
   d) $d = 105\,cm$ e) $e = 54\,cm$ f) $f = 45\,m$
   g) $g = 28\,cm$ h) $h = 29\,mm$ i) $i = 6·6\,m$
2. $73·5\,cm$ 3 $46\,m$
4. a) $a = 14\,mm$ b) $b = 6\,cm$ c) $c = 10\,m$
   d) $d = 120\,mm$ e) $e = 99\,m$ f) $f = 80\,mm$

Exercise 4C

1. a) $48\,cm^2$ b) $1800\,mm^2$ c) $248\,cm^2$
   d) $125\,cm^2$ e) $256\,cm^2$ f) $105\,cm^2$
2. a) £36   b) £81
3. a) $20\,cm$ b) $12\,m$ c) $1·5\,cm$

Exercise 4D

1. a) $810\,ml$ b) $320\,ml$ c) $312·5\ m^3$
   d) $160\,cm^3$ e) $405\,cm^3$ f) $125\,ml$
2. $20\,cm$
3. No, large bottle would cost £10·80
4. £68·60, £145·80

Exercise 4E

1. a) $270\,cm^2$ b) $1860\,mm^2$
   c) $10\,m^2$ d) $9\,m^2$
2. a) $126\,cm^2$ b) $500\,cm^2$
   c) $15750\,mm^2$ d) $6·6\,m^2$
3. a) $80\,cm^2$ b) $18\,m^2$
   c) $3150\,mm^2$ d) $124\,cm^2$
4. a) $2550\,cm^2$ b) $27\,m^2$
   c) $3300\,cm^2$ d) $94\,m^2$
5. Parallelogram $66\,m^2$, kite $16·5\,m^2$
6. a) $150\,cm^2$ b) $1050\,cm^2$ c) $600\,cm^2$

Exercise 4F

1. a) $12·57\,cm$ b) $47·12\,cm$
   c) $9·42\,m$ d) $62·83\,mm$
   e) $188·50\,cm$ f) $59·69\,in$
   g) $106·81\,mm$ h) $7·85\,m$

2. Snare $43·98\,in$, kick $69·12\,in$, tom tom $37·70\,in$
3. $39\,898·23\,km$
4. $100·53\,cm$, no not enough as $2 \times 100·53$ is greater than $2\,m$
5. a) $43·98\,cm$ b) $28·27\,m$
   c) $113·10\,cm$ d) $87·96$ inches
6. a) $462·74\,cm$ b) $67·85\,cm$
   c) $265·66\,cm$ d) $436·31\,mm$
7. a) $9·99\,cm$ b) $7·96\,cm$
   c) $19·10\,m$ d) $0·64\,km$
8. a) $12·56\,cm$ b) $47·1\,cm$ c) $9·42\,m$
   d) $62·8\,mm$, reasonable approximations but differences appearing at 4th significant figure.

Exercise 4G

1. a) $12·57\,cm^2$ b) $78·54\,cm^2$
   c) $314·16\,m^2$ d) $1256·64\,mm^2$
   e) $6361·73\,cm^2$ f) $380·13\,in^2$
   g) $907·92\,mm^2$ h) $1017·88\,cm^2$
2. a) $452·39\,cm^2$ b) $38·48\,m^2$
   c) $1885·74\,m^2$ d) $11\,309·73\,mm^2$
3. No, small $50·27\,in^2$, large $113·10\,in^2$, large is more than double ($2·25$ larger), $esf = \dfrac{3}{2}, asf = \dfrac{9}{4}$
4. a) $50·27\,ft^2$, $63·62\,ft^2$, $72\,ft^2$
   b) £110 trampoline
5. a) $2390\,cm^2$ b) $4·62\,m^2$
   c) $32\,200\,cm^2$ d) $19\,900\,cm^2$, $1·99\,m^2$
6. a) $8830\,m^2$ b) $388\,m$

Exercise 4H

1. a) $152\,cm^2$ b) $54\,cm^2$ c) $256\,cm^2$
2. a) $660\,cm^2$ b) $129·6\,m^2$ c) $1416\,m^2$
3. a) Pupils' own answers e.g. $4 \times 6 \times 3$, $2 \times 4 \times 9$
   b) Pupils' own answers e.g. $4 \times 4 \times 4·5$, $SA = 104\,cm^2$; $1 \times 1 \times 72$, $SA = 290\,cm^2$

Exercise 4I

1 a) 603 cm²   b) 86·4 m²

    c) 99 000 cm²   d) 3020 mm²

2 2850 cm²   3 154 m²

4 Cuboid $A = 2lb + 2bh + 2lh$

   Cylinder $A = 2 \times \pi r^2 + 2\pi r \times h$

Exercise 4J

1 a) 2480 cm³   b) 5250 mm³   c) 10·08 m³

    d) 1080 cm³   e) 452·7 cm³   f) 43 500 mm³

2 a) 3150 cm³   b) 1495 cm³   c) 6630 cm³

    d) 153 m³   e) 110 m³   f) 36 000 cm³

3 a) Hexagon-based prism 18 m³,
      L shaped 21 m³,   Square 24 m³

    b) L shaped 105 mins, Square 120 mins

4 a) $a = 10$ cm   b) $h = 4$ m   c) 18 cm

5 $w = 4$ m

Exercise 4K

1 a) 352 cm³   b) 88·0 m³

    c) 2 120 000 cm³   d) 13 600 mm³

2 a) 17 100 cm³   b) 11·7 m³

    c) 2 590 000 mm³   d) 2·29 m³, 2 290 000 cm³

3 Cylinder holds more, cuboid 216 000 cm³,
   cylinder 254 000 cm³

4 a) 39 800 cm³   b) 39·8 litres

5 a) 565 cm³   b) 2·01 m³

    c) 1·77 m³, 1 770 000 cm³

6 A 16π   B 18π   C 10π   D 8π

Exercise 4L

1 a) 150 cm³   b) 1050 cm³   c) 2250 mm³

    d) 144 cm³   e) 1·35 m³   f) 0·54 m³

    g) 2720 cm³   h) 0·63 m³

2 Prism: bread, paint pot, cheese wedge,
   chocolate box   pyramid: tea bag, cone

3 3 sweets

4 a) 494 cm³   b) 12 600 cm³   c) 2484 cm³

5 2700 cm³

6 a) 18 cm³ b) 48 cm c) 9216 cm³ d) 9198 cm³

Exercise 4M

1 a) 101 cm³   b) 236 cm³   c) 1880 mm³

    d) 9·68 m³   e) 3020 cm³   f) 0·679 m³

2 a) 21·13 m³   b) 2·2 m   c) 26·7 m³

3 a) Attempt 1, volume of the cone is wrong
      as the radius is 3·5 cm not 7 cm, this is the
      diameter.

    b) Attempt 2 is wrong as they have rounded
      to 1 s.f. too soon undercounting the
      volume of the cones.

    c) 115·45, 5000 ÷ 115·45 = 43·3 …
      The 5-litre bottle can fill 43 whole cups.

4 No, not quite enough. He needs 1·036… litres.

5 7·66 cm³   6 0·688 m³

7 6930 cm³   8 2610 cm³

9 a) $x = 15$ cm   b) 1740 cm³

Exercise 4N

1 a) 4190 cm³   b) 33·5 m³   c) 660 m³

    d) 1·44 m³   e) 230 000 cm³   f) 382 cm³

2 a) 2 cm, 3·35 cm, 11 cm, 11·9 cm

    b) squash 33·5 cm³   tennis 157 cm³
      football 5580 cm³   basketball 7060 cm³

3 1800 cm³   4 79 600 cm³

5 a) 12 cm   b) 1900 cm³

6 a) 2 mm   b) 9 mm   c) 146·6 mm³

7 No, he is wrong, cylinder has a greater volume of
   wax, sphere $\dfrac{4000}{3}\pi$ cm³ < cylinder 1500π cm³

8 a) 36π cm³   b) 162π cm³   c) $\dfrac{2}{3}$

9 a) $V_{\text{hemisphere}} = \dfrac{2}{3}\pi r^3$, $V_{\text{cylinder}} = \pi r^3$, Graham

   should choose the hemisphere as smaller
   volume makes more bombs

    b) 200

Exercise 4O

1 a) 5 cm   b) 10 cm   c) 4 cm

2 a) 10·61 cm   b) 1·99 cm   c) 1·07 m

3 10 cm   4 $a = 30$ cm $b = 21$ cm

5 $a = 2 \cdot 19$ cm $b = 3 \cdot 06$ cm

6 110 cm   7 23 cm   8 12 mm   9 2·5 cm

Check-up

1 a) $a = 9$ mm   b) $b = 6$ cm   c) $c = 16$ cm

2 a) $\dfrac{5}{2}$ b) 6·25 m²   c) 5000 litres

3

Circle	
Circumference	$C = \pi D$
Area	$A = \pi r^2$

Area of quadrilaterals	
Parallelogram	$A = bh$
Rhombus/kite	$A = \frac{1}{2}d_1 d_2$
Trapezium	$A = \frac{1}{2}(a + b)h$

Volume	
Prism	$V = Ah$
Cylinder	$V = \pi r^2 h$
Pyramid	$V = \frac{1}{3}Ah$
Cone	$V = \frac{1}{3}\pi r^2 h$
Sphere	$V = \frac{4}{3}\pi r^3$

4 a) 65·97 cm   b) 5·65 m   c) 62·83 cm

5 a) 346·36 cm² b) 2·54 m²   c) 314·16 cm²

6 a) 1·8 m², no   b) 2·16 m², yes c) 2·57 m², yes

7 a) 262 cm²   b) 346 cm²

8 a) caramel   b) truffle

   c) toffee   d) praline

9 3490 mm³

10 a) 60 cm   b) 170 000 cm³

11 Yes, it will overflow. Volume available in the tank = 3000 cm³. Volume displaced by sphere = 4190 cm³.

5 Algebra

Exercise 5A

1 a) $3x + 21$ b) $8x - 32$

   c) $14x + 35$ d) $24x - 88$

2 a) $2x + 6$ b) $4x + 20$ c) $7x + 42$

   d) $18x + 45$ e) $77x + 44$ f) $6x - 30$

g) $8x - 104$ h) $36x - 24$ i) $28x - 49$

j) $76x - 57$ k) $24 + 8x$ l) $28 - 70x$

m) $35 + 20y$ n) $36 - 72x$ o) $175 + 125x$

p) $6x + 9y$ q) $15x + 20y$ r) $7x - 14y$

s) $81a + 72b$ t) $48a - 84b$

3 a) $-3x - 15$ b) $-7x - 14$ c) $-5x - 5$

   d) $-18x - 117$ e) $-50x - 60$ f) $-4x + 8$

   g) $-8x + 72$ h) $-12x + 45$ i) $-24x + 128$

   j) $-39x + 78$ k) $-4x - 6y$ l) $-12x - 18y$

   m) $-2x + 4y$ n) $-15a - 35b$ o) $-44a + 36b$

   p) $-65 - 13x$ q) $-36 + 27x$ r) $-63 - 168y$

   s) $-85 + 102x$ t) $-152 - 133x$

4 a) $x^2 + 5x$ b) $2x^2 + 3x$ c) $4x^2 + 7x$

   d) $8x - 3x^2$ e) $12x - 7x^2$ f) $6x^2 + 16x$

   g) $3x^2 - 6x$ h) $12x^2 + 20x$ i) $27x^2 + 9x$

   j) $52x^2 - 32x$ k) $5x^2 + 10xy$ l) $6x^2 - xy$

   m) $30x^2 + 75xy$ n) $32a^2 - 12ab$ o) $12ab + 18a^2$

   p) $34p^2 - 85pq$ q) $96st + 40s^2$ r) $63y^2 - 147yz$

   s) $6x^3 + 28xy$ t) $35x^2 + 10x^2y$

5 a) $2x^2 + xy + 3x$ b) $5x^2 + 4xy + 2x$

   c) $4x^2 + 9xy - 11x$ d) $7x - 5x^2 - 2xy$

   e) $10x^2 + 15xy - 25x$ f) $63x^2 + 9xy - 99x$

   g) $96x - 36x^2 + 144xy$ h) $102x + 17xy - 68x^2$

   i) $70x^2 + 60xy + 90x$ j) $39x^2 - 21xy - 24x$

   k) $3x^3 + 2x^2y + 6x^2$ l) $12x^4 - 9x^3y + 24x^2y$

6 a) $2x + 19$ b) $5x + 24$

   c) $6x + 46$ d) $4x + 17$

   e) $4x^2 + 12x - 15$ f) $6x^2 + 50x$

   g) $6x + 15$ h) $32x - 14$

   i) $35x + 33$ j) $25x - 3$

   k) $20x^2 - 18x$ l) $30x + 12$

   m) $-3x + 3$ n) $-4x + 15$

   o) $-14x - 6$ p) $-36x + 33$

7 a) $6x + 21$ b) $11x + 20$

   c) $21x - 6$ d) $23x - 88$

   e) $-4x + 1$ f) $4x^2 + 28x$

   g) $-12x - 3$ h) $-8x + 15$

   i) $-12x + 59$ j) $-21x + 21$

k) $-21x + 26$

l) $26x^2 - 27x$

m) $18x^2 + 3x - 5$

n) $22x^2 - 90x + 5$

o) $-12x + 33$

p) $-72x - 27$

8 a) $6x + 30$

b) $16x + 93$

c) $9x + 9$

d) $12x + 48$

e) $6x^2 - 14x + 32$

f) $18x^2 - 4x + 3$

g) $40x - 26$

h) $x + 11$

i) $4x + 16$

j) $-3x^2 - 35x - 18$

k) $58x - 10$

l) $78x + 66$

9 A and F, B and D, C and G, E and H

10 a) $3x^2 + 10xy + 5y^2$

b) $12x^2 + 49xy + 12y^2$

c) $2x^2 - 22xy + 2y^2$

d) $4x^2 - 32xy - 6y^2$

e) $42x^2 + 22xy - 25y^2$

f) $18x^2 + 51xy + 8y^2$

g) $3x^3 + x^2y + 2y^3 - 7xy$

h) $2x^4 + 4x^2y - y^3 - 3xy$

i) $2xy^3 - y^5 - 2x^2y + 3x^3$

Exercise 5B

1 a) $x = 7$ b) $x = 9$ c) $x = 4$

d) $x = 6$ e) $x = 9$ f) $x = 11$

g) $x = 5$ h) $x = 6$ i) $x = 15$

j) $x = 12$ k) $x = 7$ l) $x = 5$

m) $x = -4$ n) $x = -3$ o) $x = -2$

p) $x = -7$

2 a) $x = 2$ b) $x = 6$ c) $x = 7$

d) $x = 8$ e) $x = 2$ f) $x = 6$

g) $x = -4$ h) $x = -4$ i) $x = 3$

j) $x = -5$ k) $x = -5$ l) $x = -9$

m) $x = 1$ n) $x = -4$ o) $x = 8$

p) $x = -12$

3 a) $x = \frac{1}{2}$ b) $x = \frac{2}{3}$ c) $x = \frac{1}{2}$

d) $x = \frac{2}{5}$ e) $x = \frac{1}{4}$ f) $x = \frac{2}{3}$

g) $x = -\frac{4}{7}$ h) $x = -\frac{1}{2}$ i) $x = \frac{3}{4}$

j) $x = \frac{2}{3}$ k) $x = \frac{1}{2}$ l) $x = \frac{1}{2}$

m) $x = -\frac{2}{5}$ n) $x = \frac{1}{3}$ o) $x = \frac{5}{7}$

p) $x = \frac{6}{13}$

4 a) Should subtract 7 from each side. $x = 2$

b) A negative has been omitted in front of 11. $x = -2$

c) A negative has been omitted in front of $7x$. $x = 6$

Exercise 5C

1 a) $x = 1$ b) $x = 6$ c) $x = 10$

d) $x = 21$ e) $x = 9$ f) $x = 2$

g) $x = -5$ h) $x = -14$

2 a) $x = 6$ b) $x = 2$ c) $x = 7$

d) $x = 14$ e) $x = 3$ f) $x = -1$

g) $x = 0$ h) $x = \frac{1}{2}$ i) $x = -8$

3 a) $x = 6$ b) $x = 2$ c) $x = 10$

d) $x = 1$ e) $x = 11$ f) $x = 7$

g) $x = -3$ h) $x = -2$ i) $x = -4$

4 a) $x = 5$ b) $x = 9$ c) $x = -3$

d) $x = -1$ e) $x = 0$ f) $x = 5$

g) $x = 12$ h) $x = 10$ i) $x = -4$

Exercise 5D

1 a) $x = 15$ b) $x = 18$ c) $x = 24$

d) $x = 32$ e) $x = 54$ f) $x = 10$

g) $x = 16$ h) $x = 12$

2 a) $x = 10$ b) $x = 9$ c) $x = 8$

d) $x = 3$ e) $x = 1$ f) $x = 4$

g) $x = 13$ h) $x = 13$

3 a) $x = 7$ b) $x = 4$ c) $x = 2$

d) $x = 5$ e) $x = 12$ f) $x = -1$

g) $x = -6$ h) $x = 9$

4 a) $x = 3$ b) $x = -4$ c) $x = 7$

Exercise 5E

1 a) 35   b) -23 c) -6 d) 51

e) -28 f) 111   g) 4   h) 15

2 a) -4 b) 5   c) -56 d) 60

e) 57   f) -8 g) 15   h) 50

3 a) -31 b) 135   c) 49   d) 16

e) 16   f) 529   g) -27 h) 125

4 a) 4   b) 5   c) 3   d) 6
   e) 5   f) 3   g) 38   h) 1

5 a) −5   b) 5   c) 2   d) 7
   e) 2   f) −2   g) 9   h) 1

6 a)
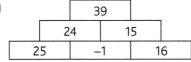

   b)
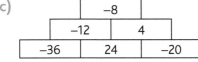

   c)

   d)

		18		
	−15		23	
−17		2		21

7 a) circle   b) square   c) algebra

8

−2	−1	0	1	2
1	2	−2	−1	0
−1	0	1	2	−2
2	−2	−1	0	1
0	1	2	−2	−1

Exercise 5F

1 a) 5 < 8   b) 13 > 9   c) 1 < 12
   d) 14 > 13   e) −6 < 4   f) 3 > −3
   g) −10 < −7   h) −25 > −30   i) 0 > −1
   j) −17 < −15   k) 0·25 < 0·5   l) −1·8 > −1·81

2 a) 5 < 10   b) 6 > 1   c) −4 < 0
   d) $\frac{1}{2} > \frac{1}{4}$ e) $x \leq 2$ f) $x > -9$
   g) $8 \geq x$ h) $46 < x$ i) $2x \leq x + 1$
   j) $10x > -\frac{2}{5}$

3 a) 3 < 5   b) 4 > 1   c) 12 < 20
      5 > 3   1 < 4   20 > 12
   d) −5 < 3   e) −1 > −7   f) −3 < 1
      3 > −5   −7 < −1   1 > −3

   g) −10 > −13 h) $x < 10$ i) $4 \geq x$
      −13 < −10 $10 > x$ $x \leq 4$

   j) $8 \leq y$
      $y \geq 8$

4 a) 4, 5, 7, 13   b) −23, −19, −5, 0
   c) 13   d) −5, 0, 4, 5   e) −23, −19

Exercise 5G

1 a) $x < 2$ b) $x > 21$ c) $x \leq 10$
   d) $x > 16$ e) $x \leq -1$ f) $x \geq -3$
   g) $x < 2$ h) $x < -6$ i) $x \geq -7$
   j) $x < -9$ k) $x < 4$ l) $x > 8$
   m) $x \leq 3$ n) $x < 7$ o) $x \geq 5$
   p) $x < 13$ q) $x < -4$ r) $x \leq 3$
   s) $x < -9$ t) $x \geq 4$

2 a) $x < 7$ b) $x > 9$ c) $x \leq 8$
   d) $x < 2$ e) $x < -2$ f) $x < 5$
   g) $x \geq 9$ h) $x \geq -1$

3 a) $x \geq 18$ b) $x > 7$ c) $x < -3$
   d) $x \geq 1$ e) $x < 3$ f) $x \geq 4$
   g) $x < -3$ h) $x < -5$

4 a) $x \geq 7$ b) $x > -2$ c) $x \leq -36$
   d) $x \geq 9$ e) $x > 9$ f) $x \leq -5$

5 a) $x \geq -55$ b) $x > 4$ c) $x < -6$
   d) $x > 2$ e) $x > 4$ f) $x \geq -8$

Exercise 5H

1 a) $x - 7$ b) $2x - 7 = 35$
   c) Sean: 21, Rowan: 14

2 a) $j + 3$ b) $2j + 6$
   c) Jenna: 7, Hughie: 10, Pauline: 20

3 a) $c + 5$ b) Child: £7, adult: £12

4 46 tickets

5 6 plants

6 a) $C = 0·15m + 3$ b) £6   c) 36 miles

7 a) $W = 40p + 20$ b) 300 kg   c) 12 pairs

8 a) $C = 30s + 50$ b) £3·20   c) 950 g

9 a) $E = 12h + P + 30$ b) £134
   c) 9 hours   d) £82·50
10 a) $t = 8j + 11s + 10$ b) 7 hours 11 minutes
   c) 14   d) 30

Exercise 5I

1 a) 26, 32, 38, $6n + 2$
   b) 45, 56, 67, $11n + 1$
   c) 15, 19, 23, $4n - 1$
   d) 27, 35, 43, $8n - 5$
   e) 3, 1, −1, $-2n + 11$
   f) −7, −11, −15, $-4n + 9$
   g) 47, 55, 63, $8n + 15$
   h) −18, −22, −26, $-4n - 2$
   i) 19, 26, 33, $7n - 9$
   j) 6, 6·5, 7, $0·5n + 4$
2 a) i) 62   ii) 242
   b) i) 111   ii) 441
   c) i) 39   ii) 159
   d) i) 75   ii) 315
   e) i) −9   ii) −69
   f) i) −31   ii) −151
   g) i) 95   ii) 335
   h) i) −42   ii) −162
   i) i) 61   ii) 271
   j) i) 9   ii) 24
3 a) Pupils' diagrams   b) 14, 17, 20, … 32
   c) $m = 3n + 2$ d) 92   e) 42
4 a) Pupils' diagrams   b) 9, 11, 13, … 21
   c) $m = 2n + 1$ d) 37   e) 48
5 a) Pupils' diagrams   b) 16, 20, 24, … 40
   c) $m = 4n$ d) 240   e) 32
6 a) Pupils' own diagrams b) 23, 29, 35 … 59
   c) $m = 6n - 1$ d) 107   e) 35
7 a) 39·2, 41, 42·8 … 50 b) $F = 1·8C + 32$
   c) 95°F   d) 20°C   e) −40°C

8 a) gap increases by 2 each time
   b) 1, 4, 9, 16, 25, 36 (3 less than values in table)
   c) $n^2 + 3$
9 a) $n^2 + 5$ b) $n^2 + 10$ c) $n^2 - 1$
   d) $n^2 - 9$ e) $-n^2$ f) $-n^2 + 2$

Exercise 5J

1 a) $l = \dfrac{A}{b}$ b) $T = \dfrac{Q}{l}$ c) $S = \dfrac{D}{T}$

   d) $a = \dfrac{F}{m}$

2 a) $m = \dfrac{W}{g}$ b) 75 kg

3 a) $c = y - 5x$ b) $q = p - 6r$ c) $k = \dfrac{m-3}{5}$

   d) $s = \dfrac{6r+1}{7}$ e) $b = \dfrac{a-d}{4}$ f) $g = \dfrac{f+h}{9}$

   g) $p = \dfrac{n+q}{10}$ h) $w = \dfrac{3v-y}{8}$ i) $f = \dfrac{d-g}{e}$

   j) $m = \dfrac{3k-n}{l}$ k) $c = \dfrac{2a+d}{b}$ l) $g = \dfrac{4e+h}{f}$

4 a) $t = \dfrac{v-u}{a}$ b) 3 seconds

5 a) $x = -y + 2$ b) $s = -r + 5$
   c) $q = -p + 1$ d) $g = -h + 4$
   e) $t = -2p + 3s$ f) $p = -5q + 6n$
   g) $l = \dfrac{-7k+mn}{2}$ h) $z = \dfrac{-xy+vw}{3}$

6 a) $q = 8p - 3$ b) $n = 9m + 4$
   c) $a = 2tb + 5$ d) $l = \dfrac{6kh+b}{2}$
   e) $b = 2a - 6$ f) $d = 30c + 5e$
   g) $q = pr - rs$ h) $h = \dfrac{3gm+mt}{2}$
   i) $b = \dfrac{21a-3c}{2}$ j) $g = \dfrac{11fh+5h}{6}$
   k) $q = \dfrac{7p-21}{2}$ l) $y = \dfrac{8-10x}{3}$

7 a) $F = \dfrac{9C+160}{5}$ or $F = \dfrac{9}{5}C + 32$ b) 86 °F

8 a) $a = \dfrac{2s - 2ut}{t^2}$ b) $0.6\,\text{m/s}^2$

9 a) $x = \dfrac{10}{y} - 4$ b) $k = \dfrac{31}{2m} + 5$

   c) $q = \dfrac{p}{8r} + 3$ d) $v = \dfrac{w}{5h} + x$

   e) $s = \dfrac{2}{3t} - \dfrac{1}{3}$ f) $a = \dfrac{7}{2b} + \dfrac{1}{2}$

   g) $p = \dfrac{4}{q} - \dfrac{5}{3}$ h) $x = \dfrac{5}{12y} - \dfrac{1}{4}$

10 a) $x = \dfrac{3b}{c} - y$ b) $p = \dfrac{4m}{n} + q$

   c) $t = \dfrac{k}{5r} - 2s$ d) $h = \dfrac{g}{fk} + 6$

   e) $c = \dfrac{a}{7b} + d$ f) $z = \dfrac{3x}{y} - 10$

   g) $s = 2r - \dfrac{p}{4q}$ h) $r = \dfrac{3p}{2q} + 5s$

11 a) $m = \sqrt{K + 1}$ b) $v = \sqrt{\dfrac{w}{8}}$ c) $r = \sqrt{\dfrac{A}{\pi}}$

   d) $v = \sqrt{\dfrac{2E}{m}}$ e) $x = \sqrt[3]{y - 4}$ f) $t = \sqrt[3]{\dfrac{r}{6}}$

   g) $f = \sqrt[3]{\dfrac{a + 1}{2}}$ h) $d = \sqrt[3]{\dfrac{c + g}{4}}$

12 $c = \sqrt{\dfrac{E}{m}}$

13 a) $x = k^2$ b) $y = (5m)^2$ or $y = 25m^2$

   c) $q = p^2 - 1$ d) $f = g\left(\dfrac{c}{\pi}\right)^2$

   e) $b = \dfrac{a^2}{2}$ f) $y = \dfrac{x^2 + 1}{3}$

   g) $q = \dfrac{p^2}{50} - \dfrac{r}{2}$ h) $g = \dfrac{25f^2}{36} + \dfrac{h}{4}$

14 a) $r = \sqrt{\dfrac{2V}{\pi h}}$ b) $r = 2\,\text{cm}$

15 a) $x = \pm\sqrt{10 - (y - 2)^2}$ b) $x = \pm 1$

   c) For a circle the two points are vertically aligned on the circumference. For the paraboloid, r is a length which cannot be negative therefore only the positive root is valid.

16 a) $x = \sqrt{5 - \dfrac{y^2}{4}}$ b) $x = \pm 2$

17 a) $L = \dfrac{T^2 g}{4\pi^2}$ b) $24.8\,\text{m}$

Check-up

1 a) $7x - 56$ b) $3x + 27$ c) $10x - 55$
   d) $-28x + 63$ e) $10x^2 + 12x$ f) $-27x + 81x^2$
   g) $48x^2 + 60x - 12$ h) $-14 + 42x - 21x^2$

2 a) $12x + 8$ b) $16x - 35$ c) $30 - 28x$
   d) $45x - 44$ e) $26x - 8$ f) $14 - 12x$
   g) $11x^2 - 14x$ h) $79 - 77x$

3 a) $8x + 7$ b) $13x + 71$ c) $36x + 51$
   d) $x - 73$ e) $33 - 78x$ f) $125x - 73$
   g) $5x^2 + 2x + 18$ h) $42x^2 - 53x + 32$
   i) $-13x^2 - 25x$

4 a) 13 b) 3 c) 7 d) 5

5 a) -25 b) 256 c) 40 d) -19

6 a) 36 b) -4 c) 10 d) 12

7 a) $x = 7$ b) $x = 4$ c) $x = -1$ d) $x = 6$
   e) $x = 7$ f) $x = 4$ g) $x = 3$ h) $x = 6$
   i) $x = 5$

8 a) $x = 6$ b) $x = -3$ c) $x = 2$ d) $x = -5$
   e) $x = 7$ f) $x = -9$

9 a) $x = 9$ b) $x = 27$ c) $x = 37$ d) $x = 10$
   e) $x = 33$ f) $x = 9$ g) $x = 5$ h) $x = -6$

10 a) $5 < 13$ b) $-1 > -4$ c) $-7 < 5$ d) $0 > -11$

11 a) $x > 3$ b) $x < 4$ c) $x \leq 4$ d) $x \geq -2$
   e) $x > -2$ f) $x \geq -5$ g) $x \leq 3$ h) $x < -4$

12 a) $x \leq 3$ b) $x \leq 3$ c) $x > 26$ d) $x > 7$
   e) $x \leq 4$ f) $x < -5$

13 a) $3a$ b) $3a - 9$ c) 14

14 a) $C = 3p + 8$ b) £29 c) 12 portions

15 a) $27, 34, 41, 7n - 1$ b) $9, 5, 1, -4n + 25$

16 a) $A = \dfrac{V}{h}$ b) $w = y - 4x$ c) $q = \dfrac{p + r}{5}$

   d) $n = \dfrac{g - 2m}{3}$ e) $b = 7a - 7c$ f) $y = \dfrac{8}{3x} + 4$

   g) $w = \sqrt{\dfrac{v}{3}}$ h) $d = c^3 + 5$

17 a) $h = \dfrac{B-T}{2}$ b) 1600 feet

18 a) $l = \sqrt[3]{\dfrac{m}{\rho}}$ b) $l = 5\,\text{cm}$

6 Coordinates and symmetry

Exercise 6A

1 a) 1 b) 2 c) 4 d) 2
 e) 1 f) 5 g) 1 h) 4
2 a) 5, 5 b) 2, 2 c) 8, 8 d) 1, 1
 e) 2, 2 f) 4, 4 g) 1, 1 h) 2, 2
3 Pupils' own diagrams
4 Pupils' own diagrams

Exercise 6B

1 a) A(3, 5), B(7, 1), C(1, –7), D(–4, –3), E(8, –3), F(–2, 3), G(–9, –1), H(3, 0), I(–7, 8), J(5, –9) K(6, 8), L(–6, –5), M(0, 7), N(–6, 2), O(0, –10), P(–7, –8)
 b) A and H, M and O, N and L, I and P
 c) I and K, D and E
2 Pupils' own diagrams
3 a) Pupils' own diagrams b) D(–4, –4)
 c) Pupils' own diagrams d) (0, –1)
4 a) Pupils' own diagrams b) D(3, 2)
 c) Pupils' own diagrams d) (0, 2)
5 a) pencil b) compass
 c) cylinder d) torus
6 a) (8, –9), (–7, –3), (4, 8), (4, 8), (–7, –3), (8, –9) / (0, 10), (9, –3), (10, 9)
 b) (2, 3), (7, 2), (1, –2), (9, –3), (–10, –7), (2, 3) / (–6, 0), (–10, –7), (3, –6), (3, –6), (–7, –3), (4, 8),(–10, –7)
 c) (–4, 6), (7, 2), (9, –3), (7, 2), (4, 8), (2, 3), (7, 2) / (–4, 6), (–7, –3), (0, 0)
 d) (5, 0), (–7, –3), (–4, 6), (8, –9), (7, 2), (9, –3), (–10, –7), (4, 8), (7, 2) / (8, 6), (0, 0), (8, –9), (4, 8), (4, –10), (0, 0), (4, 8)
 e) (–8, 5), (–7, –3), (9, –3), (10, 9), (–7, –3), (–8, 5) / (–8, 5), (–10, –7), (9, –3), (–8, 8), (–7, –3), (5, 0), (8, –9), (–7, –3), (4, 8), (–10, –7)

 f) (4, –10), (9, –3), (–10, –7), (4, 8), (–10, –7), (–6, 0), (–7, –3), (4, –10), (–7, –3) / (9, –3), (–7, –3), (–8, 5), (–7, –3), (4, 8), (–3, –3), (8, 6), (–7, –3), (4, 8)
 g) (–8, 5), (–7, –3), (9, –3), (10, 9) / (2, 3), (–7, –3), (9, –3), (–4, 6), (0, –9), (9, –3), (–10, –7), (–3, 2), (8, –9), (–4, 6)
 h) (–5, –8), (7, 2), (9, –3), (4, –10), (–10, –7) / (1, –2), (–10, –7), (–7, –3), (2, 3), (0, 0), (4, 8), (–10, –7), (4, –10)
7 a) (2, 3) b) (–4, –4)
 c) (–6, 6) d) (4, 3)

Exercise 6C

1 a) A′(1, 4), B′(8, 1), C′(1, 1)
 b) A′(–1, –2), B′(3, –2), C′(3, –5), D′(–1, –5)
 c) A′(–4, 4), B′(–3, 3), C′(–4, 0), D′(–5, 3)
 d) A′(–2, 3), B′(3, 3), C′(2, 0), D′(–1, 0)
 e) A′(–1, –1), B′(3, –1), C′(4, –3), D′(1, –5), E′(–2, –3)
 f) A′(–4, 3), B′(–2, 4), C′(–2, 0), D′(–4, –1)
2 a) right-angled triangle b) rectangle
 c) kite d) trapezium
 e) pentagon f) parallelogram
3 Add the horizontal translation to the x-coordinate and the vertical translation to the y-coordinate of each point.
4 a) A′(0, –4), B′(5, –4), C′(2, –1), D′(–3, –1)
 b) A′(5, –3), B′(2, 2), C′(–2, 2)
 c) A′(0, –4), B′(3, 3), C′(0, 1), D′(–3, 3)
5 a) A′(–3, 4), B′(–5, 2), C′(–3, –4), D′(–1, 2)
 b) A′(4, 4), B′(–1, 3), C′(–1, 0), D′(4, –2)
 c) A′(2, 2), B′(–2, 2), C′(–5, –4), D′(–1, –4)
6 a) A′(5, –5), B′(5, 2), C′(1, –5)
 b) A′(–1, 2), B′(–3, 5), C′(–5, 2), D′(–3, –1)
 c) A′(3, 0), B′(3, 2), C′(–2, 4), D′(–2, –2)
7 a) A′(4, –1), B′(4, –5), C′(1, –5), D′(1, –1)
 b) A′(2, –3), B′(1, –4), C′(0, –3), D′(1, 0)
 c) A′(4, 2), B′(4, –4), C′(2, –4), D′(2, 2)
8 a) A′(2, –4), B′(0, 3), C′(4, 3)
 b) A′(5, –5), B′(0, –5), C′(–2, –1), D′(3, –1)
 c) A′(–5, –3), B′(–5, 2), C′(1, 2)

9 a) A′(−4, 0), B′(−4, 1), C′(−3, 1), D′(−3, 4),
       E′(−1, 4), F′(−1, 0)
    b) A′(1, 3), B′(5, 5), C′(3, 0)
    c) A′(−4, −3), B′(−1, −1), C′(2, −3), D′(−1, −5)

10 a) A′(3, 9), B′(9, 3), C′(6, 0), D′(0, 6)
    b) A′(0, 8), B′(4, 8), C′(4, 4), D′(8, 4), E′(8, 0),
       F′(0, 0)
    c) A′(4, 2), B′(4, −2), C′(−4, −2), D′(−4, 2)
    d) A′(5, 9), B′(8, 6), C′(5, 3), D′(2, 6)
    e) A′(−2, 6), B′(1, 3), C′(−2, −6), D′(−5, 3)
    f) A′(1, 3), B′(2, 3), C′(2, 2), D′(3, 2), E′(3, 1),
       F′(0, 1), G′(0, 2), H′(1, 2)

Check-up

1 a) 2   b) 3   c) 1   d) 6
   e) 7   f) 1   g) 2   h) 5

2 Pupils' own diagrams

3 Pupils' own diagrams

4 A(3, 4), B(−4, 1), C(4, −2), D(0, −5), E(−2, −3),
   F(2, 0), G(−2, 4), H(−5, −2)

5 Pupils' own diagrams

6 a) Pupils' own diagrams   b) D(−5, 1)
   c) Pupils' own diagrams   d) (−1, 0)

7 a) A′(−1, −1), B′(2, −1), C′(4, −3), D′(−3, −3)
   b) A′(5, −3), B′(3, 2), C′(−4, 0)
   c) A′(4, −3), B′(4, −1), C′(1, 1), D′(1, −4)
   d) A′(4, −5), B′(−1, −4), C′(−1, −1), D′(4, −2)
   e) A′(5, 1), B′(3, −1), C′(1, 2), D′(2, 5), E′(3, 2)
   f) A′(1, 4), B′(3, 4), C′(3, 3), D′(5, 3), E′(5, 2),
      F′(3, 0), G′(1, 2)

8 a) A′(1, −2), B′(1, −5), C′(−3, −2), D′(−3, 1)
   b) A″(−2, 1), B″(−5, 1), C″(−2, −3), D″(1, −3)

9 a) reflection in the y-axis
   b) rotation 180° about the origin
   c) translation 1 unit right and 5 units down

7 Money

Exercise 7A

1 a) €110   b) 120 SFr
   c) 459 ﺏ.ﺍ   d) AU$180

2 a) $50   b) 1 192 672 ₩
   c) €3300   d) 17 646 Kč
   e) 318 SFr   f) 5875·20 ﺏ.ﺍ
   g) 100 581 ¥   h) AU$161·91
   i) 14848 Kč

3 969 046 ₩   4 54 000 SFr   5 66 920 000 ¥

6 a) £800   b) £555·56   c) £909·09
   d) £204·01   e) £7098·03   f) £100·61
   g) £479·30   h) £304·72

7 £69 818·18   8 £130 718·95 9 £100

10 £157·32   11 £708·33   12 £1000

13 €1650   14 AU$3888   15 26 768 000 ¥

16 yes, 2086·38 ﺏ.ﺍ   17 3373·65 ﺏ.ﺍ

18 US dollars

19 a) £544·03   b) 87 376 ¥

20 a) Supermarket   b) €40

21 a) Online   b) ₹1530 (rupees)

22 a) $1771·20   b) The bank, $13·80

23 Bank, 101·90 SFr

24 a) £83·57   b) £4852·94

25 a) Darren £661·16, Bob £509·55, Marc £555·56
   b) He is wrong, a lower rate is better

26 a) Supermarket   b) £338·10

Exercise 7B

1

Gross pay	£1500
Total deductions	£235
Net pay	£1265

2

Gross pay	£1542
Total deductions	£250
Net pay	£1292

3

Gross pay	£2383
Total deductions	£623
Net pay	£1760

4 Gross pay £1851, pension £148, Net pay £1425

5 Overtime £20, Income Tax £106, Net pay
   £1368

6 Overtime £138, National Insurance £275,
   Total deductions £945

Exercise 7C

1 a) Food £280, Non-food £120, Socialising £120, Total £764
   b) £36
2 a) Food £480, Household/toiletries £80, Socialising £200, Total £889·99
   b) £169·99
   c) Pupil's own, total spending no more than £720
   d) Pupil's own saving × 12.
3 a) If something needs fixed e.g. car repairs you will be able to pay for it. If you lose your job you will still have funds. You do not need to borrow money and pay interest to buy bigger items.
   b) If something unexpected happens or you lose your job you may have to go without.
4 a) Food £340, Household/toiletries £88, Clothes £30, Insurance £20, Days out £100, Total £578
   b) £43 a month
   c) Yes, but his monthly budget is very tight
5 a) Food £540, Non-food £380, Socialising £280, Total £1730
   b) £780 too much
   c) No, he cannot afford the repayments
   d) 40%
   e) Pupil's own, total spending no more than £950.
6 a) Pupil's own total spending no more than £1450
   b) Pupil's own £2500 ÷ savings amount
7 a) £988   b) £1063·97
8 a) Pupil's own
   b) Pupil's own all non-fixed costs x 5

Exercise 7D

1 None
2 a) Anna is correct   b) £12
3 a) £0, £0   b) £0, £24
   c) £0, £93·48, £2, £95·48
4 a) £44·04   b) £75·24
   c) £92·04   d) £100·24

Exercise 7E

1 No
2 a) Arjan's mum is correct   b) £665
3 £0, £396·15, £1083, total £1479·15
4 a) £10215   b) £0, £396·15, £2043, total £2439·15
5 a) £18272   b) £56570   c) £29541·57
6 £1873·70

Exercise 7F

1 a) £36·32   b) £192·56
   c) £1925·40   d) £4851·36
2 a) £801·64   b) £67·36
   c) £423·60   d) £84·12
3 4·9%, 6·8%, 17·9%, 21·8%, 316·5%, 1500%
4 a) Wow Pay has a longer repayment term (8 years > 2 years)   b) £219·12
5 a) Savings Society
   b) Alba Bank £172·96, Saving Society £78·64, Pearl Credit £408·88
   c) £330·24

Exercise 7G

1 a) £1836·01
   b) The minimum payment was less than the interest charged.
2 a) £360·25
   b) £6 charge for withdrawing cash and you will pay interest on it.
3 a) £730·35
   b) Cut down on non-essential spending.

Exercise 7H

1 Pupils' own answers. Basic athletic shoe is less expensive than branded.
2 Pupils' own answers. Second hand a big saving.
3 Pupils' own answers. Could be a good way to make healthy lunches and earn some money.
4 a) £69·17
   b) Save up; credit card will cost much more in interest and could take a long time to repay.
5 Pupils' own answers

6 a) £66·52

   b) Pupils' own answers. Second hand is a good saving.

   c) Pupils' own answers. Insurance only worthwhile for the new guitar.

7 a) £15 933·40   b) £6498·40

   c) Pupils' own answers. The 3-year old car is a good saving and has low mileage.

8 a) £343·20

   b) Total Cover Co is one payment, £134·20 less expensive but has a greater excess. Quicksure covers loss of keys and windscreen damage.

   c) Yes, £18·04

   d) Pupils' own answers. Total Cover Co is least expensive but only provides very basic cover.

9 a) Free stream £240 saving, Harmony £180 saving, pupils' own answers

   b) If Free stream chosen £292, If Harmony chosen £219

Check-up

1

Gross pay	£1958
Total deductions	£431
Net pay	£1527

2 a) £380, £80, £39, £200, £400, £75, £1174

   b) £364

   c) Pupils' own answers

3 £38·04   4 £1140

5 a) £14 842   b) £5627·57

6 £102·60

7 a) £6563·52

   b) Pupils' own answers. Simple wedding is a huge saving.

8 a) No, the APR is high and 8 years is a long repayment time. He will pay a lot of interest.

   b) Trinity Credit £2399·84   Fillan Finance £1048·16   Brendan Bank £235·28

   c) Brendan Bank is the best option

9 a) £949·44

   b) Pros, convenience, do not have to carry cash when travelling. Cons, very expensive.

10 a) $625   b) R10 550   c) R$3340

    d) $211·99   e) 576 610 Ft   f) R$1 336 000

    g) £102·70   h) £1199·16   i) £189 573·46

11 R$8016   12 $153 466·25

13 a) Online Next Day   b) €72

14 a) €8050   b) £7200

    c) £7700   d) £6382·98

8 Straight line

Exercise 8A

1 a) $\frac{1}{3}$ b) $\frac{1}{5}$ c) $\frac{5}{6}$

   d) $-\frac{1}{3}$ e) $-\frac{9}{16}$ f) $\frac{21}{23}$

2 6 m   3 8 m

4 Slope A, since its gradient is greater $\left(\frac{5}{12} > \frac{1}{3}\right)$

5 Neither; they have the same gradient

Exercise 8B

1 a) $\frac{1}{2}$ b) 3   c) $\frac{2}{5}$

   d) $-\frac{1}{2}$

2 a) 2   b) −5   c) 4   d) 12

   e) $\frac{9}{2}$ f) −7   g) 6   h) 0

   i) $\frac{1}{2}$ j) $\frac{1}{4}$ k) $\frac{2}{3}$ l) $-\frac{1}{5}$

   m) $-\frac{3}{4}$ n) $\frac{7}{8}$ o) $\frac{5}{2}$

Exercise 8C

1 Yes, since 5·5 > 5

2 No, since gradient exceeds $\frac{3}{100}$

3 a) No, since gradient exceeds $\frac{1}{12}$

   b) 420 cm

4 $t = 27$ 5 $t = -17$ 6 $t = 25$ 7 $t = 3$

Exercise 8D

1 a) $y = 3$ b) $x = 3$ c) $y = 2$

   d) $x = -4$ e) $y = -3$ f) $x = -1$

2 Pupils' own diagrams

3 a) $y = 3$ b) $y = 2$ c) $y = -7$
   d) $y = 11$ e) $y = -34$

4 a) $x = 5$ b) $x = 8$ c) $x = -7$
   d) $x = -10$ e) $x = 0$

5 a) $y = 8$ b) $x = 5$ c) $y = 6$
   d) $x = -3$ e) $y = -2$ f) $x = 0$

6 $y = 0$ 7 $x = 0$

Exercise 8E

1 a) 0, 3, 6
   b) Pupils' own diagrams   c) $(0, 0)$

2 a) 4, 5, 6
   b) Pupils' own diagrams   c) $(0, 4)$

3 a) 0, −1, −2
   b) Pupils' own diagrams   c) $(0, 0)$

4 a) 3, 1, −1
   b) Pupils' own diagrams   c) $(0, 3)$

5 a) $(0, -3)$ b) $(0, 4)$ c) $(0, 5)$
   d) $(0, -4)$ e) $(0, -1)$ f) $(0, 7)$
   g) $(0, 6)$ h) $(0, -1)$ i) $(0, 0)$
   j) $(0, 3)$ k) $(0, 2)$ l) $(0, 4)$

6 a) 1   b) 2   c) 3   d) 2
   e) −2   f) −3   g) −1   h) 6
   i) $\dfrac{1}{2}$ j) $\dfrac{1}{4}$ k) $-\dfrac{1}{5}$ l) $\dfrac{3}{2}$

7 The gradient is the coefficient of x. The y-intercept is the constant term.

Exercise 8F

1 a) $m = 2, (0, 5)$ b) $m = 6, (0, 1)$
   c) $m = 9, (0, -2)$ d) $m = -3, (0, 12)$
   e) $m = -7, (0, -1)$ f) $m = 1, (0, -6)$
   g) $m = -11, (0, -4)$ h) $m = -1, (0, 0)$
   i) $m = \dfrac{1}{2}, (0, 1)$ j) $m = \dfrac{3}{5}, \left(0, -\dfrac{1}{5}\right)$
   k) $m = -\dfrac{1}{4}, \left(0, \dfrac{2}{5}\right)$ l) $m = -\dfrac{5}{2}, \left(0, -\dfrac{3}{2}\right)$

2 a) $y = 4x + 6$ b) $y = 9x - 5$
   c) $y = -2x$ d) $y = x + 11$

   e) $y = -5x + 3$ f) $y = -x - 7$

3 a) $m = 2, (0, -3)$ b) $m = \dfrac{1}{2}, \left(0, \dfrac{1}{4}\right)$
   c) $m = -5, (0, 3)$ d) $m = 1, (0, -7)$
   e) $m = -\dfrac{3}{4}, \left(0, \dfrac{1}{2}\right)$ f) $m = \dfrac{7}{4}, \left(0, \dfrac{1}{2}\right)$
   g) $m = 6, (0, -9)$ h) $m = -11, (0, 88)$

4 $m = 0, \left(0, \dfrac{4}{5}\right)$

5 a) $y = 4x + 2$ b) $y = -2x + 12$
   c) $y = 2x - 5$ d) $y = \dfrac{1}{2}x + 13$
   e) $y = 3x + 2$ f) $y = 3x$

6 a) $y = x - 2$ b) $y = -3x + 9$
   c) $y = \dfrac{1}{2}x + 1$ d) $y = -\dfrac{1}{2}x - 4$
   e) $y = \dfrac{1}{6}x + 4$ f) $y = -\dfrac{3}{5}x - 2$

7 $y = 4x + 6$ 8 $y = \dfrac{1}{2}x - 13$

9 $y = \dfrac{1}{2}x + 5$ 10 $y = -\dfrac{7}{3}x - 12$

Exercise 8G

1 a) $y = 2x + 1$ b) $y = 4x - 19$
   c) $y = x - 5$ d) $y = 3x + 18$
   e) $y = -3x + 22$ f) $y = 5x - 13$
   g) $y = -7x + 8$ h) $y = -12x + 137$
   i) $y = 9x + 11$ j) $y = -11x + 19$
   k) $y = \dfrac{1}{2}x + 3$ l) $y = \dfrac{1}{4}x - 9$
   m) $y = \dfrac{2}{3}x - 6$ n) $y = \dfrac{3}{8}x + 9$
   o) $y = -\dfrac{1}{7}x + 4$ p) $y = \dfrac{4}{3}x + \dfrac{7}{3}$
   q) $y = -7x + 8$ r) $y = -\dfrac{1}{4}x - 2$

2 a) $y = 6x - 8$ b) $y = -2x + 7$
   c) $y = \dfrac{5}{3}x - 2$

3   a) $C = \dfrac{1}{5}d + \dfrac{13}{10}$ b) £4·30   c) 10 miles

4   a) $C = \dfrac{3}{2}d + \dfrac{1}{2}$ b) £20   c) 6 miles

Check-up

1   a) 2   b) $\dfrac{5}{7}$ c) $\dfrac{1}{3}$

2   a) 2   b) $-\dfrac{1}{2}$ c) 3

3   No, since the gradient is too steep $\left(\dfrac{2}{15} > \dfrac{1}{15}\right)$

4   $t = -12$

5   Pupils' own diagrams

6   $x = 3$ 7 $y = -5$ 8 $y = 8$

9   $x = 4$ 10 $x = 0$

11 Pupils' own diagrams

12 a) $y = 4x + 9$ b) $y = -5x + 7$

    c) $y = \dfrac{1}{4}x - 2$ d) $y = -\dfrac{5}{2}x - \dfrac{3}{8}$

13 a) $y = -2x + 19$ b) $y = -\dfrac{20}{3}x + 5$

    c) $y = \dfrac{7}{4}x - 8$

14 a) $y = 4x + 7$ b) $y = 6x + 1$

    c) $y = -2x - 11$ d) $y = -\dfrac{1}{2}x + 1$

    e) $y = -\dfrac{3}{5}x$ f) $y = -\dfrac{5}{2}x + 23$

15 $y = -3x + 12$

16 $y = 7x - 1$

17 a) $y = 4x + 5$ b) $y = -6x + 3$

    c) $y = \dfrac{3}{4}x - 2$

18 a) $y = 8x + 3$ b) $y = 13x - 15$

    c) $y = -x + 1$ d) $y = 3x - 10$

    e) $y = -4x - 9$ f) $y = \dfrac{1}{5}x - 6$

19 $y = 7x - 33$

20 $y = -\dfrac{1}{2}x + 5$ 21 $y = \dfrac{2}{5}x - \dfrac{29}{5}$

22 a) $C = 0.3n + 1.5$ b) £4·20   c) 14

23 a) $C = \dfrac{1}{25}t + 12$ b) £12

    c) 4 pence   d) £14·80

24 a) $t = 7j + 10$ b) 3 hours 40 minutes

25 a) $C = 70t + 40$ b) £40   c) £600

9 Pythagoras' theorem and trigonometry

Exercise 9A

1   13 cm

2   a) 26 cm   b) 15 cm   c) 17 cm   d) 61 m

    e) 85 mm   f) 97 cm   g) 265 m   h) 137 cm

3   a) 19·3 cm   b) 36·4 m   c) 10·2 cm

4   150 m   5 80·6 cm

6   Pythagorean triples are (5, 12, 13), (36, 77, 85) and (60, 91, 109)

Exercise 9B

1   a) 15 cm   b) 24 mm   c) 42 cm   d) 90 cm

2   a) 9·17 cm   b) 28·9 m   c) 6·23 cm

3   a) Pupils' own diagrams   b) 6·3 m

Exercise 9C

1   a) 15·2 cm   b) 14·3 mm   c) 7·81 m

    d) 17·4 m   e) 38·3 cm   f) 13·0 cm

    g) 0·964 mm   h) 3·41 km

2   66 cm   3 3·23 m

4   a) 24·1 cm   b) 2530 cm³

5   a) 514 mm

    b) No, Alison is not correct. The diagonal length of A4 is approximately 364 mm which is greater than one half of 514 mm.

6   a) 164 m   b) 1·31 m/s

Exercise 9D

1   a) Yes   b) No   c) No   d) Yes

2   No; the diagonal and edges do not form a right-angled triangle

3   Yes; the guy rope, pole and ground form a right-angled triangle

4 No; the set square is not right-angled

5 Yes; the three parts of the journey form a
right-angled triangle

Exercise 9E

1 a) 4 b) 3 c) 5

2 a) 9·85 b) 10 c) 25·2 d) 26·4

e) 10 f) 12·8 g) 7·62 h) 20

3 a) 23·7 cm b) 3 m c) 30·4 mm d) 9·84 cm

4 48·0 cm 5 10·6 cm 6 24·2 cm

Exercise 9F

1 a) i) 10 cm ii) 11·2 cm

b) i) 13 cm ii) 13·3 cm

c) i) 3·16 m ii) 5·92 m

d) i) 9·90 cm ii) 12·1 cm

2 a) 11·3 cm b) 15·1 cm

Exercise 9G

a)

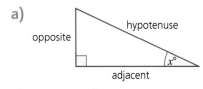

b)

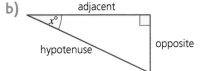

c)

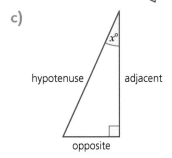

2 a) i) AC ii) AB iii) BC

b) i) PQ ii) PR iii) QR

c) i) DF ii) EF iii) DE

d) i) YZ ii) XY iii) XZ

e) i) LN ii) MN iii) LM

f) i) AC ii) AB iii) BC

Exercise 9H

1 a) 44·4° b) 60·1°

2 a) 45·6° b) 46·2° c) 55·4° d) 33·2°

e) 47·3° f) 31·3° g) 46·2° h) 55·2°

3 a) Opposite and hypotenuse substituted
wrong way round. 29·4°

b) Adjacent rather than opposite has been
substituted. 71·1°

4 a) 18·9° b) 30·5° c) 36·9°

d) 60·5° e) 36·9° f) 73·7°

5 1·47°

6 a) i) 106 m ii) 31·9°

b) i) 55 cm ii) 48·9°

c) i) 4·5 cm ii) 58·1°

Exercise 9I

1 a) 41·4° b) 48·2° c) 38·9° d) 50·5°

e) 28·1° f) 62·3° g) 37·5° h) 43·6°

2 a) 36·9° b) 58·1° c) 61·9°

3 25·9°

Exercise 9J

1 a) 50·2° b) 45° c) 17·5°

d) 56·3° e) 26·4° f) 72·6°

g) 26·0° h) 46·6°

2 a) 58·1° b) 71·1° c) 47·5°

3 323° 4 68·2°

Exercise 9K

1 a) 9·00 cm b) 10·6 m

2 a) 5·14 cm b) 5·92 cm c) 7·19 cm

d) 2·65 mm e) 1·46 m f) 13·9 cm

g) 6·02 m h) 1·98 m

3 a) 9·75 cm b) 27·4 m

4 a) 8·88 cm b) 2·56 cm c) 100 cm d) 9·55 m

5 a) The angle and side have been swapped
round in the substitution. $x = 6·15$ cm

b) Rearrangement has not been carried out
correctly. $x = 7·25$ m

6   16 m

7   No; it reaches 1·29 m which is less than 1·3 m

8   a) 5·90 cm   b) 6·79 cm

Exercise 9L

1   a) 5·44 cm   b) 13·4 cm   c) 0·303 m
     d) 2·03 mm   e) 1·01 cm   f) 0·462 m
     g) 1·49 m   h) 13·6 cm

2   a) 18·1 cm   b) 12·8 mm   c) 7·15 cm
     d) 6·87 m

3   32·9 m   4   1·97 m

Exercise 9M

1   a) 10 cm   b) 5·86 cm   c) 13·7 m
     d) 17·5 cm   e) 3·76 m   f) 13·9 mm
     g) 2·90 m   h) 3·42 cm

2   a) 3·22 cm   b) 12·4 cm   c) 37·0 mm
     d) 2·62 m

3   2·22 m

Exercise 9N

1   a) 5·82 cm   b) 65·2 mm   c) 6·47 m
     d) 4·16 cm   e) 3·77 m   f) 204 mm

2   a) 59·0°   b) 29·5°   c) 65·1°
     d) 48·8°   e) 45·9°   f) 76·1°

3   10·6 cm   4   24·0°   5   9·17 cm

Check-up

1   a) 10·0 cm   b) 22·7 cm   c) 32·8 m
     d) 6·07 mm   e) 1·20 km   f) 10·7 cm

2   No; the diagonal is only 68·4 cm which is less than 68·5 cm.

3   4·09 km

4   a) Pupils' own diagrams   b) 4·42 km

5   a) Yes   b) No   c) Yes

6   No; by the converse of Pythagoras' theorem the triangle is not right-angled.

7   Yes; the diagonal is 1·05 m long

8   a) 56·4°   b) 46·2°   c) 34·6°
     d) 54·8°   e) 65·6°   f) 33·5°

9   a) 13·4 cm   b) 10·7 m   c) 2·51 mm
     d) 5·45 m   e) 32·1 cm   f) 17·8 mm

10   14·38 km

11   a) 18·1 feet   b) 16·4 feet

12   a) 539 m   b) 539 m

13   a) 2·54 km   b) 20·9 km
     c) 1 minute 42 seconds

10 Project

Mathematics in the workplace

1   Pupils' own answers.

2

Job title	Example of Maths used
Sound engineer	The mathematical technique of Fourier analysis is used to manipulate pitch and sound.
Meteorologist	Numerical analysis and computer modelling techniques are used to make short- and long-term predictions around climate change.
Doctor	Dosages and treatment plans determined by numerical calculations and probabilities.
Architect	Building and structure safety features calculated from known and measured data.
Economist	Mathematical models are used to understand the complexities of a business cycle and the effects of inflation.
Epidemiologist	Mathematical models are used to track the progress of infectious diseases and the effectiveness of vaccines.
Pilot	Geometry is used to calculate angles and plan routes.
Estate agent	Accurate measurements are taken to produce scale drawings and plans.

11 Further algebra

Exercise 11A

1   a) $x^2 + 5x + 6$ b) $x^2 + 8x + 12$
     c) $y^2 + 8y + 7$ d) $y^2 + 18y + 81$
     e) $x^2 + 2x - 8$ f) $x^2 + x - 20$
     g) $y^2 - y - 42$ h) $y^2 - 6y - 16$

2   a) $x^2 - 7x - 8$ b) $x^2 - 3x - 28$
     c) $y^2 + 8y - 9$ d) $y^2 - 9$

e) $x^2 - 6x + 8$ f) $x^2 - 10x + 9$

g) $y^2 - 13y + 40$ h) $y^2 - 8y + 16$

i) $x^2 - 7x - 18$ j) $x^2 - 6x + 9$

k) $y^2 - 1$ l) $y^2 + 11y + 30$

3   a) $2x^2 + 11x + 12$ b) $3x^2 + 15x + 18$

    c) $4x^2 + 27x - 7$ d) $5x^2 + 28x - 12$

    e) $3y^2 - 19y - 40$ f) $2y^2 - 15y - 50$

    g) $6m^2 - 19m + 3$ h) $100h^2 - 140h + 49$

    i) $21x^2 + 43x + 20$ j) $18m^2 + 31m + 6$

    k) $20p^2 + 7p - 6$ l) $8k^2 - 47k - 6$

    m) $6x^2 + 13x - 28$ n) $45x^2 - 23x + 2$

    o) $49w^2 - 28w + 4$ p) $12a^2 - 35a + 8$

4   a) $x^3 + 8x^2 + 16x + 5$

    b) $x^3 + 9x^2 + 17x + 21$

    c) $x^3 + 13x^2 + 34x - 18$

    d) $x^3 + x^2 - 26x + 24$

    e) $2x^3 - 8x^2 + 24x - 32$

    f) $3x^3 + 26x^2 - 3x - 2$

    g) $5x^3 + 32x^2 - 26x + 3$

    h) $7x^3 - 22x^2 - 5x + 24$

    i) $4x^3 - 29x^2 + 66x - 45$

5   a) $5x^2 + 18x + 4$ b) $5x^2 + 3x - 17$

    c) $17x^2 - 12x - 57$ d) $35x^2 - 45x + 8$

    e) $4x^2 - 28x + 3$ f) $x^2 + 12x + 10$

    g) $4x^2 - 23x - 22$ h) $-8x^2 + 12x - 3$

6   a) $x^2 + 8x + 7$ b) $x^2 - 10x + 25$

    c) $x^2 + 6x - 16$ d) $\pi x^2 - 2\pi x + \pi$

7   a) $2x^2 + 21x - 36, 2x^2 + 7x + 6$

    b) 3 cm by 15 cm and 5 cm by 9 cm

Exercise 11B

1   a) $3(x + 4)$ b) $2(x + 5)$ c) $7(x + 2)$

    d) $5(1 + 3x)$ e) $9(1 + x)$ f) $2(x - 4)$

    g) $4(x - 3)$ h) $6(x - 6)$ i) $8(5 - x)$

    j) $10(3 - x)$

2   a) $2(2y + 5)$ b) $4(2y + 5)$ c) $5(4k - 3)$

    d) $3(2m + 3)$ e) $2(9w - 2)$ f) $6(2x + 7)$

    g) $7(2m + 5)$ h) $8(4x - 3)$ i) $8(5 - 2x)$

    j) $9(3 + 11x)$ k) $2(x + 4y - 2)$ l) $3(x - 3 + 6y)$

    m) $5(2y + x - 7)$ n) $7(2x - 3 - 4y)$

    o) $4(2m - 3n + 4)$

3   a) $x(y + 6)$ b) $y(5 + x)$ c) $c(9 - d)$

    d) $n(m + 2)$ e) $p(k - 1)$ f) $bc(a + 2)$

    g) $pr(q - 9)$ h) $vw(3 + x)$ i) $mn(5p - 6)$

    j) $ab(7c + 2)$

4   a) $2pqr + 5r = \boxed{r}(2pq + 5)$

    b) $9ab + 11\boxed{abc} = ab(9 + 11c)$

    c) $3x^2 + 10x = \boxed{x}(3x + 10)$

    d) $8\boxed{m^2} - 5m = m(8m - 5)$

5   a) $x(x + 6)$ b) $x(x - 8)$ c) $y(2 + y)$

    d) $p(3p + 10)$ e) $h(9 - 5h)$ f) $x^2(y - 15)$

    g) $y^2(x + 13)$ h) $m^2(8n + 1)$ i) $x^2(8x + y)$

    j) $xy^2(w + y)$

6   a) $3x(1 + 5y)$ b) $4n(m - 2)$ c) $6x(y + 9)$

    d) $5c(a - 3)$ e) $9p(2q + 1)$ f) $7x(2 - 7y)$

    g) $8v(3y - 4)$ h) $4h(7 - 10g)$ i) $20q(2p + 5)$

    j) $8n(11m + 12)$

7   a) $2x(x + 2)$ b) $5m(m - 4)$ c) $3y(3 - y)$

    d) $7k(3 + k)$ e) $10m(5m + 1)$

    f) $12k(k - 2)$ g) $9p(2 + 3p)$ h) $5w(3w + 4)$

    i) $8h(5 - 4h)$ j) $27m(m - 3)$

8   a) $5xy(w + 3)$ b) $7bc(1 - 2a)$ c) $2pq(5r - 4)$

    d) $6fh(3 + h)$ e) $2x^2(x + 6)$ f) $5h^2(1 - 9h)$

    g) $8p^2(1 + 4p^2)$ h) $10k^3(7 + 6k)$

9   a) $2p(p^2 + 3p + 4)$

    b) $3a(4a^2 + 3a - 5)$

    c) $5m(10m^2 + 2m + 3)$

    d) $4p^2(2p^2 - 3p + 9)$

10   a) $2x^2 + \boxed{18}x = \boxed{2x}(x + 9)$

    b) $20\boxed{m^2} - 5m^3 = 5m^2\left(4 - \boxed{m}\right)$

    c) $\boxed{30k^3} - 24k^2 = 6k^2\left(5k - \boxed{4}\right)$

Exercise 11C

1 a) $(x + 4)(x - 4)$ b) $(x + 5)(x - 5)$
c) $(y + 9)(y - 9)$ d) $(m + 7)(m - 7)$
e) $(p + 11)(p - 11)$ f) $(1 + x)(1 - x)$
g) $(8 + q)(8 - q)$ h) $(6 + n)(6 - n)$
i) $(10 + w)(10 - w)$ j) $(12 + y)(12 - y)$
k) $(3x + 4)(3x - 4)$ l) $(7x + 5)(7x - 5)$
m) $(8x + 7)(8x - 7)$ n) $(6y + 1)(6y - 1)$
o) $(11m + 2)(11m - 2)$ p) $(10 + 3y)(10 - 3y)$
q) $(5 + 2k)(5 - 2k)$ r) $(12 + 5r)(12 - 5r)$
s) $(20 + 3m)(20 - 3m)$ t) $(100 + 9p)(100 - 9p)$

2 a) $(x + m)(x - m)$ b) $(x + 4y)(x - 4y)$
c) $(h + 5m)(h - 5m)$ d) $(3p + k)(3p - k)$
e) $(10m + t)(10m - t)$ f) $(7x + 2y)(7x - 2y)$
g) $(5r + 6t)(5r - 6t)$
h) $(10g + 9h)(10g - 9h)$
i) $(11v + 12w)(11v - 12w)$
j) $(30x + 7y)(30x - 7y)$
k) $(15a + 13b)(15a - 13b)$
l) $(50p + 17q)(50p - 17q)$

3 Are: $81 - 25p^2$, $100p^2 - 49q^2$, $36g^2 - r^2h^2$, $196m^2 - n^2$
Not: $h^2 - 10$, $a^2b - 4c^2d$, $12x^2 - y^2$, $25a^2 + 121c^2$

4 a) $(xy + 3)(xy - 3)$
b) $(4 + mn)(4 - mn)$
c) $(5x + 7ab)(5x - 7ab)$
d) $(8pq + 9k)(8pq - 9k)$
e) $(11ab + 12cd)(11ab - 12cd)$
f) $(30xy + 13vw)(30xy - 13vw)$

5 a) $(x^2 + 3my)(x^2 - 3my)$
b) $(4x + 5w^2y^2)(4x - 5w^2y^2)$
c) $(40x^2y^2 + 9v^2w^2)(40x^2y^2 - 9v^2w^2)$

Investigation

1 a) $x^2 + 5x + 6$ b) $x^2 - 5x + 6$
c) $y^2 + 7y + 6$ d) $y^2 - 7y + 6$

2 a) You multiply together the constant terms in the bracket to get the constant term in the trinomial.

b) When the signs in the brackets are both positive, the linear term is positive. When the signs in the brackets are both negative, the linear term is negative.

c) The linear term in the trinomial comes from adding and subtracting the products of the constant terms with the variable.

3 a) $(x + 2)(x + 4)$ is correct because $+ 2x + 4x = + 6x$ which is the linear term in the trinomial.

4 a) $(x + 4)(x + 5)$ b) $(x + 4)(x + 3)$
c) $(x - 3)(x - 5)$ d) $(x - 5)(x - 2)$

5 $x^2 + 6x - 40 = (x + 10)(x - 4)$
$x^2 + 3x - 40 = (x - 5)(x + 8)$
$x^2 + 18x - 40 = (x + 20)(x - 2)$

6 a) The signs inside the brackets are different.
b) The linear term in the trinomial comes from adding and subtracting the products of the constant terms with the variable.

7 a) $(x + 10)(x - 2)$ b) $(x + 8)(x - 3)$
c) $(x + 6)(x - 2)$ d) $(x - 6)(x + 2)$

Exercise 11D

1 a) $(x + 1)(x + 5)$ b) $(x + 1)(x + 3)$
c) $(x + 1)(x + 7)$ d) $(x + 1)(x + 13)$
e) $(x + 2)(x + 4)$ f) $(x + 2)(x + 5)$
g) $(x + 3)(x + 5)$ h) $(x + 4)(x + 5)$
i) $(x + 3)(x + 4)$ j) $(x + 1)(x + 14)$
k) $(x + 2)(x + 8)$ l) $(x + 2)(x + 9)$

2 a) $(x - 1)(x - 2)$ b) $(x - 1)(x - 7)$
c) $(x - 1)^2$ d) $(x - 1)(x - 11)$
e) $(x - 2)(x - 3)$ f) $(x - 3)^2$
g) $(x - 2)(x - 12)$ h) $(x - 5)(x - 6)$
i) $(x - 1)(x - 8)$ j) $(x - 2)(x - 5)$
k) $(x - 2)(x - 6)$ l) $(x - 1)(x - 16)$

3 a) $(x + 2)(x + 3)$ b) $(x + 3)(x + 6)$
c) $(x - 3)(x - 10)$ d) $(x - 3)(x - 15)$
e) $(y - 2)(y - 18)$ f) $(n + 5)(n + 10)$
g) $(m - 4)(m - 18)$ h) $(k + 4)(k + 25)$
i) $(p + 11)^2$ j) $(q - 3)(q - 20)$
k) $(h - 3)(h - 17)$ l) $(a - 6)(a - 14)$

4 a) $(x - 1)(x + 3)$ b) $(x - 1)(x + 7)$
c) $(x - 1)(x + 5)$ d) $(x - 1)(x + 13)$

e) $(x-1)(x+9)$ f) $(x-2)(x+5)$

g) $(x-2)(x+6)$ h) $(x-2)(x+4)$

i) $(x-2)(x+7)$ j) $(x-4)(x+6)$

k) $(x-3)(x+10)$ l) $(x-3)(x+15)$

m) $(x-2)(x+16)$ n) $(x-2)(x+25)$

o) $(x-3)(x+24)$ p) $(x-3)(x+27)$

5 a) $(x+1)(x-3)$ b) $(x+1)(x-2)$

c) $(x+1)(x-11)$ d) $(x+1)(x-17)$

e) $(x+1)(x-6)$ f) $(x+3)(x-4)$

g) $(x+2)(x-8)$ h) $(x+2)(x-13)$

i) $(x+3)(x-5)$ j) $(x+3)(x-14)$

k) $(x+4)(x-9)$ l) $(x+5)(x-10)$

m) $(x+4)(x-10)$ n) $(x+7)(x-9)$

o) $(x+7)(x-8)$ p) $(x+6)(x-14)$

6 a) $(x+1)(x-5)$ b) $(x-1)(x+11)$

c) $(x+1)(x-7)$ d) $(x+2)(x-9)$

e) $(x-2)(x+8)$ f) $(x+4)(x-5)$

g) $(y+4)(y-8)$ h) $(y+4)(y-11)$

i) $(k-4)(k+14)$ j) $(a+3)(a-21)$

k) $(c-6)(c+12)$ l) $(d-4)(d+25)$

m) $(m-4)(m+21)$ n) $(q+4)(q-20)$

o) $(h-5)(h+15)$ p) $(p+3)(p-32)$

7 a) $(x-3)(x-6)$ b) $(x-3)(x+4)$

c) $(x-1)(x+14)$ d) $(x-1)(x-19)$

e) $(x+1)(x-26)$ f) $(x-2)(x-11)$

g) $(x-3)(x+10)$ h) $(x+4)(x+7)$

i) $(x+2)(x-18)$ j) $(x-5)(x+11)$

k) $(x-2)(x-23)$ l) $(x-5)(x+10)$

m) $(y+4)(y-13)$ n) $(n+4)(n+15)$

o) $(q-6)(q-9)$ p) $(p+31)(p-2)$

q) $(h-7)(h+10)$ r) $(r+8)^2$

s) $(t+4)(t-18)$ t) $(v-5)(v+17)$

u) $(w-9)^2$ v) $(g+9)(g+11)$

w) $(x-10)(x-11)$ x) $(y+8)(y-25)$

Exercise 11E

1 a) $(2x+3)(x+1)$ b) $(2x-3)(x-1)$

c) $(2x+1)(x+5)$ d) $(2x-1)(x+7)$

e) $(3x+5)(x-1)$ f) $(3x-2)(x+4)$

g) $(3x-5)(x+2)$ h) $(5x+1)(x+9)$

i) $(5x+2)(x-3)$ j) $(5x+3)(x-4)$

k) $(7x-6)(x-1)$ l) $(7x+2)(x-8)$

2 a) $(x+1)(4x+3)$ b) $(2x-1)(2x+5)$

c) $(x-1)(6x+7)$ d) $(3x-1)(2x-3)$

e) $(4x+1)(2x+1)$ f) $(8x+7)(x-1)$

g) $(2x+1)(4x-5)$ h) $(3x-1)^2$

i) $(10x-1)(x+2)$ j) $(5x-1)(2x+3)$

k) $(2x+1)(6x+5)$ l) $(3x-2)(4x-1)$

3 a) $(2x+1)(2x+9)$ b) $(2x+3)(2x+5)$

c) $(x+4)(6x-7)$ d) $(2x+1)(3x-8)$

e) $(4x-3)(2x+5)$ f) $(3x-2)(3x-4)$

g) $(3x+2)(4x-3)$ h) $(6x+5)(2x-9)$

4 a) $(1-x)(5+x)$ b) $(1+x)(3-x)$

c) $(11+x)(1-x)$ d) $(13-x)(1+x)$

e) $(4-x)(2+x)$ f) $(2-x)(5+x)$

g) $(3-x)(4+x)$ h) $(8-x)(2+x)$

i) $(4+x)(5-x)$ j) $(9-x)(2+x)$

k) $(7+x)(3-x)$ l) $(4-x)(6+x)$

5 a) $(7-x)(1+2x)$ b) $(5+2x)(1-x)$

c) $(2+x)(1-2x)$ d) $(1+5x)(1-x)$

e) $(2+x)(3-2x)$ f) $(4-x)(2+5x)$

g) $(3-2x)(2+5x)$ h) $(4+3x)(2-3x)$

i) $(3+2x)(4-3x)$ j) $(3+4x)(5-2x)$

k) $(7+6x)(9-4x)$ l) $(-3x+2)(5x+6)$

Exercise 11F

1 a) $2(a+3)(a-3)$ b) $4(2+q)(2-q)$

c) $10(x+1)(x-1)$ d) $9(y+2x)(y-2x)$

e) $2(h+11g)(h-11g)$ f) $x(x+4)(x-4)$

g) $y(7y+5)(7y-5)$ h) $x(y+2)(y-2)$

i) $3q(1+3q)(1-3q)$ j) $8m(2m+1)(2m-1)$

2 a) $5(x+1)(x+2)$ b) $2(x-3)(x-5)$

c) $4(k+1)(k-5)$ d) $3(p-2)(p-3)$

e) $x(x+4)(x+9)$ f) $q(q-3)(q-4)$

g) $h(h+7)(h-3)$ h) $2x(x-3)(x+6)$

i) $4(3x+1)(x-1)$ j) $2(5y-2)(y-3)$

k) $3(2m-1)(m-5)$ l) $5(5x+4)(x+1)$

m) $x(2x-1)(x+4)$ n) $3y(2y+1)(y+2)$

o) $4k(3k-4)(k+2)$ p) $3p(2p+1)(3p-2)$

3   a) A: She didn't take out the common factor.

B: The 3 is missing as a common factor in the final answer.

b) A: $25(x + 2y)(x - 2y)$, B: $3(a - 7)(a + 2)$

4   a) $10(3 + x)(3 - x)$ b) $2(x - 3)(x - 4)$

c) $4(5y + 4)(y + 1)$ d) $18(2k + 1)(2k - 1)$

e) $5a(a - 5)(a + 4)$ f) $3t(4t + 1)(2t + 3)$

g) $2y^2(7 + 5x)(7 - 5x)$

h) $2p^2(4p - 1)(p - 6)$

Exercise 11G

1   a) $y = x - 3$

x	0	1	2
y	−3	−2	−1

$y = -x + 7$

x	0	1	2
y	7	6	5

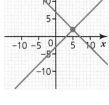

$x = 5, y = 2$ or $(5, 2)$

b) $y = 2x + 9$

x	0	1	2
y	9	11	13

$y = -2x - 7$

x	0	1	2
y	−7	−9	−11

$x = -4, y = 1$ or $(-4, 1)$

c) $y = 3x$

x	0	1	2
y	0	3	6

$y = -x - 12$

x	0	1	2
y	−12	−13	−14

$x = -3, y = -9$ or $(-3, -9)$

d) $y = x - 9$

x	0	1	2
y	−9	−8	−7

$y = -2x + 15$

x	0	1	2
y	15	13	11

$x = 8, y = -1$ or $(8, -1)$

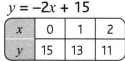

2   a) Pupil diagram leading to $(4, 0)$

b) Pupil diagram leading to $(6, 1)$

c) Pupil diagram leading to $(-7, -8)$

d) Pupil diagram leading to $\left(\dfrac{1}{2}, -2\right)$

e) Pupil diagram leading to $(5, -9)$

f) Pupil diagram leading to $(-3, -4)$

g) Pupil diagram leading to $\left(-\dfrac{1}{2}, 4\right)$

h) Pupil diagram leading to $\left(\dfrac{5}{2}, -\dfrac{3}{2}\right)$

3   They will be parallel, i.e. the lines have the same gradient.

Exercise 11H

1   a) $x = 1, y = 2$ b) $x = 2, y = 5$

c) $x = 4, y = 2$ d) $x = 8, y = 1$

2   a) $x = 3, y = 9$ b) $x = 5, y = 4$

c) $x = 2, y = 1$ d) $x = 6, y = 3$

3   a) $x = 2, y = 3$ b) $x = 3, y = 2$

c) $x = 5, y = 6$ d) $x = 7, y = 9$

e) $x = 4, y = 10$ f) $x = 3, y = 2$

g) $x = 2, y = 5$ h) $x = 7, y = 3$

i) $x = 1, y = 5$ j) $x = 2, y = 1$

k) $x = 1, y = 4$ l) $x = 4, y = 8$

m) $x = 4, y = -1$ n) $x = -2, y = 10$

o) $x = -3, y = 5$ p) $x = -1, y = -2$

4   Both adding and subtracting will give you the same solution.

5   a) Taylor

b) $-3y - (-9y) = 6y$, not $-12y$

6   a) The first equation by 3

b) $x = 2, y = 1$

Exercise 11I

1   a) $x = 4, y = 3$ b) $x = 2, y = 1$

c) $x = 3, y = 5$ d) $x = 1, y = 2$

e) $x = 5, y = 3$ f) $x = 4, y = 2$

2   a) $x = 3, y = 5$ b) $x = 2, y = 3$

c) $x = 1, y = 4$ d) $x = 5, y = 2$

Answers

e) $x = 4, y = 3$ f) $x = -1, y = 3$
g) $x = 3, y = -9$ h) $x = -2, y = -12$
i) $x = 3, y = 1$ j) $x = 7, y = 9$
k) $x = 8, y = 3$ l) $x = 6, y = 5$

Exercise 11J
1 a) $x = 1, y = 2$ b) $x = 2, y = 3$
   c) $x = 3, y = 1$ d) $x = 1, y = 4$
   e) $x = 2, y = -3$ f) $x = -2, y = 2$
   g) $x = 2, y = -1$ h) $x = 10, y = 5$
   i) $x = -3, y = -2$
2 a) $x = 3, y = 1$ b) $x = 5, y = 2$
   c) $x = 1, y = 2$ d) $x = 2, y = 9$
   e) $x = -3, y = 1$ f) $x = 6, y = -2$
   g) $x = 2, y = -1$ h) $x = 3, y = -2$
   i) $x = -3, y = 6$ j) $x = 5, y = -4$
   k) $x = -2, y = -1$ l) $x = -3, y = -4$
   m) $x = 2, y = -1$ n) $x = -3, y = 2$
   o) $x = -4, y = -5$ p) $x = -1, y = 6$

Exercise 11K
1 a) $x = 3, y = 18$ b) $x = 4, y = 9$
   c) $x = 0, y = 10$ d) $x = 2, y = 20$
2 a) $x = 24, y = 4$ b) $x = 11, y = 4$
   c) $x = 16, y = 2$ d) $x = 11, y = 3$
3 a) $x = 1, y = 8$ b) $x = 4, y = 3$
   c) $x = 3, y = 9$ d) $x = 2, y = 4$
   e) $x = 5, y = 11$ f) $x = 1, y = 5$
   g) $x = 6, y = 7$ h) $x = 5, y = 28$
4 a) $y = 2, x = 10$ b) $y = 5, x = 7$
   c) $y = 1, x = 6$ d) $y = 4, x = 16$
   e) $y = 3, x = 9$ f) $y = 6, x = 10$
   g) $y = 7, x = 8$ h) $y = 1, x = -1$
5 a) $x = 12, y = 4$ b) $x = 1, y = 7$
   c) $x = 5, y = 14$ d) $x = 2, y = 2$
   e) $x = 11, y = 51$ f) $x = 34, y = 6$
   g) $x = 11, y = 9$ h) $x = 3, y = 4$
   i) $x = -4, y = -9$ j) $x = 12, y = -6$
   k) $x = -2, y = -1$ l) $x = -2, y = -3$
6 $(-6, -17)$

Exercise 11L
1 a) $5a + c = 28$ b) $3a + 4c = 27$
   c) Entry for an adult is £5 and a child is £3.
2 a) $3s + 2w = 13 \cdot 4$ b) $5s + 3w = 21 \cdot 35$
   c) Three sandwiches and one wrap cost £10·45.
3 a) $4i + s = 95$ b) $2i + 5s = 115$
   c) Seven inner circle and four standing tickets cost £200.
4 a) Let a = adult and c = child. $2a + 7c = 110$
   b) $6a + 5c = 170$
   c) He has overestimated by £20. Nine adult bikes and 10 child bikes would cost £280 to hire.
5 a) Let c = canvas and p = photo. $3c + 5p = 17 \cdot 5$
   b) $4c + 3p = 21 \cdot 5$
   c) Eight canvases and four photos will cost £42.
6 a) Let b = badges and m = mugs. $9b + 2m = 10 \cdot 8$
   b) $4b + 3m = 11 \cdot 45$
   c) Nine badges and ten mugs cost £36.
7 a) Let x = weekday and y = weekend $20x + 13y = 454 \cdot 25$
   b) $10x + 17y = 408 \cdot 25$ c) £5·75
8 Two glasses and two bottles hold 1·6 litres.
9 Three small pots and four large pots have a mass of 6·851 kg.
10 No, he is not correct. A colour copy costs 6p and a black and white copy costs 2p so altogether it will cost David £4·04 for 49 colour and 55 black and white copies. He is 4p short in his costing.
11 42 and 25   12 5 and -3   13 48
14 There are thirty-one 10p coins and seventeen 20p coins in the jar.

Check-up
1 a) $x^2 + 5x + 4$ b) $x^2 + 9x + 14$
   c) $x^2 - 2x - 80$ d) $x^2 - 2x - 99$
   e) $x^2 + 2x - 24$ f) $x^2 + 3x - 40$
   g) $y^2 - 12y + 27$ h) $y^2 - 20y + 100$
2 a) $2x^2 + 13x + 15$ b) $6y^2 + 35y + 36$
   c) $24m^2 + 26m - 5$ d) $56k^2 + 27k - 18$

e) $15w^2 + 29w - 14$ f) $12p^2 + 28p - 5$

g) $60q^2 - 68q + 15$ h) $81x^2 - 72x + 16$

3 a) $x^3 + 7x^2 + 13x + 4$

b) $2x^3 + x^2 + 2x + 30$

c) $4x^3 + 19x^2 - 83x + 45$

d) $3x^3 - 14x^2 + 5x + 2$

4 a) $2x^2 + 4x - 24$ b) $7x^2 + 19x + 31$

c) $4x^2 - 32x - 22$ d) $-10x + 27$

5 a) $2x^2 + 13x + 21$ b) $x^2 - 2x - 3$

c) $x^2 + 8x + 16$

6 a) $4(t + 4)$ b) $6(a - 3)$

c) $2(4m + 5)$ d) $3(5 + 2k)$

e) $5(9x - 11)$ f) $x(y + 9)$

g) $p(8 - q)$ h) $n(10m + n)$

i) $ab(4c + 5)$ j) $r^2(pq - 7)$

7 a) $4x(1 + 3y)$ b) $6m(n + 5)$

c) $5q(2p - 5)$ d) $3h(6 + 7g)$

e) $7d(7c - 12)$ f) $3x(3x + 1)$

g) $8y(3 + 5y)$ h) $2y^2(4 - 3y)$

i) $5pq(r + 7)$ j) $9m^2n(2 - 3mn)$

8 a) $(y + 3)(y - 3)$ b) $(2 + x)(2 - x)$

c) $(9p + 5)(9p - 5)$ d) $(4 + 7a)(4 - 7a)$

9 a) $(x + 1)(x + 11)$ b) $(x + 2)(x + 6)$

c) $(x - 1)(x - 17)$ d) $(x - 4)(x - 5)$

e) $(x - 1)(x + 17)$ f) $(x - 2)(x + 9)$

g) $(x + 1)(x - 23)$ h) $(x - 8)(x + 2)$

10 a) $(2x - 1)(x - 3)$ b) $(5x - 1)(x + 2)$

c) $(x + 11)(7x + 1)$ d) $(7x + 4)(x - 5)$

e) $(11x - 2)(x + 9)$ f) $(2x + 1)(3x - 5)$

g) $(2x - 1)(5x + 1)$ h) $(8x - 3)(2x - 7)$

i) $(1 - x)(17 + x)$ j) $(2 - x)(3 + x)$

k) $(1 - x)(5 + 3x)$ l) $(2 - x)(2 + 3x)$

11 a) $10(m + 2n)(m - 2n)$

b) $2(11p + 4q)(11p - 4q)$

c) $8(a + 4)(a - 2)$ d) $5(x + 1)^2$

e) $4(2y - 3)(y - 1)$ f) $3k(k - 3)(k + 2)$

g) $6m(m - 1)(m + 4)$ h) $2y(2y + 1)(y + 5)$

12 a) Pupils' own diagram leading to (3, 12)

b) Pupils' own diagram leading to (5, 3)

c) Pupils' own diagram leading to (–2, –8)

d) Pupils' own diagram leading to (–2, –7)

13 a) $x = 5, y = 4$ b) $x = 2, y = 6$

c) $y = -3, x = 2$ d) $x = 1, y = -4$

14 a) $y = 2, x = 3$ b) $x = 1, y = 8$

c) $x = 5, y = 3$ d) $x = 5, y = -2$

15 a) $x = 1, y = 2$ b) $x = 2, y = 3$

c) $x = 10, y = 5$ d) $x = 2, y = 4$

e) $x = 1, y = -2$ f) $x = -1, y = 4$

g) $x = -2, y = 3$ h) $x = -3, y = -3$

16 a) $x = 4, y = 13$ b) $y = 6, x = 28$

c) $x = 1, y = 4$ d) $x = 3, y = -14$

17 a) Let m = milkshake and f = fruit smoothie
$5m + 2f = 15$

b) $6m + f = 14\cdot5$

c) The total cost of three milkshakes and four fruit smoothies is £16.

18 £18·50   19 34 points   20 10p

21 68 and 31

12 Circle

Exercise 12A

1 a) $x = 100°$ b) $y = 76°$ c) $m = 144°$

d) $n = 32°$ e) $p = 104°$ f) $q = 22°$

g) $r = 54°$ h) $w = 138°$

2 a) $x = 60°$ b) $y = 45°$ c) $r = 20°$

d) $q = 67°$ e) $p = 35°$ f) $m = 76°$

g) $n = 57\cdot5°$ h) $d = 33\cdot5°$

3 a) $x = 126°$ b) $y = 92°$ c) $p = 25°$

d) $q = 70°$ e) $r = 57°$ f) $s = 34°$

g) $t = 32°$ h) $u = 72°, v = 54°$

4 a) $x = 44°$ b) $y = 68°$

c) $w = 108°$ d) $m = 126°$

5 They add up to 90°.

Exercise 12B

1 a) $x = 65°$ b) $y = 47°$ c) $a = 19°$ d) $b = 32°$

2 a) $p = 14°$ b) $q = 28°$ c) $r = 45°$ d) $s = 81°$

3   a) $y = 8 \cdot 51\,\text{m}$ b) $q = 2 \cdot 57\,\text{m}$
    c) $x = 4 \cdot 05\,\text{cm}$ d) $p = 5 \cdot 36\,\text{cm}$
4   a) $x = 5 \cdot 83\,\text{cm}$ b) $y = 11 \cdot 3\,\text{m}$
    c) $q = 4 \cdot 47\,\text{cm}$ d) $p = 4 \cdot 80\,\text{m}$
5   a) $4\,\text{cm}$ b) $16\,\text{cm}^2$ c) $34 \cdot 3\,\text{cm}^2$
    d) $3\,\text{cm}$ e) $12\,\text{cm}^2$ f) $11 \cdot 7\%$

Exercise 12C

1   a) $x = 59°$ b) $y = 33°$ c) $m = 30°$
2   a) $y = 52°$ b) $x = 54°$ c) $k = 86°$
3   a) $a = 45°, b = 40°$ b) $p = 28°, q = 44°$
    c) $w = 85°$
4   a) $x = 31°$ b) $w = 57°$ c) $y = 124°$
5   Evana is not correct. $4^2 + 11^2 = 137$, $12^2 = 144$
    so by the converse of Pythagoras, as $4^2 + 11^2$
    $\neq 12^2$ the triangle is not right-angled and AB is
    not a tangent.
6   a) $x = 143°$ b) $y = 46°$ c) $p = 116°$
7   a) $w = 30°$ b) $p = 50°$
    c) $m = 63°, n = 46°, r = 22°$
8   $x = 29°$

Exercise 12D

1   a) $PQ = 8 \cdot 25\,\text{cm}$ b) $AB = 11 \cdot 5\,\text{cm}$
    c) $MN = 5 \cdot 85\,\text{m}$ d) $KL = 29 \cdot 3\,\text{mm}$
2   a) $x = 7 \cdot 21\,\text{cm}$ b) $y = 4 \cdot 58\,\text{cm}$
    c) $k = 7 \cdot 16\,\text{m}$ d) $p = 11 \cdot 1\,\text{cm}$
3   a) $3 \cdot 46\,\text{cm}$ b) $4 \cdot 66\,\text{cm}$
    c) $5 \cdot 74\,\text{m}$ d) $10 \cdot 7\,\text{mm}$
4   a) i) $6 \cdot 93\,\text{cm}$ ii) $14 \cdot 9\,\text{cm}$
    b) i) $12 \cdot 1\,\text{cm}$ ii) $26 \cdot 1\,\text{cm}$
    c) i) $8 \cdot 66\,\text{cm}$ ii) $18 \cdot 7\,\text{cm}$
    d) i) $19 \cdot 6\,\text{cm}$ ii) $42 \cdot 6\,\text{cm}$
5   a) i) $6 \cdot 71\,\text{cm}$ ii) $3 \cdot 71\,\text{cm}$
    b) i) $1 \cdot 3\,\text{cm}$ ii) $0 \cdot 8\,\text{cm}$
    c) i) $12 \cdot 5\,\text{cm}$ ii) $6 \cdot 53\,\text{cm}$
    d) i) $52 \cdot 9\,\text{cm}$ ii) $32 \cdot 9\,\text{cm}$
6   $0 \cdot 677\,\text{m}$
7   a) $4 \cdot 12\,\text{cm}$ b) $13 \cdot 1\,\text{cm}$
8   a) Pupils' own   b) $5 \cdot 29\,\text{cm}$
9   $10 \cdot 1\,\text{m}$

Exercise 12E

1   a) $x = 6 \cdot 81\,\text{cm}$ b) $y = 31 \cdot 8\,\text{cm}$
    c) $k = 111\,\text{mm}$ d) $p = 51 \cdot 8\,\text{cm}$
2   a) $x = 13 \cdot 8\,\text{cm}$ b) $y = 91 \cdot 4\,\text{cm}$
    c) $r = 1 \cdot 67\,\text{mm}$ d) $q = 3 \cdot 30\,\text{m}$
3   $47 \cdot 1\,\text{cm}$ 4 $r = 3 \cdot 04\,\text{cm}$

Exercise 12F

1   a) $x = 117\,\text{cm}^2$ b) $y = 163\,\text{cm}^2$
    c) $k = 42 \cdot 9\,\text{mm}^2$ d) $h = 2240\,\text{cm}^2$
2   a) $x = 54 \cdot 3\,\text{cm}^2$ b) $y = 32 \cdot 0\,\text{m}^2$
    c) $a = 159\,\text{cm}^2$ d) $b = 1 \cdot 33\,\text{cm}^2$ or
    $133\,\text{mm}^2$
3   $17312\,\text{cm}^2$ 4 $r = 5 \cdot 95\,\text{cm}$

Exercise 12G

1   a) $x = 234°$ b) $y = 123°$
    c) $w = 47 \cdot 0°$ d) $p = 39 \cdot 1°$
2   a) $x = 337°$ b) $y = 85 \cdot 6°$
    c) $p = 172°$ d) $q = 219°$
3   $80 \cdot 2°$
4   a) $x = 73 \cdot 3°$ b) $y = 64 \cdot 2\,\text{cm}$
    c) $p = 64 \cdot 0\,\text{m}^2$ d) $d = 27 \cdot 9\,\text{cm}$

Exercise 12H

1   a) $72 \cdot 6\,\text{cm}^2$ b) $618\,\text{cm}^2$ c) $52 \cdot 2\,\text{m}^2$
2   a) $62 \cdot 2\,\text{cm}$ b) $38 \cdot 0\,\text{m}$ c) $68 \cdot 0\,\text{cm}$
3   a) $155\,\text{cm}^2$ b) $58 \cdot 2\,\text{cm}$
4   a) $1 \cdot 09\,\text{m}^2$ or $10\,900\,\text{cm}^2$
5   a) $1 \cdot 75\,\text{cm}^2$ b) $4 \cdot 99\,\text{cm}^2$
6   a) $381\,\text{cm}$ b) $277\,\text{cm}^3$ or $277\,000\,\text{mm}^3$
7   a) $75\pi\,\text{cm}^2$ b) $(15\pi + 20)\,\text{cm}$
8   a) $20\pi\,\text{cm}^2$ b) $8\pi\,\text{cm}$
    c) $4\,\text{cm}$ d) $3\,\text{cm}$

Check-up

1   a) $x = 94°$ b) $y = 51°$ c) $k = 70 \cdot 5°$
2   a) $x = 31°$ b) $y = 67°$ c) $p = 36°$
3   a) $x = 9 \cdot 06\,\text{cm}$ b) $y = 6 \cdot 11\,\text{m}$ c) $p = 12 \cdot 2\,\text{m}$
4   a) $x = 51°$ b) $y = 62°$ c) $q = 32°$

5 a) $x = 7.21$ cm b) $y = 9.49$ cm c) $p = 15.5$ m

6 20 cm

7 a) $x = 20.1$ cm   b) $y = 8$ cm   c) $p = 46.7$ m

8 a) $x = 118$ cm^2 b) $y = 38$ m^2

   c) $k = 9.9$ cm^2 or 990 mm^2

9 a) $x = 177°$ b) $y = 306°$ c) $p = 231°$

10 a) $x = 130$ cm^2 b) $y = 16.1$ m   c) $q = 137°$

11 336 cm^3

12 a) 8.57 m^2 b) 16.7 m

13 Data analysis

Exercise 13A

1 a) 5   b) 105   c) 2.5   d) −1

   e) 431   f) 24.36   g) 56.92   h) 0.64

2 a) 4   b) 22   c) 2.5   d) 8

   e) 450   f) 11.9   g) 15   h) 17

3 a)

	Mean	Range
Lerwick	19·29	21
Falkirk	13·71	10

   b) Typically Lerwick was windier, the mean wind speed was higher (19·29 km/h > 13·71 km/h)

   c) Lerwick's weather was more varied, the range is higher (21 km/h > 10 km/h)

4 a)

	Mean	Range
Supermarket	£0·99	£0·65
Convenience	£1·44	£0·95

   b) A loaf is likely to be cheaper in the supermarket as the mean is lower (£0·99 < £1·44)

   c) The convenience shops have less consistent prices as the range is higher (£0·95 > £0·65)

5 a)

	Mean	Range
Trainee driver	20·25	6
Ticket examiner	30·17	46
Revenue manager	42	10

   b) Trainee driver

   c) Ticket examiner

   d) The mean age of a ticket examiner is 30·17 but most of the workforce are in their 20s and none of them are in their 30s. The two high values 59 and 65 skew the mean to the right.

6 15 kg

7 a) 720 cm   b) 185 cm

8 40 years   9 20 years

10 a) 40 years   b) 1 040 000 g, 1040 kg

   c) 13·92 µg/m^3 d) 40 reviews

Exercise 13B

1 a) mean $\bar{x} = \dfrac{\text{total}}{n}$

   $= \dfrac{912}{6}$

   $= 152$

   $s = \sqrt{\dfrac{\sum(x-\bar{x})^2}{n-1}}$

   $= \sqrt{\dfrac{614}{6-1}}$

   $= \sqrt{\dfrac{614}{5}}$

   $= \sqrt{122.8}$

   $= 11.081...$

   $= 11.08$

x	$x - \bar{x}$	$(x - \bar{x})^2$
144	−8	64
155	3	9
138	−14	196
160	8	64
168	16	256
147	−5	25
	Total	614

   b)

	mean	s
S1	152	11·08
S3	172	7·05

   c) Typically the S3s were taller as the mean is higher (172 > 152)

   d) The S1s had more varied heights as the standard deviation is larger (11·08 > 7·05)

2 a) $\bar{x} = 90$p, $s = 37.08$

   b)

	mean	s
Own-brand	90	37·08
Branded	125	15·20

   c) The own-brand beans are typically less expensive, the mean is lower (90p < £1·25). The prices of the branded beans are more consistent as the standard deviation is lower (15·2 < 37·08).

3 a) $\bar{x} = 5, s = 2$

b)

	mean	s
Teachers	5	2
Lifeguards	4	1·14

c) The teachers were typically a year older when then learned to swim (5 > 4). The lifeguards' ages were more consistent as the standard deviation is lower (1·14 < 2).

4 a) $\bar{x} = 3·4, s = 2·07$

b)

	mean	s
No spa	3·4	2·07
Spa	6·2	3·1

c) Typically guests stay for more nights after the spa is built as the mean is higher (6·2 > 3·4). The number of nights guests stay varies more after the spa is built as the standard deviation is higher (3·1 > 2·07).

5 a) Before training $\bar{x} = 29·1$ ml/kg/min, $s = 3·06$; after training programme $\bar{x} = 36·1$ ml/kg/min, $s = 4·11$

b) On average the athletes were fitter after the programme as the mean VO_2 Max is higher (36·1 > 29·1). The athletes' fitness levels were more varied after the programme as the standard deviation is higher (4·11 > 3·06).

6 $s = 6·23$ as the standard deviation is a measure of spread and the spread is identical

7 B

8 Data set 7, 4, 3, 1, 5

9 $s = 4·08$

	3	9
	4	16
	5	25
	12	144
Total	24	194

10 a) $\bar{x} = 6, s = 2·55$ b) $\bar{x} = 10, s = 5·06$

11 The price of bread is higher in convenience shops as the mean is higher (£1·44 > £0·99). The price of bread is more consistent in supermarkets as the standard deviation is lower (0·19 < 0·33).

12 Also 1·58, as standard deviation is a measure of spread and the spread is identical.

13 95% of people at the cinema are between 28 and 44

14 We know that 95% of people in that age group will score between 1·3 and 9·7, so this person's fitness is exceptionally good.

Exercise 13C

1 a) 7   b) 39   c) 20   d) 55
    e) −1   f) 0·5   g) 3147·5

2 a) 8, 10, 30, 32, 36, 45, 70 median = 32
    b) 34, 38, 41, 45, 47, 50, 80 median = 45
    c) Joseph as the median was higher (45 > 32). Joseph's maximum was also higher than Annie's (80 >70).

3 a) median = 2·9 minutes
    b) mean = 4 minutes. The median represents the typical time better as most customers completed their orders in less than 4 minutes. The mean is skewed by the two high values.

4 They are using the mean annual salary, £34 500. The median salary, £22 000, should be used as the mean is skewed by two much higher values.

Exercise 13D

1 a) $Q_1 = 59, Q_2 = 62, Q_3 = 67$
    b) $Q_1 = 6·5, Q_2 = 8, Q_3 = 10$
    c) $Q_1 = 4·5, Q_2 = 7, Q_3 = 10·5$
    d) $Q_1 = 43, Q_2 = 48, Q_3 = 54·5$
    e) $Q_1 = 412·5, Q_2 = 500, Q_3 = 527$
    f) $Q_1 = 6·7, Q_2 = 7·5, Q_3 = 8·1$
    g) $Q_1 = 22 000, Q_2 = 26 500, Q_3 = 35 000$

2 a) IQR = 8   b) IQR = 3·5   c) IQR = 6
    d) IQR = 11·5 e) IQR = 114·5 f) IQR = 1·4
    g) IQR = 13 000

3 a) The Call Centre: $Q_1 = 18·5, Q_2 = 21, Q_3 = 22·5$
    Monday Morning: $Q_1 = 5·5, Q_2 = 26·5,$
    $Q_3 = 29$
    b) The Call Centre IQR = 4
    Monday Morning IQR = 23·5

c) The audience switched off The Call Centre faster as the median time watched is lower (21 < 26·5). The audience reaction was more consistent for The Call Centre as the IQR is smaller (4 < 23·5).

4 a) Original: $Q_1 = 18$, $Q_2 = 60$, $Q_3 = 105$;
New: $Q_1 = 14$, $Q_2 = 36$, $Q_3 = 85$

b) Original IQR = 87, New IQR = 71

c) Typically, the new material dissolves faster as the median is lower (36 < 60). The times taken for the new material to dissolve are more consistent as the IQR is smaller (71 < 87).

5 a) Bethton: $Q_1 = 3$, $Q_2 = 5$, $Q_3 = 8$, L = 1, H = 10
Billford: $Q_1 = 5$, $Q_2 = 7$, $Q_3 = 11·5$, L = 3, H = 15

b)

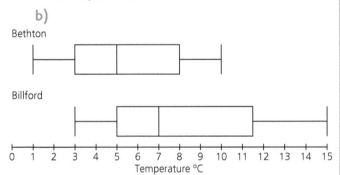

Bethton

Billford

c) Bethton IQR = 5, Billford IQR = 6·5

d) Typically the temperatures were warmer in Billford as the median is higher (7 > 5). The temperatures in Bethton were more consistent as the IQR (box) is smaller (5 < 6·5).

Exercise 13E

1 a) Sour   b) R   c) M and XL

d) 25   e) There is no mode

2

Data	Measure of location	Measure of spread
Numerical data, no unusually high or low values	Mean	Standard deviation
Numerical data, unusually high or low values	Median	IQR
Categorical data	Mode	

3 a) Median and IQR

b) Mode with no measure of spread

c) Mean and standard deviation

4 a) Mean £105, median £103·50, mode £95. The mean and the median both represent the data well. The mode is lower than most of the prices and is not the best choice.

b) Mean £229 900, median £111 000, there is no mode. The median as the mean is higher than most of the data (skewed by the high values).

c) Mean = 6·17, median = 5·5, mode = 5. If you owned this shop the mode would be the most useful.

5 Pupils' own answers e.g. 3, 5, 5, 9, 10, 11, 13; check most common = 5, 4th term = 9, total = 56

6 We need two 4s for mode, total of 30 for the mean. 6 options in total e.g. 4, 4, 5, 6, 11

7 Need at least two 4s for mode. Must add to 25 for mean. Can't be done with a median of 3.

Exercise 13F

1 a)

Sponsor money
0	5
1	0 3 5 5 9
2	1 3 8
3	0 0 4 5 9
4	1 8

$n = 16$ 2 | 3 = £23

b) £25·50

2 a)

Noise level
7	0 9
8	0 5 6 8
9	0 2 5 5 8
10	0

$n = 12$ 7 | 0 = 70 decibels

b) 95 Db   c) 89 Db   d) $\dfrac{5}{12}$

3 a)

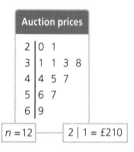

Auction prices	
2	0 1
3	1 1 3 8
4	4 5 7
5	6 7
6	9

$n = 12$ —— $2 \mid 1 = £210$

   b) £410   c) 25%

4 a) 4   b) $\dfrac{1}{4}$

   c) 11   d) $\dfrac{11}{16}$

5 a) peat 2·6 cm, clay 1·9 cm

   b) peat 3·3 cm, clay 2·0 cm

   c) The typical sapling grows taller in peat soil (3·3 cm > 2·0 cm). Saplings grown in clay have more consistent heights as the range is lower (1·9 < 2·6).

6 a)

	200 m sprint times	
World Championship	6 5	28	2 8	Rio Paralympics
	9	29	4	
		30		
		31		
	9 7 0	32	7	
	7	33	1 9	

$n = 7$ —— $n = 6$ $29 \mid 4 = 29{\cdot}4$ seconds

   b) Rio 31·05, World 32·0, Rio Paralympics faster by 0·95 seconds

7 a)

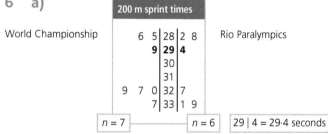

				Cost of projects						
		Education				Clean water				
		5 4	4	0						
7 7	5	3	1	1	8 9					
		4	2	0 1	4 5	8 9				
		8	3	0 0	1 4	6 6	9			
		0	4	2						
6	3 0	0	5	0						
		2	0	6	0 7					
		8	7							

$n = 18$ $n = 19$ $1 \mid 9 = £1{\cdot}9$ million

   b) Clean water IQR £1·5million, Education IQR £4 million

   c) Clean water (1·5 < 4)

Exercise 13G

1 a) $m = 4$ b) (0, 10)   c) $y = 4x + 10$
   d) $T = 4D + 10$ e) $T = 42$ minutes

2 a)

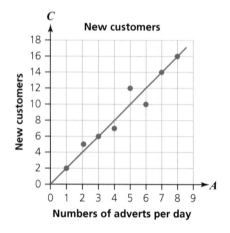

   b) Pupil's own answers, example above
   c) To match pupil's own, example above, $m = 2$
   d) To match pupil's own, example above, $C = 2A$
   e) To match pupil's own, example above, 30 customers

3 a)

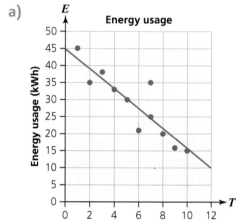

   b) Pupil's own, example above
   c) To match pupil's own, example above, $E = -3T + 45$
   d) To match pupil's own, example above, 51 kWh

4 a)

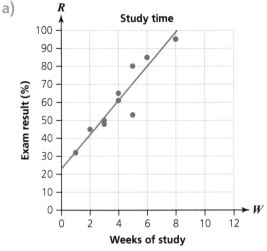

b) Pupil's own, example above, $R = 9.5W + 23$

c) 5 weeks

Exercise 13H

1 a) Great Britain   b) 19   c) 10 more

d) 64   e) China   f) $\frac{5}{14}$

2 a) 13 °C   b) 4 hours   c) 7 am–9 am

d) 15 °C   e) 4 °C

f) The temperature in Glasgow rose steadily from a low 3 °C at 4 am to a peak of 18 °C late afternoon. Following the peak the Glasgow temperature dropped back down after 8 pm, falling as low as 5 °C again at night-time. The Moscow temperature does not rise and fall so dramatically but stays warmer throughout the day and night, it drops only a little at night to 14 °C. Dawson's Creek has a similar pattern to Glasgow but is overall colder at this time of year.

3 a) 108   b) 432   c) 1080

d) 108°   e) 54°   f) 28   g) 45

h) No, this is true for the songs Charlie streamed but she streamed a lot more R&B songs (432) than Country (54).

4 a) 10   b) $\frac{1}{6}$ c) 19 g per 100 g

d) 19 g per 100 g   e) 19 g per 100 g

f) On average the cold sandwiches contain less fat per 100 g as the mean is lower (5·65 < 19). The amount of fat in the sandwiches is more varied as the standard deviation is higher (2·99 > 2·04).

Exercise 13I

1 a) pie chart

b) compound bar graph

c) compound line graph

d) scatter graph

e) stem and leaf.

2 a)

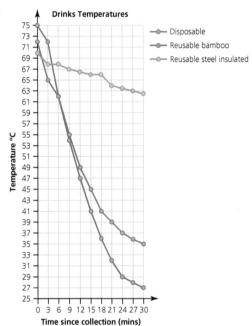

b) Disposable

c) Disposable 40°C   reusable bamboo 33 °C reusable steel 64·5 °C

d) The disposable cup has a larger heat loss (48 °C, compared to 8°C).

e) No

3 a) $\frac{1}{3}$

b)

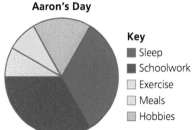

Aaron's Day

Key
■ Sleep
■ Schoolwork
☐ Exercise
☐ Meals
☐ Hobbies

4   a) England

b)

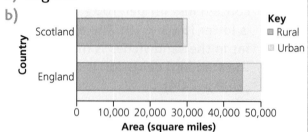

c) The proportion of land designated as urban is very small. Scotland is almost $\frac{2}{3}$ of the size of England.

5   a) 6·5, 0·5

b)

Cybersecurity scores	
0	5
1	0 1 2 3 5 5 8 9 9
2	0 0 0 1 2 5 5
3	0 5
4	0
5	5
6	5

$n = 22$ 6|5   2|1 = 2·1

c) Median = 2·0

6   a)

Call volume

Number of calls vs *Time*

b) The volume of calls is high at 9:00 and peaks at 9:20 it then falls gradually with slight upticks at 9:50 and 10:50.

c)

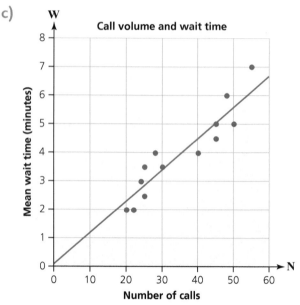

There is a positive correlation, the more calls there are the longer the wait time.

d) To match pupils' own, example line of best fit above $W = \frac{1}{9}N$

e) To match pupils' own, example above 6 minutes 40 seconds

Exercise 13J

1   a) $\frac{5}{7}$ b) $\frac{2}{7}$ c) $\frac{2}{7}$ d) $\frac{1}{7}$ e) $\frac{6}{7}$

2   a) $\frac{1}{50}$ b) $\frac{2}{25}$ c) $\frac{1}{10}$ d) $\frac{9}{10}$

3   a) $P(6) = \frac{1}{6}$, $P(H) = \frac{1}{2}$, Heads more likely

   b) $P(\text{even}) = \frac{1}{2}$, $P(J) = \frac{1}{4}$, Even more likely

   c) $P(\text{Win 1}) = \frac{1}{16}$, $P(\text{Win 2}) = \frac{1}{30}$, more likely to win raffle 1

4   a) $\frac{1}{8}$ b) $\frac{1}{4}$ c) $\frac{1}{4}$

   d) $\frac{3}{8}$ e) $\frac{5}{8}$ f) $\frac{7}{8}$

5 a) 4   b) 10   c) 64   d) 125   e) 225

6 a) 30   b) 35   c) 17   d) 13

   e) Helen as she did the most rolls of the dice

7 a) 300 seconds, 5 minutes

   b) $\dfrac{2}{5}$ c) $\dfrac{1}{2}$ d) $\dfrac{1}{30}$ e) $\dfrac{1}{10}$ f) 360

8 a) $\dfrac{7}{20}$ b) $\dfrac{1}{3}$

   c) Raffle is more likely, $\dfrac{7}{20} = 35\% > \dfrac{1}{3} = 33\dfrac{1}{3}\%$

9 a) Blue marble is more likely,

      $\dfrac{4}{7} = 57\cdot1\% > \dfrac{1}{2} = 50\%$

   b) Red or green marble is more likely,

      $\dfrac{3}{7} = 42\cdot9\% > \dfrac{3}{8} = 37\cdot5\%$

10 a) A multiple of 6, $\dfrac{8}{50} > \dfrac{7}{50}$

   b) A prime number, $\dfrac{15}{50} > \dfrac{12}{50}$

Exercise 13K

1 a) $\dfrac{2}{25}$ b) 200

2 a) $\dfrac{1}{30}$ b) 180 zips

3 a) Minicar   b) 7
   c) 1   d) Yes, 700 expected

4 20 sets   5 275 000 people

Exercise 13L

1 a)

	1	2	3	4	5	6
1	2	3	4	5	6	7
2	3	4	5	6	7	8
3	4	5	6	7	8	9
4	5	6	7	8	9	10
5	6	7	8	9	10	11
6	7	8	9	10	11	12

   b) $\dfrac{5}{18}$ c) $\dfrac{13}{18}$ d) 2, 12   e) 7

f) Easy eight 6, 2 or 5, 3; hard eight 4,4
   (less likely)

g) 100

2 a) $\dfrac{1}{12}$ b) $\dfrac{1}{100}$ c) $\dfrac{1}{21}$ d) $\dfrac{1}{8}$

   e) 0·00002, $\dfrac{1}{50000}$, assuming that having red
      hair does not make a mother more likely
      to have triplets!

   f) 0·000 125, $\dfrac{1}{8000}$

Check-up

1 a) 22   b) 76

2 a) 11   b) 40

3 a) 21   b) 75·5

4 a) $\bar{x} = 7, s = 1\cdot41$

   b) The software is not understanding female
      voices as well, the mean is lower (5 < 7).
      The response to female voices is less
      consistent as the standard deviation is
      higher (1·96 > 1·14).

5 a) $Q_1 = 115$ $Q_2 = 130$ $Q_3 = 147$
   b) $Q_1 = -2\cdot5$ $Q_2 = 5$ $Q_3 = 12$
   c) $Q_1 = 71\,250$ $Q_2 = 80\,625$ $Q_3 = 86\,000$

6 a) 32   b) 14·5   c) 14 750

7 a) 3   b) C and H

8 a)

Party guests	
1	5 8 8 9
2	0 0 0 0 0 1 1 1 1
3	
4	8 9
5	
6	7 9

n = 17   1|3 = 13 years

   b) 69, 15; range = 54   c) 20

   d) 20   e) 28·65   f) $\dfrac{9}{17}$

   g) The mode and median are the same and
      both represent the data better than the
      mean which is skewed by the few older
      people.

9   a) $\bar{x} = 25, s = 2$

10 a) $\dfrac{1}{20}$ b) $\dfrac{1}{25}$ c) $\dfrac{1}{500}$

11 a) $\dfrac{3}{8}$ b) 111

12 a) 2   b) $s = 1\cdot41$

   c) On average, the programme did help as the mean wellbeing score was higher on day 30 (5 > 2). There was less variation amongst the wellbeing scores on day 30 as the standard deviation is lower $\left(1 < \sqrt{2}\right)$

13 a) This is a grouped scattergraph, it shows the participants twice, once on day 1 and once on day 30. So there are 2 × 8 = 16 data points.

   b) 3·5 hours   c) 2 hours   d) 5

   e)

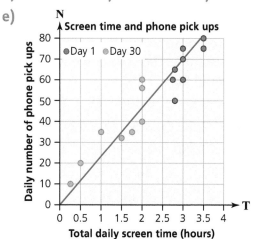

e) Pupils' lines will vary, the example shown has $m = \dfrac{70}{3}$

f) To match pupil's line, example shown $N = \dfrac{70T}{3}$

g) To match pupil's own line, example shown approx 117 times

h) Yes, the cluster of points for day 1 shows higher screen time and more phone pick-ups. At day 30 the participants were using their phone less.

14 The participants are using social media less. They are also spending less time on work email. There is a new category of wellbeing apps on the day 30 graph, perhaps they have found new more positive uses for their phones.